Spectroelectrochemistry

THEORY AND PRACTICE

Spectroelectrochemistry
THEORY AND PRACTICE

Edited by
Robert James Gale

Louisiana State University
Baton Rouge, Louisiana

Plenum Press • New York and London

CHEMISTRY

Library of Congress Cataloging in Publication Data

Spectroelectrochemistry: theory and practice / edited by Robert James Gale.
   p.   cm.
   Includes bibliographies and index.
   Contents: Introduction / Robert J. Gale—X-ray techniques / James Robinson—
Photoemission phenomena at metallic and semiconducting electrodes / Ricardo Borjas Severeyn and Robert J. Gale—UV-visible reflectance spectroscopy / Dieter M.
Kolb—Infrared reflectance spectroscopy / Bernard Beden and Claude Lamy—Surface
enhanced Raman scattering / Ronald L. Birke and John R. Lombardi—ESR spectroscopy of electrode processes / Richard G. Compton and Andrew M. Waller—Moss-
bauer spectroscopy / Daniel A. Scherson.
   ISBN 0-306-42855-5
   1. Electrochemistry. 2. Spectrum analysis. I. Gale, Robert J., 1942–
QD555.6.S65S66   1988   88-15431
541.3′7—dc19   CIP

© 1988 Plenum Press, New York
A Division of Plenum Publishing Corporation
233 Spring Street, New York, N.Y. 10013

Printed in the United States of America

Contributors

Bernard Beden • Laboratory of Chemistry I—Electrochemistry and Interactions, University of Poitiers, U.A. CNRS No. 350, 86022 Poitiers, France

Ronald L. Birke • Department of Chemistry, City College, City University of New York, New York, New York 10031

Ricardo Borjas Severeyn • Chemistry Department, University of Wyoming, Laramie, Wyoming 82071

Richard G. Compton • Physical Chemistry Laboratory, Oxford University, Oxford OX1 3QZ, United Kingdom

Robert J. Gale • Chemistry Department, Louisiana State University, Baton Rouge, Louisiana 70803

Dieter M. Kolb • Fritz Haber Institute of Max Planck Gesellschaft, D-1000 Berlin 33, Federal Republic of Germany

Claude Lamy • Laboratory of Chemistry I—Electrochemistry and Interactions, University of Poitiers, U.A. CNRS No. 350, 86022 Poitiers, France

John R. Lombardi • Department of Chemistry, City College, City University of New York, New York, New York 10031

James Robinson • Department of Physics, University of Warwick, Coventry CV4 7AL, United Kingdom

Daniel A. Scherson • Case Center for Electrochemical Sciences and the Department of Chemistry, Case Western Reserve University, Cleveland, Ohio 44106

Andrew M. Waller • Physical Chemistry Laboratory, Oxford University, Oxford OX1 3QZ, United Kingdom

Preface

The intention of this monograph has been to assimilate key practical and theoretical aspects of those spectroelectrochemical techniques likely to become routine aids to electrochemical research and analysis. Many new methods for interphasial studies have been and are being developed. Accordingly, this book is restricted in scope primarily to *in situ* methods for studying metal/electrolyte or semiconductor/electrolyte systems; moreover, it is far from inclusive of the spectroelectrochemical techniques that have been devised. However, it is hoped that the practical descriptions provided are sufficiently explicit to encourage and enable the newcomer to establish the experimental facilities needed for a particular problem.

The chapters in this text have been written by international authorities in their particular specialties. Each chapter is broadly organized to review the origins and historical background of the field, to provide sufficiently detailed theory for graduate student comprehension, to describe the practical design and experimental methodology, and to detail some representative application examples. Since publication of Volume 9 of the *Advances in Electrochemistry and Electrochemical Engineering* series (1973), a volume devoted specifically to spectroelectrochemistry, there has been unabated growth of these fields. A number of international symposia—such as those held at Snowmass, Colorado, in 1978, the proceedings of which were published by North-Holland (1980); at Logan, Utah in 1982, published by Elsevier (1983); or at the Fritz Haber Institute in 1986—have served as forums for the discussion of nontraditional methods to study interphases and as means for the dissemination of a diversity of specialist research papers.

Many thanks are extended to those who have facilitated the publication of this volume. The patient and competent secretarial help of Juanita Miller and the office staff of the LSU Chemistry Department is gratefully acknowledged. The efforts of the editorial staff at Plenum—Mary Phillips Born, Amelia McNamara, and Jeanne Libby—and particularly of those who helped in the manuscript review stages are greatly appreciated. Finally, this text is dedicated to the memory of my father, who instilled in me his love for books and knowledge, and to that of my Ph.D. research director and mentor, Carl A. Winkler.

Baton Rouge   Robert J. Gale

Contents

Chapter 5.   Infrared Reflectance Spectroscopy
             Bernard Beden and Claude Lamy

Chapter 7. ESR Spectroscopy of Electrode Processes

Richard G. Compton and Andrew M. Waller

Chapter 8. Mössbauer Spectroscopy

Daniel A. Scherson

Introduction

Robert J. Gale

1. Motivations for Spectroelectrochemistry

It has long been recognized by electrochemists that measurements of electrical currents, voltages, charges, or capacitances do not always provide unequivocal identification of electroactive molecules, i.e., although a diffusion current might be correlated to a particular species, with its peak or half-wave potentials for reduction or oxidation and a diffusion coefficient appropriate to the media, the molecular identity has to be inferred from the measured physical properties of standard systems. In more complex (multilayer) or natural (biochemical or environmental) systems, these properties may not always be resolvable. The ability, therefore, to utilize additional, perhaps more specific, physical characteristics of molecules to monitor electrode processes, in either dynamic or equilibrium conditions, would be immensely valuable. In the past several decades in particular, there have been considerable efforts expended to develop spectroelectrochemical techniques to aid electrochemical research. Molecular properties such as molar absorptivities, vibrational absorption frequencies, and electronic or magnetic resonance frequencies, in addition to the traditional electrical parameters, now are being used routinely to better our understanding of electron transfer reaction pathways and the fundamental molecular states at interfaces.

Since bulk properties, of either solid or liquid phases, clearly are not always appropriate to describe these reactions and molecular states at electrode/electrolyte phase boundaries, it is evident that these unique systems require innovative spectroscopic techniques and theoretical advancements to enable us to understand fully the origins of their spectra. Development of *in situ* techniques is especially important because crystalline electrode surfaces can restructure when removed from an electrolyte to *vacuo*, e.g., Ref. 1, or by

Robert J. Gale • Chemistry Department, Louisiana State University, Baton Rouge, Louisiana 70803.

internal thermodynamic rearrangements due to impurities or adsorbates, e.g., Refs. 2 and 3. While the joint application of electrochemical and spectrographic techniques is admittedly more involved than either individual method of measurement, the complications are worthwhile because the outcome provides advantages to each branch of chemistry; i.e., electrochemical science benefits from a more complete description of the molecular aspects of charge transfer mechanisms and surface equilibria, whereas the field of spectroscopy gains by reaping new information concerning molecules that comprise interfaces and their vicinities. Such combinations of electrochemistry and spectroscopy, in fact, have proved to be a powerful approach to exploring redox and interfacial phenomena.

2. Methodologies Available

A wide variety of spectroelectrochemical methods is available and developments continue apace. Figure 1 contains a classification of methods by spectroscopic mode. Some techniques involve one or more modes of absorption, luminescence, or emission, e.g., X-ray or Mössbauer, so a rigorous and comprehensive classification has not been attempted. Mass spectrometry

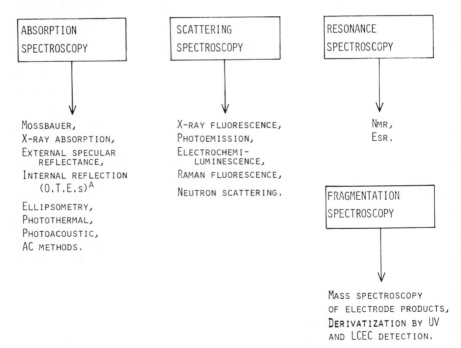

FIGURE 1. Range of spectroscopic techniques applied to the study of electrode processes. (^Optically transparent electrodes.)

is included, too, even though it is not easily categorized in terms of the energy of radiation. The choice of a technique depends on the particular problem, be it related to biochemistry, corrosion processes, battery or fuel cell performance, etc. The nature of the electroactive species (neutral molecule, ion, radical, etc.) and the substrate electrode's homogeneities and morphologies may dictate the selection or restriction of methodologies. General references to spectroelectrochemical techniques are provided in recent standard electrochemical textbooks, e.g., Refs. 4–6, journal articles, e.g., Refs. 7 and 8, and literature reviews, e.g., Refs. 9–11. In addition, various spectroelectrochemical topics have been given in-depth converage in the electrochemistry series edited by Bard, e.g., Ref. 12, and Bockris and coworkers, e.g. Ref. 13.

Seven chapters, written by specialists, follow in this book. In Chapter 2, Robinson outlines the uses of X-ray scattering and absorption techniques in electrochemistry. Powerful synchrotron radiation sources are finding new applications to the study of electrode surfaces in this emerging field.[14] In Chapter 3, Severeyn and Gale have reviewed the theoretical background of metal and semiconductor photoemission. Many photoemission theories have been formulated, some still controversial and lacking complete physical understanding. Further experimental and theoretical advances are needed in this area, in particular to clarify the roles of electron–hole recombination and the electrostatic screening influences on electrons photoemitted from semiconductor electrodes.[15] Kolb expertly discusses current knowledge in UV–visible reflectance spectroscopy in Chapter 4. Fundamental progress is impressive in this field, despite the complexity of the systems which have to be modeled. Even the simplifications of single-crystal electrodes and polarized light do not eliminate all of the difficulties. Here, both classical and modern quantum physics overlap with phenomenological observations to unnerve all but the most tenacious of researchers. Specular reflectance refers to the macroscopic optical reflections from "mirrorlike" electrode surfaces, and, of course, many practical electrodes permit only diffuse reflections to occur. However, this field remains at the forefront of those needed for the serious student of spectroelectrochemistry. Impressive advances have occurred also in the description of the vibrational properties of adsorbed molecules at electrodes. Infrared reflectance spectroscopy is ably reviewed and its practice detailed by Beden and Lamy in Chapter 5. Further evidence of the progress in this latter spectroscopic area is provided by the development of surface-enhanced Raman scattering (SERS) at electrodes. The in-depth contribution of Birke and Lombardi (Chapter 6) underscores the advances that have been achieved in understanding SERS over a surprisingly short time. Radical chemistry studies have diverse interest, especially in the life sciences, and Compton and Waller have outlined in Chapter 7 the best approaches for ESR spectroscopy of electrode products. Electrochemistry provides a controlled source of radicals for kinetic studies. One can anticipate that combined optical/electrochemical methods of radical generation at semiconductor or novel organic polymer electrodes may be the focus of future work in this area. Finally, the field of Mössbauer spectroscopy is discussed by Scherson in Chapter 8. This application is limited to selected nuclei, but the diversity of problems associated with the electrochemistry of these elements has barely been touched upon. We may certainly expect, therefore, many more applications for Mössbauer spectroscopy to practical problems in the future.

Not every spectroelectrochemical technique is provided detailed coverage in this text and not all of the topics that are provided could be covered from every viewpoint. Two techniques that should be singled out because of their omission are nuclear magnetic resonance spectroscopy and mass spectrometry of electrode products (NMR studies of electrolyte structure are excluded because they are not an *in situ* electrode technique). Unlike ESR, the former suffers from a low sensitivity, long periods can be needed to acquire spectra, and the degradation of the magnetic field homogeneity can diminish further the sensitivity and resolution; however, electrogenerated species have been detected by NMR in certain configurations, e.g., Refs. 16 and 17. Mass spectrometry has required considerable ingenuity in cell design but the technique, which has been pioneered by Bruckenstein and coworkers,[18,19] is viable using porous Teflon plug electrodes for volatile products.

3. Computer-Based Data Processing

Development of spectroscopic techniques to elucidate electrochemical phenomena has necessitated considerable innovation in cell designs, in modes of excitation, and in response detection and interpretation. The number of molecules reacting at a heterogeneous interface is usually orders of magnitude below that for homogeneous reactions. This latter constraint has necessitated the widespread adoption of modulation and transform techniques for data enhancement and processing. Utilization of on-line computer data collection increases significantly the quantity of data available, improves its quality, and permits direct statistical analyses or data manipulation by transforms. Rediscovery of the Fast Fourier Transform (FFT) by Cooley and Tukey in 1965[20,21] has meant that extremely rapid software and/or hardware computations of powerful mathematical integrals are feasible. The wide applicability of the FFT algorithm to physical problems perhaps is not as generally appreciated as it might be. For example, the Kramers–Kronig relation (mathematically the Hilbert transform) between absorption $A(\omega')$ and dispersion $D(\omega)$ spectra,

$$D(\omega) = -\frac{1}{\pi} \int_{-\infty}^{+\infty} \frac{A(\omega')}{\omega - \omega'} \, d\omega'$$

can be implemented efficiently using the much more rapid FFT algorithm.[22] Discrete convolution and correlation may be achieved using the convolution theorem,

$$h(t) * x(t) \Longleftrightarrow H(f)X(f)$$

where the convolution integral is given by

$$y(t) = \int_{-\infty}^{+\infty} x(\tau)h(t - \tau) \, d\tau \text{ or } h(t) * x(t)$$

and $h(t)$ has the Fourier transform $H(f)$ and $x(t)$ has the Fourier transform $X(f)$; or, using the similar correlation theorem,

$$\int_{-\infty}^{+\infty} h(\tau)x(t + \tau) \, d\tau \Longleftrightarrow H(f)X^*(f)$$

where $X^*(f)$ is the conjugate. If $x(t)$ is the same function as $h(t)$, the term autocorrelation is used, whereas if $x(t)$ and $h(t)$ are different functions, the term cross-correlation is used. Thus, high-speed convolution and correlation can be achieved efficiently and complex integrals can be evaluated using an inexpensive microcomputer.[23] In fact, the use of AC techniques (termed admittance or, conversely, impedance spectroscopy) to study electrochemical phenomena, e.g. corrosion processes,[24,25] may be included as a spectroelectrochemical technique insomuch as a sinusoidal electrical perturbation represents an electromagnetic wave excitation. Hence, using time-to-frequency domain methods to analyze transient responses, the definition of what constitutes a spectroelectrochemical technique becomes fuzzy (the prefix spectrorefers to responses to radiation when the parts are arranged according to wavelength). Carney and Gale[26] and Gabrielli and coworkers[27] have shown that, contrary to earlier thoughts,[28] equally good precision can be obtained from Fourier analyses of pulse transient (PT) data as from pseudorandom white noise (PRWN), if the signals are appropriately weighted for their mean amplitudes in Fourier space. To be explicit, for a given input power (in time), PRWN gives a flatter precision versus frequency curve (ignoring nonideality

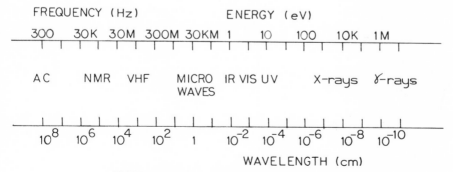

FIGURE 2. The spectroelectrochemical spectrum.

effects), while PT gives better precision at low frequencies than at high frequencies. The signal degradation at high frequencies with pulses is due simply to the fact that less power is available from the sinc function at high frequencies, and the smaller signal amplitudes lead to lower precision. The significance for spectroelectrochemistry using optical excitations is that it is not always easy experimentally to produce complex waveforms such as PRWN. Figure 2 shows that a broad energy range of the electromagnetic radiation spectrum is accessible and has been utilized to probe electrode reactions and interfaces.

4. The Future

The exploration of electrode interphasial regions by electromagnetic radiation is an ongoing adventure; experimentalists in the spectroelectrochemical field increasingly are able to describe the molecular properties and mechanisms within two-dimensional atomic geometries. With the development of these new techniques, novel theories have been required to explain and model the often unexpected complexity of many of the data. The terms "interface" and "interphase" often are used interchangeably, with some electrochemists preferring the latter because they feel it conveys a better sense of the uniqueness and dimensionality of the electrode surface regions.

Conceptually at least, other interfaces exist. In the United States, electrochemists wear either of two hats—that of physical or analytical chemist. Analytical chemists originally made routine measurements and their research focused on applying methods developed by physical chemists. More recently, the "electroanalytical" chemist has been involved with providing new methodologies and mathematical treatments to conduct superior measurements at electrodes—qualitative and quantitative. The goals, by and large, are to solve applied problems but the results assist in adding to fundamental knowledge of physical interfacial equilibria and electron transfer processes. New means are available to make measurements and to conduct experiments. The study of semiconductor electrodes—in the mainstay, applied research—has expanded enormously in the past decade among not only the electroanalytical fraternity but among physical, inorganic, and organic chemists as well as solid-state physicists and materials scientists. This research has heightened interest in the internal (surface) electronic structure and conduction and recombination mechanisms as well as the mechanisms of interactions of optical radiation with the electrode. Novel conducting polymers and inorganic superconductors require electrical and optical analyses and may provide new electrode materials. Many materials have to be studied in nonaqueous or molten salt media. Spectroelectrochemistry in molten salts, which experimentally requires elaborate cell and instrument design, has been reviewed by Norvell and Mamantov.[29]

Fortunately, the rewards of research are complementary, and whatever direction or thrust a particular spectroelectrochemical project may take, several fields of science ultimately may benefit. The interchange of knowledge and ideas in electrochemistry and surface science at the analytical–physical interface, for example, is notable. Such progress offers a better understanding of electrochemical processes through theoretical advances and experimental discovery or validation, by both pure and applied motivation. These new electroanalytical techniques permit reevaluation of important practical electrochemical systems such as corrosion mechanisms, biochemical redox intermediates, kinetic and catalytic processes of analytical and chemicals production importance, and optical devices. Spectroelectrochemistry continues to be an exciting and challenging field in which to work.

References

1. E. Yeager, in: *Non-Traditional Approaches to the Study of the Solid–Electrolyte Interface* (T. E. Furtak, K. L. Kliewer, and D. W. Lynch, eds.), p. 1, North-Holland Publishing Co., Amsterdam (1980).
2. R. Parsons, in: *Electronic and Molecular Structures of Electrode–Electrolyte Interfaces* (W. N. Hansen, D. M. Kolb, and D. W. Lynch, eds.), p. 51, Elsevier, Amsterdam (1983).
3. E. Schmidt and H. Siegenthaler, in: *Electronic and Molecular Structure of Electrode-Electrolyte Interfaces* (W. N. Hansen, D. M. Kolb, and D. W. Lynch, eds.), p. 59, Elsevier, Amsterdam (1983).
4. A. J. Bard and L. R. Faulkner, *Electrochemical Methods,* John Wiley and Sons, New York (1980), p. 577.
5. P. T. Kissinger and W. R. Heineman (eds.), *Laboratory Techniques in Electroanalytical Chemistry,* Marcel Dekker, New York (1984), Chapters 12, 19, 23, and 24.
6. Southampton Electrochemistry Group, *Instrumental Methods in Electrochemistry,* Ellis Horwood Ltd., Chichester (1985), Chapter 10.
7. W. R. Heinemann, *Anal. Chem.* **50,** 390A (1978).
8. J. K. Foley and S. Pons, *Anal. Chem.* **57,** 945A (1985).
9. M. D. Ryan and G. S. Wilson, *Anal. Chem.* **54,** 24R (1982).
10. D. C. Johnson, M. D. Ryan, and G. S. Wilson, *Anal. Chem.* **56,** 14R (1984); **58,** 42R (1986).
11. J. Robinson, *Electrochemistry* (Specialist Periodical Report) (D. Pletcher, ed.), Vol. 9, p. 101, The Royal Society of Chemistry, London (1984).
12. W. R. Heineman, F. M. Hawkridge, and H. N. Blount, in: *Electroanalytical Chemistry* (A. J. Bard, ed.), Vol. 13, p. 1, Marcel Dekker, New York (1984).
13. R. E. White, J. O.'M. Bockris, B. E. Conway, and E. Yeager (eds.), *Comprehensive Treatise of Electrochemistry,* Vol. 8, Plenum Press, New York (1984).
14. News Review, *Chemistry in Britain* **23,** 102 (1987).
15. R. B. Severeyn and R. J. Gale, *J. Phys. Chem.* **90,** 4187 (1986).
16. J. A. Richards and D. H. Evans, *Anal. Chem.* **47,** 964 (1975).
17. D. W. Mincey, Ph.D. thesis, Univ. of Cincinnati (1978).
18. S. Bruckenstein and R. Rao Gadde, *J. Am. Chem. Soc.* **93,** 793 (1971).
19. S. Bruckenstein and J. Comeau, *Faraday Discuss. Chem. Soc.* **56,** 285 (1974).
20. N. Metropolis, J. Howlett, and Gian-Carlo Rota, *A History of Computing in the Twentieth Century,* Academic Press, Orlando (1980), p. 4.

21. E. Oran Brigham, *The Fast Fourier Transform*, Prentice-Hall, Englewood Cliffs, New Jersey (1974), p. 8.

22. A. G. Marshall (ed.), *Fourier, Hadamard, and Hilbert Transforms in Chemistry*, Plenum Press, New York (1982), p. 108.

23. Ref. 21, Chapters. 4, 7, and 13.

24. C. Gabrielli, F. Huet, M. Keddam, and H. Takenouti, in: *Proc. Symp. on Computer Aided Acquisition and Analysis of Corrosion Data* (M. W. Kendig, U. Bertocci, and J. E. Strutt, eds.), p. 1, The Electrochemical Society, Inc., Pennington, New Jersey (1985).

25. D. D. MacDonald and M. Urquidi-MacDonald, *J. Electrochem. Soc.* **132**, 2316 (1985).

26. K. R. Carney and R. J. Gale, Electrochemical Society 166th Meeting, New Orleans, Louisiana (Oct. 10, 1984), Abstract No. 426.

27. C. Gabrielli, M. Keddam, and J. F. Lizee, *J. Electroanal. Chem.* **205**, 59 (1986).

28. Ref. 22, p. 475.

29. V. E. Norvell and G. Mamantov, in: *Molten Salt Techniques* (D. G. Lovering and R. J. Gale, eds.), Vol. 1, p. 151, New York (1983).

X-Ray Techniques

James Robinson

1. Historical Background

The development of spectroelectrochemical techniques has been a major field of research for electrochemists since the late 1960s.[1] As can be seen from the other chapters in this book, there are now many such techniques being routinely applied in electrochemical laboratories around the world, while further developments are appearing all the time. One of the most significant successes of spectroelectrochemistry has been the elucidation of structure at the electrode/electrolyte solution interface. In particular, it is possible to investigate the electronic structure of the electrode surface, and of species near it, using UV–visible light as a probe, while vibrational spectroscopy can be used to study the molecular structure and local environment. Unfortunately, these optical techniques do not provide any direct information about either long- or short-range order parameters. Such information would clearly be very valuable both from a fundamental point of view as well as from a technological one. For example, it would be very interesting in metal deposition studies to be able to monitor the evolution of structure as nuclei are formed and as they subsequently grow into thick deposits. Similarly, the battery scientist would welcome the opportunity to follow structural changes occurring during the charge and discharge cycles of a battery system. While the electrochemist's need for a technique, or techniques, capable of providing this type of information is readily apparent, it is not immediately clear what is the best approach to follow when attempting to develop one.

1.1. Ultrahigh Vacuum Techniques

In recent years surface scientists have made considerable advances in their knowledge of clean solid surfaces (particularly of metals and

James Robinson • Department of Physics, University of Warwick, Coventry CV4 7AL, United Kingdom.

semiconductors) and of adsorbate-covered ones, through the application of techniques such as low-energy electron diffraction (LEED),[2] X-ray photoelectron spectroscopy (XPS),[3] surface extended X-ray absorption fine structure (SEXAFS),[4] and, most recently, scanning tunneling microscopy (STM).[5] These techniques all either use electron probes or involve the detection of emitted electrons and therefore necessitate the use of at least high vacuum (HV), and generally ultrahigh vacuum (UHV), conditions. Clearly, they cannot be used to study the electrode/electrolyte solution interface *in situ* (STM may be possible in a modified form). However, in view of their obvious power, a number of electrochemists have investigated the possibility of using them to study electrochemical systems *ex situ*.[6] In particular, much effort has been expended on trying to ascertain to what extent the surface studied under UHV conditions corresponds to that existing in the electrochemical cell. Certainly, some restructuring will occur on transfer from the electrochemical environment to the UHV chamber (at the very least, the bulk of the water will be lost), but efforts have been made to develop transfer techniques that minimize these difficulties.

One approach that can be used with any electrode material, with the exception of high-vapor-pressure ones such as Hg, is to perform the electrochemical experiments in a side chamber to the main UHV system and to use a totally volatile solvent/electrolyte system (HF/H_2O is commonly used).[7] By using a simple two-electrode cell with only a small solution volume (one drop on the working electrode), it is possible to rapidly pump off the electrolyte after performing an electrochemical experiment and to transfer the electrode to UHV in less than five minutes. The restrictions placed on the solvent/electrolyte combinations that can be used do limit, however, the range of systems that can be studied in this way. For metal electrodes with low heats of melting, e.g., gold, an alternative approach has been proposed that places no such restrictions on the solvent system. It is claimed[8] that if the electrode is removed from the cell while polarized, the double layer is removed intact. The evidence to support this claim has been obtained from a variety of investigations including reflectance spectroscopic studies[9] and ESCA studies of the ions present on the electrode after emersion.[10] Recently, however, on the basis of photoemission studies in the presence of the water left on the electrode after emersion, it has been suggested[11] that the emersed electrode under UHV conditions is not after all a good model for the immersed electrode, at least with regard to the water in the double-layer region, as the nature of this water changes on transfer. In this respect, of course, the emersed electrode is no different from electrodes that have been transferred to UHV conditions using other techniques.

Despite the problems involved in transfer, a number of interesting observations have been made using these UHV techniques. Investigations have been concentrated largely on studies of strongly adsorbed species, in particular, of underpotential metal deposits, since these are the systems least likely to be affected by transfer. In conventional LEED measurements with solid surfaces, the standard procedure is to clean/renew the surface using argon-ion bombardment until sharp diffraction spots are obtained. Without this pretreatment, results are irreproducible. Clearly, this cannot be applied to an electrode after removal from the cell as the electrochemically interesting layer will be lost. Thus, in these investigations, it is usual to apply this treatment prior to immersion, i.e., the electrode is cleaned in a UHV chamber, transferred to the cell where it is polarized, and then transferred back to the UHV chamber for study. Interestingly, the electrochemical behavior of a surface prepared in this way is frequently quite different from that of a conventionally prepared electrode, though it usually slowly reverts to conventional behavior after repeated potential cycling[7] (presumably this cycling facilitates surface restructuring). Provided that this restructuring is avoided, satisfactory LEED patterns can be obtained when the electrode is transferred back to the UHV system[12] (for example, of Cu underpotential layers on Au), and the structure of these surface layers can be studied.

The difficulties encountered in having to use Ar-ion bombardment can be avoided by using reflectance high-energy electron diffraction (RHEED) instead of LEED, with the added advantage that only HV conditions are required. Here, the surface sensitivity is obtained through using a glancing angle of incidence while the beam is sufficiently energetic to penetrate surface impurities. Studies of Cu monolayers on Au using RHEED reach essentially the same conclusions as LEED studies.[13]

In the last few years, a number of studies of electrode systems have been made using these vacuum chamber techniques, and undoubtedly some progress has been made in our understanding of interfacial structure. The fact remains, however, that most of the region of interest is lost on transfer to the vacuum chamber. It is also still uncertain whether or not even strongly adsorbed species, which survive the transfer, restructure. It is therefore highly desirable that *in situ* techniques capable of providing similar information be developed. The only probes likely to have this capability are X-rays and neutrons, and it is techniques based on these that will be discussed in this chapter. The bulk of the discussion will be related to the development of surface-sensitive X-ray diffraction systems, but the application of other techniques such as neutron scattering and extended X-ray absorption fine structure to electrochemical systems will also be considered. Whereas the techniques discussed elsewhere in this book have now become quite widely applied, those discussed in this chapter are still very much in the development stage, and this will be reflected in this discussion. In particular, it is hoped to identify possible future developments in this field.

1.2. X-Ray Techniques for Surface Study

X-ray diffraction is, of course, the most widely used technique for the elucidation of the structure of crystalline solids. Since the X-rays are scattered by all atoms in their path, they are not inherently surface sensitive. Therefore, until recently, X-ray-based techniques have not been widely used in the study of surfaces. However, in parallel with the further development of techniques based on either electron emission or scattering, surface scientists have recently turned their attention to the possible use of X-ray-based methods. This new area of study has arisen largely because of the construction around the world of a number of electron storage rings dedicated to the production of synchrotron radiation.[14] This radiation is tunable over a wide energy range from hard X-rays to the infrared, and is also very intense. It is these properties that have caused surface scientists to consider possible applications.

1.2.1. Scattering Methods

We saw earlier that high-energy electron diffraction could be made surface sensitive if a total reflection configuration was used. It had been realized for many years that the same ought also to be true of X-ray diffraction, but since the critical angle for X-rays at the air/metal interface is typically less than 1°, it was felt that the experiment was too difficult to perform. Marra *et al.*[15] were the first to show that with careful cell design, this experiment was indeed possible, even when using a rotating anode source. The main requirement for the experiment, which is known as total reflection Bragg diffraction (TRBD), is a highly collimated intense X-ray beam, and this is most readily achieved with synchrotron radiation. The penetration depth of the X-ray beam is typically in the range of 100 to 1000 Å, and since this beam couples with all reciprocal lattice vectors in the scattering plane, the study of long-range order in the interfacial region is possible. Applications to date appear to have been restricted largely to multilayer and similar systems, under vacuum. In theory, there is nothing to prevent this technique being applied to electrochemical systems, but this has not yet been attempted. One of the main difficulties is that the glancing angle of incidence will generally lead to long solution pathlengths and the consequent absorption of both incident and scattered X-rays. Careful cell design will reduce this problem, but, as we will see later, it is often very difficult to reconcile the requirements of an X-ray experiment (short pathlengths in solution) with a good electrochemical configuration. The use of hard radiation may also help to reduce absorption but may lead to electrode damage. It is very likely that TRBD will be applied to electrochemical systems in due course, but since this has not yet happened, the technique will not be considered further here.

At much the same time that TRBD was developed, another surface-sensitive X-ray technique was also reported. This relied on the well-known fact that the dynamical diffraction of an X-ray plane wave at a perfect crystal leads to the formation of an interference field on either side of the surface. As the angle of incidence is scanned through the Bragg angle, so this interference field moves. By monitoring the X-ray fluorescent yield as the beam is scanned, atomic positions relative to the diffraction plane can be determined (the fluorescent yield for any atom will peak when the interference field has an antinode at the atomic position).[16] This technique has been used to study a number of surface films and, most recently, an electrochemical system, namely, thallium underpotential deposition on Cu(III).[17] This was an *ex situ* study, but, in principle, *in situ* investigations should be possible. The major drawback with this technique is that the crystal has to be of very high quality in order to set up the standing waves, and therefore it will never be possible to study systems of technological importance.

The surface sensitivity of both TRBD and the standing-wave technique has been achieved by the use of special experimental configurations and high-flux sources. The final X-ray diffraction technique to be mentioned here, and the one that will be considered at greater length in the rest of this chapter, is in many senses much simpler and is not dependent on synchrotron radiation (though the use of such a source might be beneficial in reducing the duration of experiments). Briefly, this technique, which was developed by Fleischmann and coworkers,[18-21] utilizes a simple laboratory X-ray source and conventional geometry. The surface sensitivity is achieved through the use of position-sensitive X-ray detection in conjunction with electrode potential modulation. Thus, it is changes in the X-ray diffractogram that are recorded as the potential is modulated and scattering from bulk materials is subtracted out. The approach, therefore, has much in common with that used in potential-modulated reflectance methods. The details of this technique are discussed later.

1.2.2. Absorption Techniques

While X-ray diffraction is well suited to the study of the structure of crystalline samples, it is less well suited to amorphous or disordered ones. In this area, the availability of synchrotron radiation has again led to the development of new techniques, in particular, extended X-ray absorption fine structure (EXAFS).[22] In EXAFS, structural information is derived from an analysis of the oscillations that appear above an X-ray absorption edge, and this technique is now widely applied. The conventional EXAFS experiment is not surface sensitive but some variants are. Under HV conditions, the electron yield is a measure of the absorption, and since these electrons can only escape from a thin layer at the surface, this detection method is surface sensitive. This is the basis of SEXAFS.[4] Clearly, this cannot be used for *in situ* studies, but in these cases the fluorescent X-ray yield (which is also a measure of the absorption) can be measured and, particularly when this is coupled with a glancing incidence geometry, reasonable surface sensitivity is achieved. Some initial studies of this type on electrochemical systems have now been made, and these will be described in greater detail later.

1.3. Neutron Scattering

It should, of course, be possible to perform elastic neutron scattering experiments in a very similar way to X-ray ones; indeed, when studying low-atomic-number samples this may be advantageous. However, the difficulty in getting regular access to neutron sources has meant that little progress has been made in this direction with electrochemical systems. The only such study appears to be that of Bomchil and Rekel[23] on nickel oxide formation. Since the availability of sources is unlikely to improve significantly, it is unlikely that this will ever become a widely applied technique.

2. Theory—The Interaction of X-Rays with Matter

When an X-ray beam falls on an atom, it may either pass through unaffected or interact with the electrons. This interaction takes one of two forms; the X-ray photon is absorbed, with electrons and photons being emitted by the atom, or alternatively the beam is scattered. Scattering cross sections are generally rather small, while the absorption cross section is very dependent on the X-ray wavelength.

Classical theory provides a simple physical description of the scattering process. However, electrons do not scatter X-rays exactly according to this simple approach, and a full quantum mechanical description shows that in addition to the elastic scattering predicted by classical theory, a small amount of inelastic scattering (Compton effect) also occurs. Fortunately, this inelastic scattering can be easily differentiated from the elastic scattering since it is incoherent and cannot therefore give rise to interference. It is consequently present only as a diffuse background and, anyway, for most materials is of low amplitude. It is unlikely that inelastic scattering will affect the X-ray diffraction experiments to be described here, other than perhaps to increase the background and thus reduce the signal-to-noise ratio. The simple classical description of X-ray scattering will therefore be adequate here.

The remainder of this section is concerned firstly with a brief outline of the classical description of X-ray scattering by electrons and of how this scattering leads to the production of X-ray diffraction patterns. The second part presents a description of X-ray absorption and in particular of absorption fine structure.

2.1. X-Ray Scattering

Let us consider an X-ray beam propagating along the x-axis and encountering an electron at the experimental origin, O. If this electron has a velocity small compared with the velocity of light, c, it will be accelerated by interaction with the oscillating electric field of the X-ray beam and will become a source of radiation at the same frequency as the X-rays. This effectively gives rise to a scattered wave. At a point P a distance R from the origin and at an angle θ with respect to the incident beam, the intensity of this scattered wave, I_e, is given by the Thomson equation,

$$I_e = \left(\frac{\mu_0}{4\pi}\right)^2 I_0 \frac{e^4}{R^2 m^2 c^4} \left(1 + \frac{\cos^2 \theta}{2}\right) \tag{1}$$

When considering scattering by a group of electrons in an atom, it is useful to introduce the concept of the electron scattering factor, f_e, given by

$$f_e = \int_0^\infty 4\pi r^2 \rho(r) \left(\frac{\sin(4\pi r \sin \theta / \lambda)}{(4\pi r \sin \theta / \lambda)}\right) dr \tag{2}$$

where $\rho(r)$ is the electron density, assuming spherical symmetry. The scattering factor for an atom, f_i, is then given by the sum of all the electron scattering factors, and therefore

$$f_i = \sum_n \int_0^\infty 4\pi r^2 \rho_n(r) \left(\frac{\sin(4\pi r \sin \theta / \lambda)}{(4\pi r \sin \theta / \lambda)}\right) dr \tag{3}$$

It follows from Eq. (3) that for small values of $\sin \theta / \lambda$

$$f_i \to \sum_0^\infty 4\pi r^2 \rho_n(r) \, dr \tag{4}$$

which is equal to z, the number of electrons in the atom. Accurate tables of f_i and f_e as a function of $\sin \theta / \lambda$ are available in the literature.[24]

Here, we are not interested in scattering by isolated atoms but rather by bulk and, in particular, crystalline materials. A suitable starting point for a discussion of this is the familiar Bragg equation:

$$n\lambda = 2d \sin \theta \tag{5}$$

In his early work, Bragg showed that the angular distribution of the scattered radiation could be understood by considering that the diffracted beams behaved as if they were reflected from planes in the crystal lattice. For this reason, diffraction peaks are often called Bragg reflections. The Bragg law simply states that constructive interference will only occur when the pathlengths of reflected wavefronts are out of phase by a whole number of wavelengths. This law successfully predicts scattering angles but tells us nothing about scattering intensities. This information can be determined by considering the scattering factors of all the atoms in the unit cell. It can be shown that the scattered amplitude of a wave from the (hkl) plane of the three-dimensional lattice, F_{hkl} (the structure factor), is given by

$$F_{hkl} = \sum_i f_i \exp[2\pi i(hx_i + ky_i + lz_i)] \tag{6}$$

unit cell

and the intensity of any diffraction spot, or peak, is proportional to the square of this factor.

A detailed discussion of the shape of the diffraction peak is beyond the scope of this chapter; however, it is worth remembering that with high-quality data, useful information about the degree of order can be obtained from this. There is one feature that should be mentioned and that is the different lineshapes for two- and three-dimensional lattices. While diffraction peaks for three-dimensional lattices have the familiar Lorentzian shape, this is not true of peaks for two-dimensional systems. For these, Warren[25] has shown that the peak shape is that shown in Figure 1. This is therefore a good test for scattering from a two-dimensional lattice. Further information is obtained from the widths of the peak at half-height, as this is determined by the extent of the two-dimensional lattice and therefore provides particle, or domain, size information.

Finally, before leaving this discussion of scattering, we must remember that X-rays also will be scattered by amorphous and liquid phases, the most important of which in the present context is the electrolyte solution. In these cases, the diffraction pattern will not consist of sharp lines or spots but will instead resemble a radial distribution function with broad peaks. Even when the scattering from the solution is not directly of interest, it must be borne in mind that this will be a significant contribution to the total scattering and will often be the factor that limits the sensitivity of the technique.

This discussion of X-ray scattering has been very brief, but further information is readily available in any book on X-ray crystallography. The application of X-ray scattering techniques to electrochemical systems introduces no novel theoretical concepts, and as we will see later, it is only in the experimental design that new approaches are used.

2.2. X-Ray Absorption

As was stated earlier, while an X-ray photon may be scattered when interacting with matter, a more likely fate is that it will be absorbed. The relationship between the transmitted and incident X-ray intensities (I and I_0)

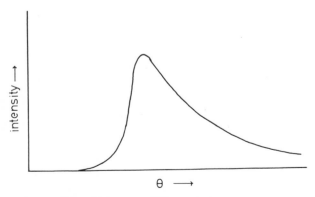

FIGURE 1. A schematic representation of the lineshape for scattering from a two-dimensional lattice of a powder sample.

is given by Lambert's law

$$I = I_0 \exp\left(-\mu_m m\right) \tag{7}$$

where μ_m is the mass absorption coefficient and m is the mass per unit area. For any element, the value of μ_m increases with increase of wavelength, but there is no simple relationship to predict the form of this dependence. This is because of the existence of sharp absorption edges, which arise when the X-ray energy is just sufficient to eject a core level electron. The shape of these absorption edges is shown in Figure 2, and it can be seen that with increasing energy, the absorption coefficient increases sharply and then falls in an oscillatory manner. The position of the absorption edge is characteristic of the absorbing atom, while fine structure on the absorption edge and the precise position of the edge can provide additional information. For example, the edge energy for a cationic absorber shifts to higher energies with increasing valence of the cation (the chemical shift). A full discussion of these properties is inappropriate here; the topic is thoroughly dealt with in a book by Agarwal.[26] Our interest is with the oscillations above the edge, known as extended X-ray absorption fine structure (EXAFS). This does not mean however that X-ray absorption edge data may not be useful in electrochemical studies, particularly as the data can be obtained at the same time as the EXAFS.

The EXAFS oscillations extend from the absorption edge to energies of between 500 and 1000 eV above this point, and arise because of interference effects causing oscillations in the absorption cross section of the absorbing atom. Briefly, the explanation for this is as follows; if we consider an atom surrounded by several others as shown in Figure 3, then if the central atom

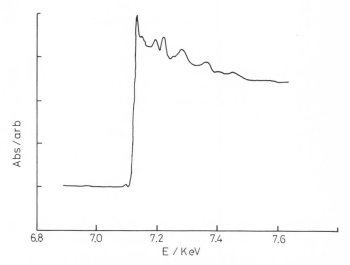

FIGURE 2. The X-ray absorption spectrum for a sample of iron.

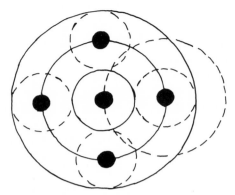

FIGURE 3. A representation of the interference between outgoing and backscattered photoelectron waves. The solid curves represent the outgoing waves and the dashed curves the backscattered ones.

absorbs an X-ray photon, a photoelectron is emitted. This photoelectron wave (represented by the solid curves) is then backscattered by the surrounding atoms and the backscattered wave (dashed curves) interferes with the outgoing wave. This interference, which can be either destructive or constructive, affects the absorption cross section of the central atom and thus gives rise to the oscillations above the absorption edge. When the waves interfere constructively, there is an absorption maximum, and when they interfere destructively there is a minimum.

It can be shown[27] that the normalized oscillatory part of the absorption rate in momentum space, $\chi(k)$, is given by

$$\chi(k) = \frac{m}{4\pi h^2 k} \sum_j \frac{N_j}{r_j^2} S_j(k) \sin\left[2kr_j + 2\delta_j(k)\right] e^{-2r_j/\lambda} e^{-2k^2\delta_j^2} \qquad (8)$$

where $S_j(k)$ is the backscattering amplitude for each of the N_j neighboring atoms in the jth shell, λ is the mean free path of the electron, $\exp(-2k^2\delta^2)$ is a Debye–Waller factor to account for thermal vibrations, and k is determined from the kinetic energy of the excited photoelectron, i.e., the energy above the absorption edge. Thus $\chi(k)$ is a superposition of terms of the form $\sin(2kr_j)$ where r_j is the distance to the neighbors around the central atom. The Fourier transform given in Eq. (9) should therefore peak at the shell distances $r = r_j$ and thus provide information about the structure of the material:

$$F(r) = \frac{1}{2\pi} \int e^{-2kr} \chi(k) \, dk \qquad (9)$$

This is the basis of the standard method of analyzing EXAFS data. The EXAFS oscillations are extracted from the absorption data and they are then transformed (possibly after weighting the data). Unfortunately, the peaks in the transform are not exactly at the positions of the various shells but are shifted due to the unknown phase shift $\delta(k)$. This is usually determined by using EXAFS data from model compounds (containing the same atoms) for which structural data have been obtained by other methods. Recently, more detailed data analysis techniques which rely on complex computational algorithms have been developed.[28,29] Using the EXAFS technique, excellent structural information can be obtained, and the major application (as far as solids are concerned) has been for the study of amorphous systems where X-ray diffraction is not particularly well suited. Many electrochemical systems are of this type and therefore the technique has much promise in this area.

3. Experimental Details

Probably the main difficulties associated with *in situ* X-ray techniques are reconciling the cell requirements of the electrochemistry with a suitable configuration for the X-ray technique and detecting the inevitably small signals. This is true of both diffraction and absorption measurements. This section will be concerned with how these problems are overcome and how the relevant experiments are performed.

3.1. In Situ X-Ray Diffraction

3.1.1. X-Ray Detection Methods

Conventional X-ray techniques generally use one of two detection methods, photographic film or scanning proportional counters. The former has the advantage of being an area detector but has poor contrast, while the latter has excellent contrast but is slow in recording a diffraction pattern over a wide angular range. In the electrochemical experiment, the phase of interest will probably be fairly thin and therefore will scatter X-rays only weakly, whereas there will also be a number of other materials present that will also scatter (e.g., the solution). The phase of interest may therefore only contribute a small percentage of the total scattered photons. It is important, therefore, to use a detection system with good contrast which at the same time makes optimum use of all the photons. This requirement is satisfied by using a position-sensitive detector.

Position-sensitive X-ray detection systems, which were originally introduced for imaging purposes, particularly in medical applications, have been used in X-ray crystallography for about 20 years. Their virtue is that they count all the electrons, all the time, either along a line or over an area.

There are three principal types of position-sensitive X-ray detectors of use for crystallography—the vidicon, solid-state devices (largely charge-coupled devices), and proportional counters. Vidicons are area detectors which use image intensifier techniques to produce TV-type images. They are not count rate limited and are therefore well suited for use with high-flux sources such as storage rings,[30] but their small size limits their suitability for the present problem. Charge-coupled devices are also very small but have exceptionally good spatial resolution and are therefore ideal for lineshape studies. By far the most suitable detectors for the present work, however, are position-sensitive proportional counters (PSPC), either linear or multiwire area types.

As with most proportional counters, the linear PSPC consists of a central anode wire between one or two cathodes. The space between the anode and cathode is filled with a conventional counter gas consisting of an ionizable gas, frequently Ar or Xe, and a quencher, e.g., CH_4, typically in the ratio 9:1 and often at pressures of a few atmospheres. A large voltage (~ 1.5 kV) is maintained between the anode and cathode. When an X-ray photon enters the counter, there is a finite probability that it will ionize a gas molecule and the emitted electron then accelerates towards the anode under the influence of the field. This electron may then collide with further gas atoms, resulting in the formation of an electron avalanche near the central wire and thus the formation of a charge pulse on the anode. The total charge in this pulse is proportional to the energy of the incident photon. Thus far, this description is that of a conventional proportional counter. The position sensitivity can be achieved in several ways.

The simplest and probably most widely used is that first proposed by Borkowskii and Kopp,[31] which relies on using a high-resistance anode wire. Since the cathode and anode are capacitively coupled, the detector behaves like a distributed R.C. delay line, and therefore the charge pulse takes a finite time to propagate down the anode. By measuring the difference between the arrival times of the pulses at each end of the wire, the location of the ionizing event can be determined. A schematic diagram of a detector of this type is shown in Figure 4, while Figure 5 is a diagram of the signal processing electronics. In order to determine the location of the X-ray event, the charge pulses are first amplified and then differentiated. Zero-crossing detectors are then used to detect the zero-crossing points of the two signals, and outputs from these are used to start and stop a time-to-amplitude converter (TAC). The TAC output is fed to a multichannel analyzer (MCA), which sorts the pulses by amplitude (the amplitude relates to the location of the event), or to a computer.

One of the disadvantages of this type of detector is that it requires a very-high-resistance anode. Carbon-coated quartz (25-μm radius and 8-kΩ mm^{-1} resistance) is frequently used, but this suffers X-ray damage. One way of overcoming this is to use lower-resistance wire, e.g., nichrome, and to add extra capacitance to each end of the wire.[32] The maximum count rates of both these detectors are typically several tens of kilohertz, and they may be made to almost any required dimensions, typical values being 5 to 20 cm. The criteria for detector design are given in a review by Radeka,[33] and detector systems of this type are also available commercially for those who do not wish to construct them.

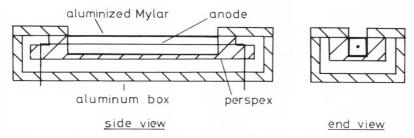

aluminized Mylar   anode

aluminum box   perspex

side view   end view

FIGURE 4. A schematic representation of a resistive anode position-sensitive proportional counter.

An alternative readout system is to use the induced charge on the cathode and to either make the cathode a delay line[34] or use a segmented cathode connected to a tapped delay line.[35] Since this approach does not rely on a resistive anode, conventional materials such as gold-plated tungsten may be used, thus overcoming the radiation damage problem. The signal processing for these detectors is essentially the same as for resistive anode ones, though they can operate at higher data rates. Therefore, it may be beneficial to replace the TAC and MCA by a time-to-digital converter (TDC) coupled to histogramming memory, as this is potentially faster. Again, commercial systems of this type are available, while design criteria have been discussed by Boie *et al.*[36]

All the diffraction results reported in this chapter were obtained using a resistive anode type detector. For those considering new installations, a cathode delay line readout system may be preferable in view of its greater speed and freedom from radiation damage. The detectors discussed so far have all been linear, and this design results in parallax errors with obliquely incident X-ray beams. A better configuration is a detector in the form of an arc of a circle,[37] and these are beginning to be available commercially.

The one-dimensional PSPC is well suited to the study of powder-type systems (though wasteful of photons) where the diffraction pattern consists of rings. For single-crystal materials with spot patterns, a two-dimensional area detector is required. In these cases, a multiwire proportional counter

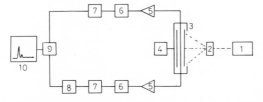

FIGURE 5. A schematic diagram of the electronic circuit required for a PSPC: (1) source, (2) sample, (3) detector, (4) high-voltage supply, (5) preamplifiers, (6) shaping amplifiers, (7) crossover detectors, (8) variable delay, (9) TAC, (10) MCA.

(MWPC) should be used. A full discussion of the design of these is beyond the scope of this chapter, but essentially they consist of orthogonal arrays of anode and cathode wires and the readout is again by delay lines.[38] They are capable of fairly high count rates (2 MHz) and resolution comparable with the wire spacing (~1 mm). As yet, a detector of this type has not been used in an electrochemical study.

3.1.2. X-Ray Sources

For laboratory-based applications, fixed anode sources (1–1.5 kW) will undoubtedly be the most widely available, though some may have access to rotating anode systems. With such a fixed anode source, it is probably prefer-able to use simple filters for wavelength selection, as crystal monochromators would reduce the flux too severely. This will result in part of the bremsstrahlung being present in addition to the desired line, but this will only contribute to the background of the diffraction pattern, and as we will see later, this gets subtracted out in data processing. With a rotating anode source, the use of a crystal monochromator may be beneficial.

The choice of source material is a difficult problem. Hard radiation is required (to achieve high penetration through the electrolyte solution) but hard X-ray photons are difficult to detect in a proportional counter, requiring the use of high pressure and expensive xenon gas. This is acceptable with a sealed detector but it means that the construction materials must be carefully chosen to be "clean," or the detector gas will soon become contaminated. Softer radiation is much easier to detect, and, for example, for Cu Kα photons reasonable detection efficiencies are obtained with Ar at atmospheric pressure. All the results reported here were obtained with a Cu Kα source, which limits the solution pathlength to a few tenths of a millimeter. This clearly makes cell design critical. For Mo Kα radiation, considerably thicker solution layers would be acceptable.

In addition to laboratory sources, synchrotron sources are becoming increasingly available. These have the advantage of higher flux and wider choice of wavelength and thus may be beneficial in the study of real technologi-cal systems. It is uncertain whether it will be possible to take full advantage of the higher flux as the detectors may become saturated. However, since synchrotron radiation is white, it should be possible to conduct the experiment in a different manner, that is, in the energy-dispersive mode. In this configur-ation, an energy-sensitive solid-state detector is used at a fixed scattering angle and white radiation is used. This is essentially analogous to the use of the PSPC, in that photons corresponding to a large section of the diffraction pattern are counted all the time. No reports of this type of study have yet appeared in the literature.

Finally, while discussing X-ray sources, the whole question of safety must be considered. The electrochemical application does not however introduce any new problems, and therefore the steps that must be taken for safe operation are identical to those for conventional crystallographic studies.

3.1.3. Cell Design

As has already been indicated, the design of the cell is critical to the success of *in situ* X-ray scattering experiments. In deciding on an acceptable design, the following points have to be considered:

1. The provision of uninterrupted paths for the incident and scattered photons.
2. The minimization of scatter and absorption of the X-ray photons by matter other than the phase of interest.
3. The provision of an even current/potential distribution across the working electrode.

X-ray diffraction experiments can be performed in one of two configurations—the reflection (Bragg) mode and the transmission (Laue) mode. In the present context, the Laue mode is best suited to studies of correlations parallel to the electrode surface, e.g., adsorbates, while the Bragg mode is better for thicker films and systems of a more technological nature such as passive layers. Different cell designs are required for each type of experiment but they do have a common feature in requiring windows. These need to be strong and should not decompose in common electrolytes but they must also have a high transmission coefficient for X-rays. Materials that have been used for this type of application are generally polymers. Mylar or Melinex are the most successful, and thicknesses of between 5 and 25 μm have been found to be appropriate.

In the Laue configuration, the X-ray beam must pass through the working electrode, the surface layer of interest, and a layer of electrolyte. Therefore, in order to have a reasonable number of scattered photons to detect, the electrode and electrolyte layer must be kept thin. Figure 6 shows a design of a typical cell of this type. It is made of glass, except for the Mylar windows, which are fixed with epoxy resin over the end of the cell and the end of the hollow syringe barrel. The working electrode is prepared by evaporating or sputtering a thin layer of the required material onto the fixed window (this should be possible with most commonly used electrode materials, e.g., Ag, Au, and Pt). In order to define the thin electrolyte layer, small spacers (typically 100 μm thick) are placed on the end of the syringe barrel and this is then pushed against the fixed window. The secondary electrode is a platinum ring placed around the syringe barrel. The tip of the Luggin capillary to the reference electrode is placed near the working electrode. The incident X-ray beam passes down the syringe barrel, while the scattered photons exit through the fixed window.

A design for Bragg mode experiments is shown in Figure 7. This cell is again made of glass and is fairly similar to the transmission mode one, except that the incident and scattered X-rays pass through the same window and the

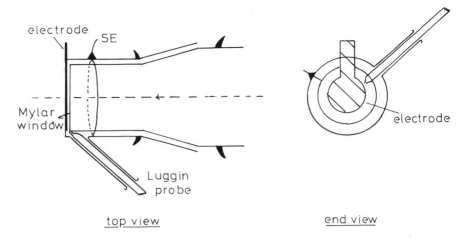

top view   end view

FIGURE 6. Cell design for transmission mode measurements.

working electrode can therefore be made of bulk materials sealed in glass or polymer tubes. Again spacers are used to define the electrolyte layer.

Neither of these designs permits replenishment of the electrolyte layer (other than by natural convection and diffusion), and therefore they cannot be used when large amounts of material are transformed, e.g., in metal deposition, though they are ideal for studies of adsorption and processes involving restructuring. When solution replacement is required, a flow system must be used, and a possible cell design is shown schematically in Figure 8. Here, a narrow band electrode is used which defines the thickness of the electrolyte layer since the windows are fixed on either side of it.

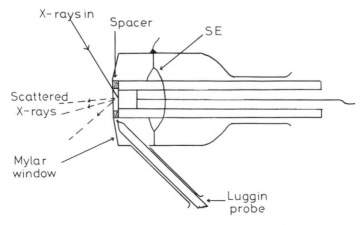

FIGURE 7. Cell design for reflection mode measurements.

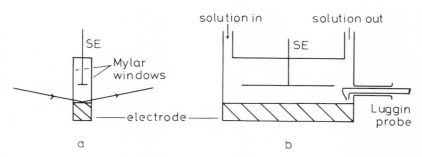

FIGURE 8. A schematic diagram of a flow cell for reflection mode measurements: (a) end view, (b) side view.

These cell designs are, of course, only examples and others are possible, though it will be found to be difficult to reconcile the various design criteria as the needs of the X-ray technique are in many senses contrary to those of a good electrochemical system.

3.1.4. The Experiment

Even when great care is taken in cell design, it will be found that for most systems, any diffraction pattern is dominated by scattering from the solution, windows, and the bulk electrode, and that it will probably be impossible to identify directly any scattering arising from a surface film. Clearly, something has to be done to improve the surface sensitivity, and as already stated in the introduction, that something is to use potential modulation techniques and to acquire difference diffractograms, i.e., representations of the changes that occur in the diffraction pattern as a result of the potential modulation. It might appear to be simpler to acquire diffraction patterns over a long period of time at each potential and then to difference these. X-ray sources, however, are not stable over long periods, and fluctuations in output would mean that background subtraction would be inexact and that little improvement in surface specificity would be achieved. The essential feature of the modulation technique is that only changes are recorded in the difference diffractogram, and scattering from the windows, bulk electrolyte, and bulk electrode is subtracted out.

Figure 9 is a schematic diagram of the apparatus for a transmission experiment. A typical experimental study proceeds as follows. Conventional electrochemical techniques (usually cyclic voltammetry) are used to establish the potentials where the processes of interest occur, e.g., adsorption and desorption potentials. The working electrode potential is then modulated between these values at a low frequency (typically 10^{-1} to 10^{-2} Hz) using a computer-controlled potentiostat. The choice of modulation frequency is not critical except, of course, it should be considerably less than the time constant

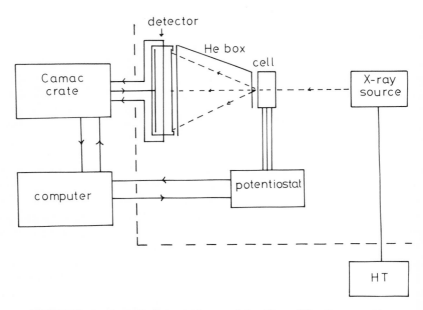

FIGURE 9. A schematic diagram of transmission X-ray diffraction apparatus.

of the cell. The computer used to control the potential modulation is also used to control the routing of the diffraction data, the diffraction pattern at each potential being stored in a separate part of the MCA or histogramming memory. It is considered advisable to acquire the diffractograms and only to subtract them at the end of the experiment, rather than to acquire the running difference, as it is easier to detect any instrument malfunctions in this way. For related reasons, it is also preferable, if possible, to write the data stored in the MCA or histogramming memory to a disk frequently to avoid corruption of the data if there is a breakdown during a long experimental run. A further refinement is to prevent data acquisition for a short period after each potential change until all structural changes have taken place. In this way, diffractograms correspond to final and not changing states.

In view of the long duration of some of these experiments, it is advisable to monitor the electrochemical behavior by sampling the current after each pulse and to save these data in the computer. For this, an A/D converter connected to the current follower of the potentiostat is required. As a further check on the validity of any data, the electrochemical behavior of the system should be checked at the end of the experimental run, probably with cyclic voltammetry again, to ensure that no significant changes have occurred. The duration of any experiment will depend on the system being studied, the requirement, of course, being that there are significant features above the noise level. For thick films, an hour may be sufficient, whereas for adsorbates on low-surface-area substrates, times in excess of 100 hours may be required.

Degradation of diffraction patterns by air scatter is reduced by placing a He-filled bag between the cell and detector.

The procedure described above involves acquiring diffraction patterns at two potentials; there is no reason, however, why this approach cannot be extended to situations where the electrode potential is repeatedly pulsed up a potential staircase with the diffraction pattern at each potential being recorded. All that is required are further MCA or histogramming memory channels.

No details as yet have been given on the requirements of the computer controlling the experiment, and clearly many strategies are possible. In view of the high data rates, it is probably preferable not to require the computer to do the data acquisition directly (i.e., not to use it as an MCA) but rather to interpose local electronics and memory to do this. The computer is then used only for overall control and as a file store. All the required components for this approach are available as modules to the CAMAC standard, and therefore this is a convenient way to proceed. These modules are housed in a CAMAC crate which is controlled by a computer through an appropriate interface (a wide range of computer specific interfaces are available, in addition to one which uses the IEEE-488 standard). After initialization by the computer, the CAMAC system performs the required data acquisition and storage until the computer halts the process and reads the data into its own memory. Commercially available X-ray detection systems based on PSPCs are frequently constructed in this modular form. If a system to be purchased is not, it should be ascertained whether it is versatile enough to permit the required data partitioning and synchronization to the electrochemistry.

3.2. In Situ X-Ray Absorption Studies

As we have seen, EXAFS studies almost invariably will be made using synchrotron radiation as the X-ray source. The use of such a central facility inevitably means that, as a user, one has little control over the choice of much of the experimental equipment. Typically, the beam line will already be equipped with an appropriate monochromator, the software to drive it, as well as detectors and data acquisition software. In addition, all the safety features will be in place. One does, however, have to decide on an appropriate experimental configuration and the type of detector to be used.

Before going on to consider experimental design, we must first be aware that the EXAFS technique will not be applicable to every *in situ* electrochemical system. For nuclei lighter than chromium, the absorption edge is at such low energies that the X-ray photons are unable to penetrate significantly the electrolyte solution. Indeed, for very light nuclei, e.g., Al, high-vacuum conditions are required to record the EXAFS.

For nonelectrochemical, solid-state EXAFS studies of the heavier nuclei, the most generally used experimental configuration is transmission, i.e., the incident and transmitted photon fluxes are measured with ionization chambers, as a function of X-ray energy, and hence the absorption spectrum is obtained. An electrochemical cell of the type described earlier for transmission X-ray diffraction experiments in theory could be used in this configuration, though this does not as yet appear to have been tried. Instead, the most widely used approach seems to have been to take advantage of the fact that one of the consequences of X-ray absorption is the emission of fluorescent X-ray photons. The yield of these fluorescent X-rays is a measure of the X-ray absorption cross section and hence can be used to record the absorption spectrum. A schematic diagram of the apparatus required for this experiment is shown in Figure 10. The cell, which would be of similar design to that used for Bragg mode diffraction experiments, is mounted so that the electrode is inclined at about 45° to the incident beam and a series of scintillation detectors are mounted around the cell to detect the fluorescent photons. To minimize the detection of scattered photons, appropriate filters (made of material containing nuclei one or two atomic numbers lighter than the test nuclei) are placed in front of some, or all, of the detectors. The optimum thickness of these filters can be found by experiment. An ionization chamber again records the incident photon flux. To record the EXAFS simply involves scanning through the energy range of interest (typically from 100 eV below the edge to 500 to 1000 eV above it) and acquiring data. Resolution of 2 to 5 eV is generally acceptable for EXAFS studies, but for near-edge work, better resolution is desirable. Typical energy scans may take 30 minutes and perhaps 10 such scans will need to be averaged to obtain satisfactory signal-to-noise ratios with low-concentration/thin samples.

A major difficulty with this configuration arises from the fact that all nuclei of a particular type in the beam will contribute to the X-ray absorption spectrum. Thus, for a system where there is a thin surface layer on a bulk substrate (e.g., an iron oxide on iron) the observed spectrum will arise from

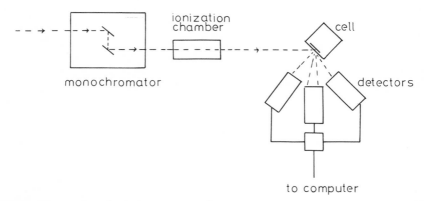

FIGURE 10. A schematic diagram of the apparatus for EXAFS studies of electrode surfaces using fluorescence detection.

both the surface film and the bulk material. For thin surface layers, the spectra may even be dominated by the bulk material. To overcome this, thin layers of bulk material can be used or it may even be possible to use a technique analogous to the modulation method used for *in situ* diffraction studies. For EXAFS work, however, it would probably be preferable not to difference the EXAFS signals directly but rather their Fourier transforms, as this information will be easier to interpret. Experiments of this type have not yet been attempted but are likely to be in the near future.

The approach for improving surface sensitivity that has already been tried is to use glancing angles of incidence (either sub- or near-critical angle). This experimental configuration is shown in Figure 11. The determination of the absorption spectrum can be either by direct monitoring of the reflected beam (probably with an ionization chamber) or of the fluorescence yield (with either an ionization chamber or scintillation detectors). The glancing angle of incidence inevitably leads to long solution pathlengths for the X-rays, and therefore this technique may be restricted to higher-energy absorption edges (higher-atomic-number nuclei). Again, the acquisition of the absorption spectrum requires scanning over the required energy range and averaging of a series of spectra, if possible. Detection of the reflected beam can of course only be achieved for subcritical angles of incidence, and under these conditions, the surface sensitivity is very high. At larger angles of incidence, the fluorescence detection is used, and if the yield is measured as a function of angle of incidence, then profiling of the surface structure can be achieved since the thickness of the surface layer that is being probed increases with the angle of incidence.

The two examples of EXAFS experiments described above have both used a spectrometric configuration, and therefore absorption spectra have been acquired by scanning the energy. An alternative is to use an energy-dispersive system, and potentially this has many advantages. Figure 12 shows a schematic diagram of the apparatus with the X-ray beam in glancing incidence at the electrode surface (for thin samples, the transmission mode could be used). In this system, a broad band (500–1000 eV) of X-ray radiation is focused at the sample by a bent crystal monochromator. The bandpass is angularly dispersed and can be resolved geometrically. Thus, the absorption spectrum can be obtained by placing a linear position-sensitive detector, as shown. This technique has much in common with the use of position-sensitive detection in

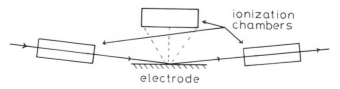

FIGURE 11. A schematic diagram of the reflEXAFS configuration.

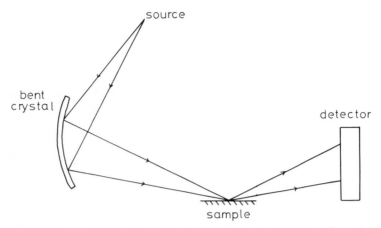

FIGURE 12. A schematic diagram of the energy-dispersive EXAFS configuration.

X-ray diffraction studies, in that the entire spectrum is collected simultaneously. Therefore, for a given signal-to-noise ratio, data collection times are reduced, time-resolved studies become possible, and potential modulation techniques are facilitated. Electrochemical studies using this configuration are only just beginning but early results appear very promising.

4. Applications

4.1. In Situ X-Ray Diffraction

As discussed earlier, the Bragg configuration is ideally suited to X-ray scattering studies of films on bulk substrates. A typical example of a study using this configuration is that of nickel hydroxide electrodes by Fleischmann et al.[21] These electrodes are, of course, of considerable technological importance in view of their use as the positive plate in the nickel–cadmium battery, and their efficiency in this application is known to be affected by the precise structure of the hydroxide.

Electrochemically precipitated $Ni(OH)_2$ consists primarily of the α form[39] and is rather amorphous. On exposure to the concentrated KOH electrolyte, it becomes more crystalline, and also it slowly converts to the β form. The aim of the study was to investigate the structural changes accompanying the charging of these two phases.

Figure 13 shows a series of cyclic voltammograms for a 7500-Å-thick layer of $Ni(OH)_2$ electrodeposited on a nickel electrode in 8 M KOH. The effect of the aging process is clearly seen. Aging occurs less rapidly in more dilute KOH, and therefore, in order to study the structural effects for the $\alpha Ni(OH)_2$ phase, the electrolyte concentration was reduced to 1 M. For these structural

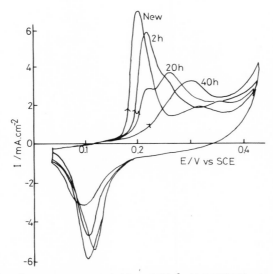

FIGURE 13. Cyclic voltammograms recorded for a 7500-Å-thick Ni(OH)$_2$ electrode in 8 M KOH at a sweep rate of 1 mV s^{-1}. The periods of aging are shown.

studies, the thickness of the hydroxide was limited to 1500 Å (for thicker films, the charge/discharge time was too long) and a cell of the type shown in Figure 7 was used. The electrode potential was modulated at 5×10^{-1} Hz for six hours between the potentials of +400 and +100 mV (corresponding to the charged and discharged states) and the relevant diffractograms were acquired as described earlier. Figure 14 shows the accumulated diffractogram in the discharged state. The two narrow peaks were assigned to the [111] and [200] diffractions of the nickel substrate while the broad peak arises from scattering by the hydrogen-bonded water structure. It is clear that no peaks are visible attributable to scattering by the hydroxide. However, on subtracting the diffractogram obtained at +400 mV, some clear features become apparent, as shown

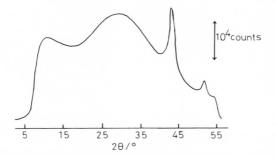

FIGURE 14. *In situ* X-ray diffractogram for an unaged Ni(OH)$_2$ electrode held at 100 mV vs. SCE in 1 M KOH. Data were acquired for three hours while using Cu Kα radiation.

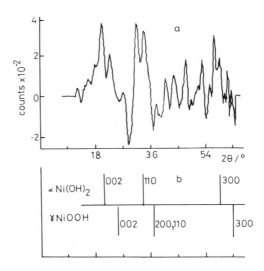

FIGURE 15. Difference diffractogram for the $Ni(OH)_2$-NiOOH system in 1 M KOH for an unaged electrode. The total data acquisition time was six hours at a modulation frequency of 5×10^{-3} Hz. The data are displayed as the diffractogram at +100 mV minus that at +400 mV. (a) Experimental data, (b) assignment to reflections of $\alpha Ni(OH)_2$ and $\gamma NiOOH$.

in Figure 15. As further indicated on this figure, these can be assigned to major reflections of $\alpha Ni(OH)_2$ in the discharged state and $\gamma NiOOH$ in the charged state. The efficiency of the differencing procedure is demonstrated by the absence in the difference diffractogram of anything that can be attributed to the nickel substrate or water. A similar experiment was performed for an aged electrode, and the difference diffractogram is shown in Figure 16. Here, the main features were attributed to $\beta Ni(OH)$ in the discharged state and $\beta NiOOH$ in the charged state. Studies of incompletely aged electrodes indicated that mixtures of phases were present.

This study clearly demonstrates that by using the differencing technique, sufficient sensitivity is achieved to permit the study of electrochemically induced structural changes in thin surface films. The results obtained indicate that it should be possible to study films of about 100-Å thickness fairly easily with a small, fixed anode X-ray source, with such a system probably requiring less than 24 hours of data acquisition time. The study of adsorbed monolayers clearly will be more difficult. Such investigations will benefit from the use of high-surface-area electrodes, e.g., platinum black, and/or the use of the Laue geometry.

The first *in situ* X-ray diffraction study of adsorption from an aqueous solution was the investigation by Fleischmann and coworkers[18,21] of iodine adsorption on graphite. To increase the surface sensitivity in this study, a high-surface-area compressed exfoliated graphite (Papyex) was used as the substrate. This material has a surface area of between 20 and 30 $m^2 g^{-1}$,[40]

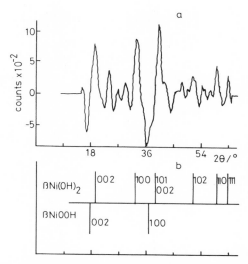

FIGURE 16. As Figure 15 for aged electrode. (a) Experimental data, (b) assignment to reflections of βNi(OH)$_2$ and βNiOOH.

and the small crystallites are partially aligned so that a high proportion of them are oriented with their basal planes parallel to the surface of the bulk material. Since it is these basal planes that are the adsorption surface, it is possible to perform experiments where the orientation of the scattering vector with respect to the two-dimensional adsorbate phase can be controlled.

This investigation of iodine adsorption was nonelectrochemical and simply involved equilibrating samples of Papyex with KI$_3$ solutions for several days and then performing X-ray diffraction experiments in the Laue configuration. Figure 17 contains some diffractograms for this system under various experimental conditions. Figure 17a shows the pattern obtained in the absence of any iodine when using the in-plane geometry (scattering vector parallel to basal planes). The peaks at $2\theta = 27°$ and $44°$ were assigned to the (002) and unresolved (100) and (101) diffractions of the graphite, while that at $16°$ and the shoulder on the low-angle side of the substrate (002) peak were due to the Mylar windows. Comparison with Figure 17b, which is for the same geometry but with adsorbed iodine present, shows that in this latter case, there is an additional peak. This was attributed to the (10) diffraction of a two-dimensional array of adsorbed I$_2$ molecules, in view of the absence of the peak in the out-of-plane diffraction pattern (Figure 17c). Further support for the assignment was obtained from the shape of the peak after background subtraction, which, as shown in Figure 18, is typical of that of a two-dimensional lattice. Comparison of the area under the (10) diffraction peak and that under the Mylar window peak (used as an internal standard) for various concentrations of iodine in solution showed that the X-ray diffraction

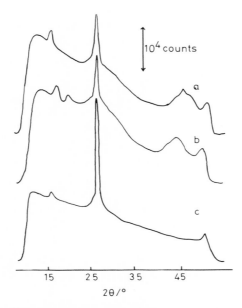

FIGURE 17. *In situ* diffraction patterns obtained for a 0.3 mm sample of Papyex with an acquisition time of 4000 s in (a) 0.1 M KI, in-plane geometry; (b) 0.02 M I_2 in 0.1 M KI, in-plane geometry; (c) 0.02 M I_2 in 0.1 M KI, out-of-plane geometry.

data agreed very closely with coverage data obtained by classical wet analytical techniques and fitted a Langmuir isotherm.

The position of the (10) diffraction peak was found to be independent of the I_2 coverage, which implied that the forces between the adsorbed I_2 molecules were attractive and that these molecules associated to form islands.

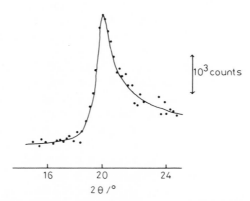

FIGURE 18. Difference diffractogram [(absorbate + adsorbent) − absorbent] for condition of Figure 17b.

We saw earlier that the width of a diffraction peak for a two-dimensional lattice can be used to obtain the coherence length. In this case, the value was estimated to be 110 ± 20 Å (independent of coverage). This is a similar value to that obtained for gas adsorption studies on Papyex, and it has been attributed to the graphite particle size. It was thus concluded that even though the system obeyed the Langmuir isotherm, it might perhaps be better to consider the adsorption to be a higher-order phase transition.

By using the high-surface-area graphite substrate, the number of surfaces probed by the X-ray beam was increased to a few hundred. Despite this, however, the total count rate under the (10) diffraction peak was still only about ten counts per second. Clearly, measurements on a single layer of adsorbate will be much more demanding. One such system has been investigated in an attempt to determine the sensitivity of the technique and that is lead underpotential deposition on silver. Using a cell such as that shown in Figure 6 with an evaporated silver electrode, Fleischmann *et al.*[21] acquired data over a period of 100 hours, while modulating the electrode potential at 10^{-2} Hz between values corresponding to a monolayer-covered surface and an adsorbate-free one. Figure 19a shows the diffraction pattern obtained for the adsorbate-covered electrode. The broad peak is due to scattering by the water, while those at $2\theta = 38°$ and $44°$ were attributed to the Ag(111) and (200) reflections for the substrate. After subtracting the diffraction pattern obtained for an adsorbate-free electrode, the difference diffractogram shown in Figure 19b was obtained. The main conclusions were that while no structure directly related to the adsorbate could be determined with any confidence, it was clear that the adsorption of the lead caused some changes—most notably, the appearance of the differential peak around 26°, which was attributed to the lengthening of the O—O distance in hydrogen-bonded water near the electrode when the lead adsorbed. It was felt that the loss of structure around 17° was probably related to changes in ion distribution.

These results were quite interesting but it is clear that a better signal-to-noise ratio is required if the surface structure of an adsorbate is to be investigated. The acquisition time at 100 hours is already long and cannot be realistically increased. Other possible ways of improving the situation include the use of a single-crystal substrate with an area detector (the counts will then be concentrated in spots rather than being distributed around a circle) and increasing the photon flux (e.g., using a synchrotron source). It is likely that in the next few years both these possibilities will be investigated.

4.2. EXAFS Studies

X-ray absorption studies of systems of electrochemical interest have been made by several groups of workers. The principal field of interest has been passive layers, in particular those on iron and its alloys, and to a lesser extent those on nickel. While early work in this area was performed *ex situ*, generally

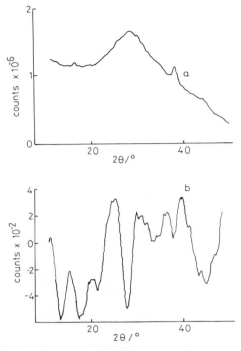

FIGURE 19. *In situ* diffractograms for Ag electrode in 0.1 *M* Pb(ClO$_4$)$_2$, 0.5 *M* NaClO$_4$, 1 m*M* HClO$_4$: (a) after 50 hours at -400 mV, (b) difference diffractogram (that at -400 mV minus that at -100 mV) after 100 hours of acquisition while modulating the electrode potential at 10^{-2} Hz.

using electron yield detection, recently some *in situ* studies have been reported, though in only a few cases has the electrode been polarized during the measurements.

An example of the use of fluorescent X-ray yield measurements is the study by Forty *et al.*[41] of the passive films formed on Fe/Cr alloys in NaNO$_2$ solutions. These samples were prepared by evaporation onto glass substrates, and the film thicknesses were restricted to 20 Å to ensure that all the metal was converted to passive layers on immersion in the solution. Figure 20 shows the absorption spectrum obtained for an Fe/13% Cr alloy at the iron edge, whereas Figure 21 shows the Fourier transform of the oscillatory part of the absorption spectrum. For comparison, the transform of the EXAFS of γFeOOH also is shown. The main conclusions of this study were that the film *in situ* had a structure similar to that of γFeOOH but that *ex situ* it dehydrated to a structure similar to γFe$_2$O$_3$. It was also found that on increasing the Cr content of the alloy, the passive layer became more amorphous and the Fe—Fe and Fe—O bond lengths increased. EXAFS at the Cr edge also were reported by this group. Similar investigations have been made by Long *et al.*,[42] while

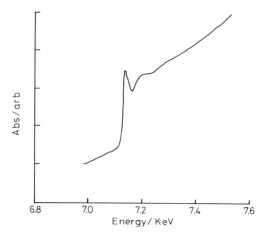

FIGURE 20. In situ Fe K edge absorption spectrum for Fe/13% Cr film passivated in 0.1 M $NaNO_2$.

Kordesch and Hoffman[43] have used a novel cell design that incorporates an emersed electrode and permits the use of electron yield techniques, remaining essentially *in situ.*

Most of the work to date on the use of the reflection geometry has been of an exploratory nature, and the data often do not appear to have been fully analyzed. As an example, Figure 22 contains some results obtained by Bosio *et al.*[44] for passive films on nickel. This figure shows k-weighted $\chi(k)$ data for the electrode before and after passivation and, for comparison, curves for Ni and NiO. An example of the use of the energy-dispersive mode is the study

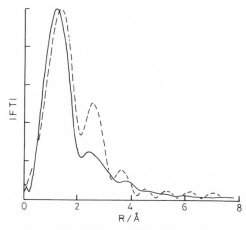

FIGURE 21. Fourier transform of data in Figure 20 (solid curve). The dashed curve shows data for γFeOOH.

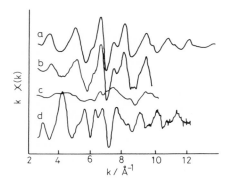

FIGURE 22. ReflEXAFS results at an angle of incidence of 3.4 mrad for nickel electrode passivated in 1 N H_2SO_4: (a) Ni, (b) Ni electrode *in situ*, (c) electrode after passivation, (d) NiO (after Ref. 44).

by Dartyge *et al.*[45] of the electrochemical inclusion of Cu into a conducting polymer, poly(methyl thiophene), which was formed as a thin film on a platinum substrate. Their results clearly demonstrate the existence of Cu^+ ions in the film.

It can be seen that as yet only some very preliminary studies have been made using EXAFS. It is clear, however, that the technique has great potential in this area, and it can be anticipated that there will be a rapid increase in reports of this type of work.

List of Symbols

c	velocity of light
d	interlayer spacing
e	electronic charge
f_e	electron scattering factor
f_i	atomic scattering factor
F_{hkl}	structure factor for *hkl* planes
$F(r)$	Fourier transform of $\chi(k)$
h	Planck's constant
hkl	Miller indices
I_e	scattered X-ray intensity
I_0	incident X-ray intensity
I	transmitted X-ray intensity
k	electron momentum
m	electronic mass or mass per unit area
N_j	number of atoms in *j*th shell
R	distance from scattering electron
r_j	distance to *j*th shell

r	radial distance
$S_j(k)$	backscattering amplitude of atoms in jth shell
x_i, y_i, z_i	atomic positions in unit cell
$\delta(k)$	phase shift
θ	scattering angle
λ	X-ray wavelength or electron mean free path
μ_m	mass absorption coefficient
μ_0	vacuum permeability
$\rho(r)$	electron density
$\chi(k)$	oscillatory part of absorption rate

References

1. J. Robinson, in: *Electrochemistry* (D. Pletcher, ed.), Vol. 9, p. 101, Royal Society of Chemistry, London (1984).
2. J. B. Pendry, *Low Energy Electron Diffraction*, Academic Press, New York (1974).
3. T. A. Carlson, *Photoelectron and Auger Spectroscopy*, Plenum Press, New York (1974).
4. E. A. Stern, *J. Vac. Sci. Technol.* **14**, 461 (1977).
5. G. Binning, H. Rohrer, C. Gerber, and E. Weibel, *Phys. Rev. Lett.* **49**, 57 (1982).
6. P. N. Ross and F. T. Wagner, in: *Advances in Electrochemistry and Electrochemical Engineering* (H. Gerischer and C. Tobias, eds.), Wiley-Interscience, New York (1985).
7. E. Yeager, A. Homan, B. D. Cahan, and D. Scherson, *J. Vac. Sci. Technol.* **20**, 628 (1982).
8. W. N. Hansen, *Surface Sci.* **101**, 109 (1980).
9. D. M. Kolb and W. N. Hansen, *Surface Sci.* **79**, 205 (1979).
10. D. M. Kolb, D. L. Rath, R. Wille, and W. N. Hansen, *Ber. Bunsenges. Phys. Chem.* **87**, 1108 (1983).
11. N. Neff and R. Koetz, *J. Electroanal. Chem.* **151**, 305 (1983).
12. U. Nakai, M. S. Zei, D. M. Kolb, and G. Lehmpfuhl, *Ber. Bunsenges. Phys. Chem.* **88**, 340 (1984).
13. H. O. Beckmann, H. Gerischer, D. M. Kolb, and G. Lehmpfuhl, *Faraday Symp. Chem. Soc.* **12**, 51 (1977).
14. C. Kunz (ed.), *Synchrotron Radiation*, Springer-Verlag, Berlin (1979).
15. W. C. Marra, P. Eisenberger, and A. Y. Cho., *J. Appl. Phys.* **50**, 6927 (1979).
16. P. L. Cowan, J. A. Golovchenko, and M. F. Robbins, *Phys. Rev. Lett.* **44**, 1680 (1980).
17. G. Materlik, J. Zegenhagen, and W. Uelhoff, *Phys. Rev. B* **32**, 5502 (1985).
18. M. Fleischmann, P. J. Hendra, and J. Robinson, *Nature* **288**, 152 (1980).
19. A. Bewick, M. Fleischmann, and J. Robinson, *Dechema Monographie* **90**, 87 (1980).
20. M. Fleischmann, P. R. Graves, I. R. Hill, A. Oliver, and J. Robinson, *J. Electroanal. Chem.* **150**, 33 (1983).
21. M. Fleischmann, A. Oliver, and J. Robinson, *Electrochim. Acta* **31**, 899 (1986).
22. P. A. Lee and J. B. Pendry, *Phys. Rev. B* **11**, 2795 (1975).
23. G. Bomchil and C. J. Rekel, *J. Electroanal. Chem.* **101**, 133 (1979).
24. *International Tables for X-ray Crystallography*, Vol. III, Kynoch Press, Birmingham (1962).
25. B. G. Warren, *Phys. Rev.* **59**, 693 (1941).
26. B. K. Agarwal, *X-ray Spectroscopy—An Introduction*, Springer-Verlag, Berlin (1979).
27. F. W. Lytle, D. E. Sayers, and E. A. Stern, *Phys. Rev. B* **11**, 4825 (1975).
28. S. J. Gurman, N. Binstead, and I. Ross, *J. Phys. C* **17**, 143 (1984).
29. N. Binstead, S. J. Gurman, and I. Ross, *J. Phys. C* **19**, 1845 (1986).
30. G. T. Reynolds, J. R. Milch, and S. M. Gruner, *Rev. Sci. Instrum.* **49**, 1241 (1979).
31. C. J. Borkowskii and M. K. Kopp, *Rev. Sci. Instrum.* **39**, 1515 (1968).

32. M. K. Kopp, *Rev. Sci. Instrum.* **50**, 382 (1979).
33. V. Radeka, *IEEE Trans. Nucl. Sci.* **21**, 51 (1974).
34. P. Lecomte, V. Perez-Mendez, and G. Stokes, *Nucl. Instrum. Meth.* **153**, 543 (1978).
35. A. Gabriel, *Rev. Sci. Instrum.* **48**, 1303 (1977).
36. R. A. Boie, J. Fischer, Y. Inagaki, F. C. Merritt, V. Radeka, L. C. Rogers, and D. M. Xi, *Nucl. Instrum. Meth.* **201**, 93 (1982).
37. E. R. Wolfel, *J. Appl. Cryst.* **16**, 341 (1983).
38. C. Cork, D. Fehr, R. Hamlin, W. Vernon, N. H. Xuong, and V. Perez-Mendez, *J. Appl. Cryst.* **7**, 319 (1973).
39. G. W. D. Briggs and W. F. K. Wynne Jones, *Electrochim. Acta* **7**, 241 (1962).
40. C. Marti and P. Thorel, *J. Phys. (Paris)* C4 **38**, 26 (1977).
41. A. J. Forty, M. Kerkar, J. Robinson, and M. Ward, *J. Phys. (Paris)* C8 **47**, 1077 (1986).
42. G. G. Long, J. Kruger, D. R. Black, and M. Kuriyama, *J. Electroanal. Chem.* **150**, 603 (1983).
43. M. E. Kordesch and R. W. Hoffman, *Nucl. Instrum. Meth.* **222**, 347 (1984).
44. L. Bosio, R. Cortez, A. Defrain, and M. Froment, *J. Electroanal. Chem.* **180**, 265 (1984).
45. E. Dartyge, A. Fontaine, G. Tourillon, and A. Jucka, *J. Phys. (Paris)* C8 **47**, 607 (1986).

3

Photoemission Phenomena at Metallic and Semiconducting Electrodes

Ricardo Borjas Severeyn and Robert J. Gale

1. Introduction

The original photoelectric effect was discovered by Hertz[1] in 1887 while he was investigating the properties of electromagnetic waves. It was observed that a spark would jump a gap between two electrodes more readily when the electrodes were illuminated with light, ultraviolet light having a greater effectiveness than light from the visible region of the spectrum. This simple observation, however, was to have profound impact on our conception of how radiation and matter interact.

In 1905, while an examiner at the Swiss Patent Office, Einstein[2] in his spare time devised a theory to explain photoemission. His theory adopted Planck's quantum theory of a blackbody radiator and assumed that radiation itself was quantized. When a quantum of energy ($h\nu$) falls on a metal surface, its entire energy may be used to eject an electron from an atom. Because of the interaction of the ejected electron with surrounding atoms (their electronic distributions), a certain minimum energy is required for the electron to escape from the surface. The minimum energy to escape (ϕ) depends on the metal and is called the work function. The maximum kinetic energy of an emitted photoelectron is given by Einstein's equation

$$\tfrac{1}{2}mv^2 = h\nu - \phi \tag{1}$$

The threshold frequency for emission (ν_0) can be determined from measured work functions, and at absolute zero, the photoelectric threshold is equal to the work function,

$$h\nu_0 = \phi^0 \tag{2}$$

Ricardo Borjas Severeyn • Chemistry Department, University of Wyoming, Laramie, Wyoming 82071. *Robert J. Gale* • Chemistry Department, Louisiana State University, Baton Rouge, Louisiana 70803.

Although Einstein's theory explained the basic phenomenon, it did not provide a complete description of photoelectronic emission. To have a more complete view, it is necessary to specify not only the maximum velocity of the emitted electrons but their number, their velocity distribution, and the way these factors depend on the frequency and the state of polarization of the light.

Photoelectric effects have been researched at all types of interfaces— solid/vacuum (gas), solid/liquid, solid/solid, liquid/vacuum (gas), and liquid/liquid systems—and a comprehensive theory would be extremely complicated. Nevertheless, considerable knowledge has been gained and many theories have been proposed which either partially or almost completely describe aspects of the processes. For the metal/vacuum case, one may consider the works of Richardson,[3] Thomson,[4] Uspensky,[5] Wentzel,[6] and Fowler[7] as having greatly aided the theoretical description of this effect (Table I). A treatise on metal emission spectroscopy is available.[8] The purpose of this chapter is to provide an introduction to the topic of photoelectric emission at electrodes. Detailed discussion of all of the metal/vacuum and semiconductor/vacuum theories is not possible here but it is helpful, nonetheless, to be cognizant of these as they provide a foundation and, in some instances, a means of expanding the scope and applicability of current theories of the electrode/condensed media cases that are considered.

1.1. Some General Features of Photoelectronic Emission

As early as 1839, Becquerel[9] had detected electric currents when one of two immersed electrodes in dilute acid solutions was illuminated. These effects were found to depend on the pH and they were larger when the violet part of the optical spectrum was used. Since then, considerable research has been devoted to the study of such processes. An outline of the historical developments of this field is available in a monograph,[10] together with details of the major theoretical and experimental studies of photoemission at metallic and semiconductor electrodes. This chapter, therefore, will be restricted in scope

TABLE I. Historical Development of Semiclassical Theories for the Metal/Vacuum Photoelectric Effect

Author	Year	Description
Richardson	1912	Classical: Thermodynamic electron "evaporation"[a]
Thomson[b]	1926	Classical: Quantum/electron collision[a]
Uspensky	1926	Classical: Electrons impeded by viscous force[a]
Wentzel	1928	Wave mechanical
Fowler[c]	1931	Free-electron model with temperature dependency

[a] Spectral distribution function found to be unsatisfactory.
[b] cf. Becker (1928).
[c] Matthews and Khan modification (1975).

to phenomenological and quantum mechanical descriptions of photoemission processes and to a brief coverage of those theoretical approaches which are tested more readily by experiment. In the practical section, an emphasis is placed upon semiconductor electrode studies because of the authors' experiences; however, methodology is similar regardless of the nature of the electrode.

Various theories of photoemission have been developed since Becquerel discovered photocurrents at electrodes in liquid electrolytes. To cover all of these is beyond the scope of this chapter; however, two important early contributions will be mentioned briefly. Berg and Reissman[11,12] studied the action of light on dropping mercury electrodes using conventional polarographic techniques. They concluded that the photocurrents were due mainly to the absorption of light by the metal—an increase in the energy of electrons inside the metal was thought to improve the possibility of electrode reactions. Heyrovsky[13,14] explained such photoeffects by the decomposition of a surface charge transfer complex formed between the electrode metal and a species in solution, or the solvent. The polarization of the chemical bond in such a complex will depend on the electron donor or acceptor nature of the adsorbate. Upon light absorption by the metal, this bond is ruptured and the bond electrons migrate to the electrode (anodic photocurrent) or to the adsorbed molecule which leaves the electrode surface (cathodic photocurrent). Charge transfer reactions of surface complexes are important in certain systems of practical interest and may be distinguished from substrate emission by their insensitivity to polarized light.[15] Homogeneous photochemical reactions whose photocurrents are produced as a result of light absorption by the solution leading to products which react heterogeneously will not be treated here, nor will photoreactions in which the metal is covered with a solid electrolyte layer (e.g., an oxide film) or a photoactive surface-absorbed species.

There are two important differences between photoemission into vacuum and into condensed media. Firstly, the work function in the latter case usually (but not necessarily) is lowered because of an interaction of the photoelectron with the solvent. Further, chemical reactions are possible with deliberately introduced scavengers or trace impurities present in an electrolyte. Secondly, the presence of an electrical double layer introduces a potential drop across the Helmholtz plane and diffuse layer. In some earlier theories, the electrical double layer in the presence of solute molecules was thought to screen effectively the positive charge remaining after the electron is emitted (unlike the case of emission into vacuum), thereby reducing image effects and changing the theoretical form of the photocurrent rate expressions.

1.2. Reaction Step Models for Photoemission

A simple, mechanistic description of the overall photoprocess has evolved from a series of studies by Barker and coworkers.[16-18] These pioneering studies have been able to relate the photoemission interfacial chemistry to radiolysis knowledge. It was proposed that the solvated electron might be an intermediate of a multiple-step process, and this was confirmed using electron scavengers such as N_2O, NO_3^-, NO_2^-, and H^+ ions, whose homogeneous reactions had been characterized through radiation studies of liquids. Elementary reaction steps are illustrated in Figure 1 using the example of N_2O. Nitrous oxide gas is soluble in aqueous media to about 0.025 mol liter^{-1} at 25°C. Its reduction occurs by the homogeneous steps

$$N_2O + e_{aq}^- \rightarrow N_2O^- \tag{3}$$

$$N_2O^- + H_2O \rightarrow N_2 + OH^- + OH\cdot \tag{4}$$

The OH· radical then can be reduced heterogeneously at the electrode to form a hydroxyl ion,

$$OH\cdot + e^- \rightarrow OH^- \tag{5}$$

Note above that the reaction between N_2O molecules and the solvated electrons is described as homogeneous. This is because the solvated electrons are considered to be generated from "dry" (delocalized) electrons that are deposited in a zone of electrolyte, perhaps 10 Å thick, adjacent to the electrode.

Photoemission, in this instance, can be considered to involve two main stages; firstly, the photoexcitation mechanism within the electrode material followed by the injection of a hot electron into the liquid phase, and secondly, the subsequent interactions of the emitted electrons with the solvent and dissolved solutes. A physical description of the excitation event itself is complex and may involve collective vibration of conduction electrons, called plasmons. In chemical terms, to successfully model absorption of a light quantum by the solid it is necessary to have a detailed knowledge of the matrix elements for transitions between states and of the fate of an electron that crosses the interface. Once ejected into the electrolyte, a series of reaction steps can be postulated for the electron. The overall process may comprise:

1. transition of hot electrons through the interface after absorption of light quanta such that both energy and momentum are conserved;

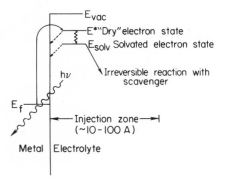

FIGURE 1. Elementary electronic photoemission into aqueous media.

2. reduction of the initial energy of electrons to the level of thermal kinetic energy of the solvent by a series of elastic scattering interactions to form solvated electrons in about 10^{-12} s (thermalization and hydration); and

3. reaction of solvated electrons (e_{aq}^-) with the scavenger, e.g., N_2O, O_2, CO_2, H^+.

If no efficient scavenger is present, the solvated and dry electrons presumably are able to return to the electrode at a very rapid rate, with the consequence that the stationary photocurent will be zero in the system.† Some slight irreversible decomposition of aqueous solvent is possible, and the lifetime of the unstable solvated electron in water is of the order of one microsecond.

Barker[18] largely was able to substantiate this mechanism using the organic additives ethanol and methanol, which react with the OH· radical. As a consequence of the reaction between OH· and the alcohol, a decrease in the N_2O photoemission current occurs. This competitive reaction involves the abstraction of the α-hydrogen from the alcohol,

$$OH· + RCH_2OH \rightarrow RCHOH· + H_2O \tag{6}$$

and the photocurrent decrease occurs because of diminished reaction (5) and the possible oxidation of the product RCHOH· radical to aldehyde at certain potentials,

$$H_2O + RCHOH· \rightarrow RCHO + H_3O^+ + e^- \tag{7}$$

At negative potential $(< -1.4$ V, for example, for methanol additive), the $CH_2OH·$ radicals are reduced to the alcohol.

$$CH_2OH· + H_2O + e^- \rightarrow CH_3OH + OH^- \tag{8}$$

From kinetic analyses of these systems, Barker was able to determine the rate constants for the reactions of ethanol and methanol with both OH· and H· radicals. Values of k were equal for each alcohol, with magnitudes of 2.4×10^7 mol^{-1} liter s^{-1} for the reactions with atomic hydrogen and 1.0×10^9 mol^{-1} liter s^{-1} for those with the OH· radicals. The constants agree well with pulse radiolysis data, and these chemical reactivity studies provide some of the best evidence for the presence of the solvated electron. The original references should be consulted for a detailed description of these results.

2. Theoretical: Metals

Firstly, this section describes Fowler's "square law" for thermionic emission from clean metals into vacuum, and some shortcomings of the existing

---

† Another way to conceive this process for dry electrons is that a standing wave, perpendicular to the electrode, penetrated the surface zone of the electrolyte.

theories are noted. Next, the quantum mechanical threshold theory of Brodskii and Gurevich[19] for photoelectronic emission into electrolyte solutions is outlined. Each of these theories has been successful in treating certain experimental data. A similar format is followed to introduce photoemission from semiconductors. Extensive reviews of photoemission theories are available elsewhere, e.g., Refs. 10, 20, and 21.

2.1. Fowler's Theory for Metal/Vacuum Interfaces

Fowler proposed a theory in 1931 which showed that the photoelectric current variation with light frequency could be accounted for by the effect of temperature on the number of electrons available for emission, in accordance with the distribution law of Sommerfeld's theory of metals. Sommerfeld's theory (1928) had resolved some of the problems surrounding the original models for electrons in metals. In classical Drude theory, a metal had been envisaged as a three-dimensional potential well (or box) containing a gas of freely mobile electrons. This adequately explained their high electrical and thermal conductivities. However, because experimentally it is found that metallic electrons do not show a gaslike heat capacity, the Boltzman distribution law is inappropriate. A Fermi–Dirac distribution function is required, consistent with the need that the electrons obey the Pauli exclusion principle, and this distribution function has the form

$$P(E) = 1/\{1 + \exp\left[(E - E_F)/kT\right]\} \qquad (9)$$

At ordinary temperatures, the Fermi energy $E_F \gg kT$. Fowler's law, referred to above as the "square law," is readily tested in practice since it predicts at $T = 0 \text{ K}$ [22,23]

$$I \propto (\nu - \nu_0)^2 \qquad (10)$$

where ν_0 is the photoelectric threshold frequency and ν refers to a range of energies from threshold to a few kT above threshold.

The theory begins by considering that an electron may be emitted when the energy of its motion normal to the surface exceeds the work function ϕ (other velocity components are, to a first approximation, irrelevant). The electron emission rate, therefore, may be expected to be proportional to the intensity of the light, the number of suitable electrons impinging per unit area of the surface in unit time, the probability that they will acquire the quantum energy $h\nu$ in the proper velocity component, and the chance that they then will be transmitted through the potential at the interface (the boundary field). By using the model of free electrons inside the metal, the whole excitation takes place in the surface field of the metal, and for the case when the boundary field is well represented by the image field (i.e., clean metals), the last probability hardly varies and may be taken as unity. Additionally, the probability of absorbing the quantum will vary with ν, as in other absorption phenomena, but this variation is not significant near the threshold frequency ν_0. Thus, to a good approximation, the photoelectric yield per unit light intensity near ν_0 is proportional simply to the number of electrons incident per unit time. The energy condition can be written, $(1/2m)p^2 + h\nu > \phi$, where p is the initial momentum of the electron normal to the surface. The escaping electrons have been termed the "available electrons."

Fowler's derivation for a single photon does not explicitly involve the quantum mechanical form of current; instead, a semiclassical flux of electrons arriving at the metal surface is used.[10] The electron gas in a metal will obey Fermi–Dirac statistics, and the number of electrons per unit volume having velocity components in the ranges $u, u + du, v, v + dv, w$, and $w + dw$ is given by the formula

$$n(u, v, w)\, du\, dv\, dw = 2\left(\frac{m}{h}\right)^2 \frac{du\, dv\, dw}{1 + \exp\left[\frac{1}{2}m(u^2 + v^2 + w^2) - E_F\right]/kT} \quad (11)$$

in which u is the velocity normal to the surface, E_F is the Fermi level energy for the metal, and m is the electron mass. E_F is constant to a good approximation and equals the highest-energy occupied electronic level which is filled at absolute zero. The number of electrons per unit volume, $n(u)\, du$, with their velocity component normal to the surface in the range $u, d + du$ is then given by:

$$n(u)\, du = 2\left(\frac{m}{h}\right)^3 \int_0^\infty \int_0^{2\pi} \frac{\rho\, d\rho\, d\theta}{(1 + \exp\left[\frac{1}{2}m(u^2 + p^2) - E_F\right]/kT}$$

$$= \frac{4\pi kT}{m}\left(\frac{m}{h}\right)^3 \log\left[1 + \exp\left(E_F - \tfrac{1}{2}mu^2\right)/kT\right] du \quad (12)$$

Using the above hypothesis, namely, that those electrons can escape which have sufficient energy when their kinetic energy normal to the surface is augmented by $h\nu$, it is possible to solve the integral for their number,

$$\int_{\frac{1}{2}mu^2}^\infty n(u)\, du$$

$$= \frac{4\pi kT}{m}\left(\frac{m}{h}\right)^3 \int_{\frac{1}{2}mu^2 = \phi_0 - h\nu}^\infty \ln\left\{1 + \exp\left[(h\nu - \phi) - \tfrac{1}{2}mu^2\right]/kT\right\} du \quad (13)$$

in which the potential step at the boundary is defined as $\phi_0 = \phi + E_F$ and ϕ is approximated to the thermionic work function.

The solution for the photocurrent has the form[22,23]

$$I \propto \frac{T^2 f(\mu)}{(\phi_0 - h\nu)^{1/2}} \quad (14)$$

where $f(\mu)$ is a series function resulting from the logarithmic expansion necessary to solve the integral. For a small energy range at the threshold, the approximation holds that $(\phi_0 - h\nu)^{1/2}$ is constant, and taking logarithms of Eq. (13) with the approximation $f(\mu) = (h\nu - \phi)/kT$ yields

$$\log \frac{I}{T^2} \approx \text{const.} + F\left(\frac{h\nu - \phi}{kT}\right) \tag{15}$$

If the variable $\log(I/T^2)$ is plotted versus the variable $h\nu/kT$, each metal provides a curve shifted from the theoretical curve by its threshold energy, ϕ/h or ν_0. The law has been well supported by experimental results, and the works of DuBridge and Roehr[24,25] in the early 1930s contributed appreciably to its acceptance.

A modification to this theory has been made by Matthews and Khan.[26,27] They pointed out that there was a discrepancy in considering the photocurrent as proportional to the number of available electrons per unit volume instead of the number of electrons available per unit area of surface per unit time. After analytical integration of the corrected form, they obtained the following expression for the photocurrent:

$$I \propto e_0 \frac{2\pi m}{h^3}(h\nu - \phi)^2 \tag{16}$$

where e_0 is the charge on the electron. This expression showed a better linear relation between the square root of the experimental photocurrent in vacuum with light frequency.

A better model for photoemission at the metal/vacuum interface requires a quantum mechanical account of the interaction of the photon with electrons. This was recognized by Wentzel, and such an approach was attempted by Fowler in his original paper.[22] The solution for the photoelectric current resulted in the form

$$I \propto \frac{(h\nu - \phi)^{3/2}}{(\phi_0 - h\nu)^{1/2}} \tag{17}$$

and when $h\nu = \phi$, $I \propto T^{3/2}$. This result did not fit the experimental data quite as well as the square-law relation in Eq. (10). However Fowler viewed this theoretical approach as equally plausible.

2.2. Tunneling through the Potential Barrier

In the presence of an interfacial electric field, the barrier height and width may be modified, which favours a tunneling process (Figure 2). Normally, when no electrical field is applied, the electron must overcome the barrier to be photoemitted, but in this case the combination of image forces and the applied field diminishes the barrier.[27] The expression for the photoemission current in vacuum and in the presence of applied field V can written as

$$I = e_0 \int n(E)P_T(E)\,dE \tag{18}$$

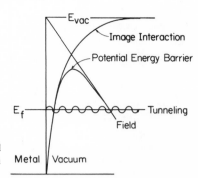

FIGURE 2. Electron tunneling through barrier formed by field and image potential at the metal/vacuum interface.[27]

where $n(E)$ is an expression similar to that given in Eq. (11), and $P_T(E)$ is the tunneling probability through a barrier of height E_m and width d. This probability can be written

$$P_T(E) = \exp\left[-\frac{\pi^2 d}{h}(2m(E_m - E))^{1/2} \right] \qquad (19)$$

This expression is the Wentzel-Kramer-Brillouin (WKB) tunneling probability for a parabolic barrier[27] and $n(E)\,dE$ is the Matthews and Khan modified form of the expression for available electrons. The complete photocurrent must include contributions for tunneling, i_t, as well as over-the-barrier currents, i_c;

$$I = i_c + i_t = e_0 \int_{E_m}^{\infty} n(E)\,dE + e_0 \int_0^{E_m} P(E)n(E)\,dE \qquad (20)$$

The barrier height E_m with respect to the bottom of the conduction band in the presence of an applied field is shown to be

$$E_m = E_F + \phi - h\nu - e^{3/2}V_0^{1/2} \qquad (21)$$

where V is the field in volts per centimeter, and $E_F + \phi = \alpha$. The barrier width d at different energy levels can be obtained from the relation

$$E = \alpha - \frac{e}{4d} - e_0 Vd \qquad (22)$$

where d is the perpendicular distance from the metal surface. This equation, after rearrangement, results in a quadratic in d as a function of E and yields two values of d, d_1 and d_2. Matthews and Khan[26] used the results from this approach to explain photoemission for the cases of medium and high fields. They observed that the square law was followed but the threshold energy differed in each case. Whereas the medium field gave a work function of 405 kJ/mol, the high field gave a value of 67 kJ/mol. This latter value is quite small and it is thought to be a consequence of the increased tunneling contribution due to the high field. In the case where the field changed and the light energy was constant, the plot of I^2 versus V was observed not to follow the square law. Matthews and Khan have explained this observation by considering that the field changes both the barrier width and height as one passes from a region of negligible tunneling to one of significant tunneling.

Photoemission from metals into vacuum under applied electrical field should resemble the electrochemical system better than the simple Fowler's law does. However, tunneling processes have not always been invoked to explain photoinjection from metals into electrolytes. In fact, one of the most accepted theories is the "$\frac{5}{2}$ rate law," to be described below. This theory does not take into account tunneling effects. So, although Fowler's theory has been accepted universally for photoemission into vacuum in the threshold frequency range, the improvement in the wave mechanical approach has been to solve for the partial flux more precisely, using methods for threshold creation phenomena.[28,29]

2.3 Quantum Mechanical Photoemission Theories for the Metal/Vacuum and Metal/Electrolyte Interfaces

This section summarizes the quantum mechanical threshold theories of Brodskii and Gurevich[10,19] for photoelectronic emission from metal electrodes into vacuum and into electrolyte solutions.

Metal/Vacuum Case

Details of the mathematical solution to this problem and the origins of the underlying physical assumptions are provided in the excellent monograph of Gurevich *et al.*[10] and the original literature.[19,30-33] It is helpful to compare this approach with Fowler's semiclassical theory. As before, the optical energy range of interest contains frequencies close to ν_0 (the near-threshold region) and the electrons considered photoemitted are those which can escape the electrode phase without undergoing additional inelastic (i.e., many particle) interactions. Many electrons fail to become emitted, and these can dissipate their energies internally in the metal in the order of picoseconds. The mathematical solution for the photocurrent involves calculating the first factor in the integral of Eq. (23) using self-consistent quantum mechanical methods,

$$I = e \int J_u(E, P_\parallel, \nu)\{1 + \exp\left[(E - E_F)/kT\right]\}^{-1}\rho(E, P_\parallel)\, dE\, dP_\parallel \quad (23)$$

in which E and $P_\parallel$ variables are the energy and the component of the momentum parallel to the surface of the initial electrons within the metal. The second factor describes the Fermi–Dirac distribution of electrons within the metal, as similarly used in Eq. (11). The third factor is the probability density function of initial states, which usually cannot be defined.

The solution for J_u uses the quantum mechanical current operator,

$$J_u[\Psi_f] = \frac{hi}{2m}\left(\Psi_f\frac{\partial\Psi_f^*}{\partial u} - \Psi_f^*\frac{\partial\Psi}{\partial u}\right) \quad (24)$$

in which Ψ_f is the time-independent wave function of the electron in its final state far from the emitter surface, Ψ_f^* is its complex conjugate, and $i = \sqrt{-1}$. As previously, the integration range is determined by the assumption that the tangential momentum components (p_v, p_w) are conserved when the electron crosses the boundary. In other words, there is no energy transfer from the component of the momentum vertical to the surface to the horizontal (parallel) plane components when the electron traverses the surface plane. Boundary fields affecting electron motion at the interface are assumed to be restricted to a layer thickness δ. At distances from the surface $u > \delta$, the electron can be treated as if it experiences a homogeneous potential $V(u)$, which becomes zero as $u \to \infty$. Thus, assuming $P_\parallel$ is conserved and choosing that the dependence of the wave function Ψ_f on v and w components is exponential,

$$\Psi_f = \Psi_u \exp\left[i(p_v v + p_w w)/h\right] \quad (25)$$

the basic one-dimensional Schrödinger equation becomes

$$\left[\frac{\partial}{\partial u^2} + \frac{P^2}{h^2} - \frac{2m}{h^2}V(u)\right]\Psi(u) = 0 \quad (26)$$

The potential due to image forces arising from the surface charge produced by the emitted electron is given by $V(u) = -e_0^2/4\varepsilon u$, where ε is the dielectric constant of the contiguous phase ($\varepsilon = 1$ for vacuum). Equation (26) can thus be written

$$\left[\frac{\partial}{\partial u^2} + \frac{P^2}{h^2} + \frac{me_0^2}{2h^2\varepsilon u}\right]\Psi(u) = 0 \quad (27)$$

The solution of Eq. (27) for $\Psi(u)$ is known as the Jost solution, and this can be substituted into Eq. (24) to obtain an expression for J_u. For the case of photoemission into vacuum, it is found that, indeed, the relation $J_u = \text{const.}$ holds, as originally postulated phenomenologically by Fowler. The magnitude of J_u reflects the presence of image forces and is characteristic of the external media (if other than vacuum). Physically, this result has been derived without any assumptions concerning the electron dispersion law for metals.[10]

As a consequence of this quantum mechanical derivation, it is remarkable that the general solution for the photocurrent predicts the same relation as Fowler's semiclassical model, namely,

$$I = \text{const.}\,(\nu - \nu_0)^2 \quad (28)$$

for ν in the threshold range of the final energies for emitted electrons. This result should apply to both bulk (volume) and surface transitions from metals into vacuum, provided that the transition is an interband one.

Metal/Electrolyte Case

The Brodskii–Gurevich theory for photoemission into an electrolyte solution differs from the vacuum case since it is assumed that for $u > h\nu_0$ the potential $V(u) = 0$ and the image forces are effectively screened. Gurevich and coworkers[10] have argued that the screening time, which is due to collective internal motion of electrons in the metal, is less than the time the wave packet requires to cross the interface. The precise, physical nature of this mechanism is unclear and, presumably, this type of relaxation could not be measured in practice. Any role of the ions or dipoles in the electrolyte in screening has been questioned[27,34] because their relaxation times would be too long. With the assumption of full screening, the appropriate form of the Schrödinger equation becomes

$$\left[\frac{\partial}{\partial u^2} + \frac{p^2}{h^2}\right] \Psi(u, p) = 0 \qquad (29)$$

Following a similar method of calculation to that above, the relation for the total photoemission current for concentrated electrolytes becomes

$$I \propto [h\nu - h\nu_0(0) - e\Psi]^{5/2} \qquad (30)$$

in which $\nu_0(0)$ is the photoelectric threshold at the potential of zero charge, and ϕ is the electrode potential on this scale (the choice of zero potential, in fact, can be arbitrary). This result, known as the "$\frac{5}{2}$-power law," has been tested by many investigators for photoemission into electrolytes and has given better results than Fowler's square law.[32]

Korshunov *et al.*[35] were the first to verify the $\frac{5}{2}$ law experimentally. Subsequently, evaluations were made by Rotenberg and Pleskov,[36] as well as DeLevie and Kreuser[37] and a number of other groups. In the DeLevie and Kreuser study, careful measurements of the photoemission current at a mercury electrode were made with modulated light ($\lambda = 436$ nm) and N_2O as electron scavenger. Fowler's square law, the $\frac{5}{2}$-power law and the logarithmic form were compared, and it was concluded that the $\frac{5}{2}$-power law gave a better linear fit of the experimental data. The photoemission current, which is the difference between the light and dark currents, usually is raised to the appropriate power (0.4 in this instance) and plotted versus the energy functions of either potential (E) or optical energy ($h\nu$) (Figures 3 and 4). Strictly, the potential should be corrected for the double layer capacity, but in concentrated electrolytes this contribution can be ignored (Section 2.5; Refs. 32, 38, and 39). Many of the verifications of this law have been made at mercury, an amorphous liquid electrode. However, the law appears to hold at a variety of polycrystalline metal electrodes which have a broad region of nearly ideal polarizability. Some systems that have been interpreted in this manner are collected in Table II.

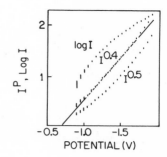

FIGURE 3. Experimental photocurrents (I) raised to power p, and log I versus potential; Hg/0.1 M KCl interface, 436 nm, N_2O scavenger.[37]

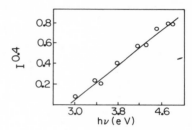

FIGURE 4. Experimental photocurrent fitted to $\frac{5}{2}$ law as function of optical energy.[35]

Before leaving this section, mention should be made of attempts to devise quantum mechanical theories for a one-step model of photoemission, i.e., treating the excitation of the photoelectron, its transmission within the metal and through the interface, and its capture in the condensed phase as a single event. Grider has described this approach in some detail.[20] The escape

TABLE II. Photoemission Metal/Electrolyte Studies Supporting the $\frac{5}{2}$ Law

Substrate(s)	System	Reference
Hg	H_3O^+ scavenger in $HClO_4$, H_2SO_4, H_3PO_4, and their salts	35
Hg, Tl(Hg), In(Hg)	H_3O^+ scavenger in H_2SO_4/Na_2SO_4	36
Hg	N_2O (saturated) in NaF	37
Hg, Pb, Tl(Hg), In(Hg)	N_2O in KF; H_3O^+ in Na_2SO_4 and KCl	32
Hg	N_2O in KF; H_3O^+ in KCl[a]	38
Zn, Cd, Pb, Tl, Ga, Sn, In, Sb, Bi, Al, Fe, Hg, Ni, Co,[b] Cu, Ag, Au, Pt	N_2O in aqueous NaF, $LiClO_4$, and H_2SO_4	40
Hg	N_2O in $LiClO_4$/DMF nonaqueous system	
Hg	N_2O, H_3O^+ in C_2H_5OH electrolytes	
Pb	H_3O^+ in KCl	41
Pt	N_2O, NO_3^- scavengers in KOH	42

[a] Modified law with inner Helmholtz potential correction for diluted electrolytes.
[b] Limited polarizability range prevented satisfactory test of law.

condition imposed by parallel momentum conservation creates the threshold for escape in the usual manner but the initial and final state wave functions inside the solid are written in the form of two-dimensional Bloch functions. These form the basis of many quantum mechanical calculations of crystalline solid and have the form

$$\Psi(x) = \Psi_0(x) U(x - a) \qquad (31)$$

where the solution of the one-dimensional Schrödinger equation for a free electron in a potential has the same period a as the periodic potential, $U(x)$. With Fourier series expansion, it can be shown that the solution for electronic energy in a one-dimensional periodic potential is parabolic. This methodology does permit account of the substrate epitaxy but it is difficult to describe the final state wave function and take account of surface effects. Several authors have equated this problem to that encountered in low-energy electron diffraction (LEED) of surfaces.[43-46]

2.4. Optical Polarization and Crystal Epitaxy Effects

A dependence of the photocurrent of the polarization state of the incident light is not accounted for by the $\frac{5}{2}$ law for bulk excitation. However, for both the metal/vacuum and metal/electrolyte cases, it is found experimentally that photocurrents do vary with the light polarization. In the models discussed thus far, free-electron behavior within the metal has been assumed and any electronic structure influences have been neglected (apart from the one-step model which has only been alluded to). Studies as a function of the light polarization and of crystal orientation thus are proving especially valuable to probe the angular dependency of electronic emission. Photoemission yield spectroscopy, a technique which uses polarized light and single-crystal substrates, has been described by Sass and Lewerenz.[47]

The problem, in essence, is to provide for an angular interaction by an accurate quantum mechanical description of the transition matrix for absorbance and reflectance. In other words, the field of the electromagnetic wave which penetrates the bulk metal will determine the photocurrent, provided that a correction is made for the reflected light. It has been shown that, approximately,[48]

$$I \propto \nu^{-4}(1 - R)g(\theta, \beta) \qquad (32)$$

in which ν^{-4} is a term arising from the frequency dependence on the matrix element of the one-photon photoeffect and R is the specularly reflected loss calculable from the appropriate classical, macroscopic Fresnel formulas, e.g., Ref. 49. Here θ is the angle of incidence and β is the angle between the plane of incidence and the plane of vibrations of the electric vector of the polarized wave. It is assumed that the field of the metal is constant within the photoemission zone. The function $g(\theta, \beta)$ describes the angular relation of the probability for photoexcitation. Usually, the term $g(\theta, \beta)$ will be expected to vary with the substrate epitaxy. On the other hand, if $g(\theta, \beta)$ can be regarded as approximately constant, the angular variation of the photoemission current will vary with the difference between the total incident light intensity and the reflected portion, i.e., with this approximation, it is possible to correlate the magnitudes of the photocurrents with the extent of optical energy dissipated by the substrate.

An earlier account of the angular dependency of the surface photoeffect was given by Korshunov *et al.*,[50] in which the assumption is made that only the lightwave component with its electric field vector polarized perpendicular to the metal surface (i.e., parallel to the plane of incidence) would contribute to the photocurrent (Figure 5). This condition arises because of the postulation of parallel momentum conservation of the emitted electrons which cross the interphasial boundary. Since the current, I, is proportional to the mean square field strength, $\langle E_{\parallel} \rangle^2$, and the relation

$$E_{\parallel} = E \sin \theta \cos \beta \tag{33}$$

holds, we have

$$I \propto \sin^2 \theta \cos^2 \beta \tag{34}$$

Subsequently, objections to this model were raised by Gurevich *et al.*,[10,48] who questioned the validity of inserting matrix elements of the type given in Eq. (34) into macroscopic (classical) formulas. An alternative physical model accounts for the real part of the permittivity of the metal becoming zero near to the surface. At this point, the field strength exhibits a sharp maximum and the functional dependence of the photocurrent becomes

$$I \propto \nu^{-2} \cos^2 \beta \tag{35}$$

Obviously, a more comprehensive treatment is needed for crystalline materials. In addition, experimental studies interpreted using this type of analysis have revealed that "surface" contributions must be included to supplement the bulk effects.

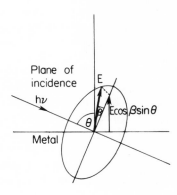

FIGURE 5. Macroscopic interpretation of polarized light vector interaction.[50]

Vectorial Photoeffects

A particularly helpful discussion of what they call the "vectorial photo-effect" by polarized light has been given by Sass and Gerischer.[51] The optical transition matrix element, in restricted form,[52] is given by three terms:

$$M_{fi} = -\frac{i}{\nu}\left[\langle f|A \cdot \nabla V_B|i\rangle + (f|\langle A \cdot \nabla V_s|i\rangle) - ih\left\langle f\left|\frac{\partial A}{\partial z}\right|i\right\rangle\right] \qquad (36)$$

These terms represent a bulk effect, a classical surface potential effect, and a surface field effect, respectively. Selection rules are stated through which these various contributions to the net photocurrent can be resolved. A complete theoretical description of photoemission, therefore, will enable resolution of each of these components as functions of the angles (θ and β), the available light intensity with depth, and the substrate epitaxy. This theory is discussed in detail in Ref. 51; note also Chapters 4 and 6 in this book.

2.5. Role of the Electrical Double Layer

Two questions can be asked and will be examined in this section; firstly, how does the electrical double layer at an electrode affect the photoemission process, and secondly, as one of the key aims of electrochemical experimenta-tion is to clarify our understanding of the structure and function of the double layer, can photoemission studies assist with determining its role in heterogeneous equilibrium and dynamic processes? Experiments have been performed from each of these viewpoints and these topics have been represen-ted in detail elsewhere.[10,39]

The Double-Layer Influence on Photoemission

As relatively less is known concerning electrical double layers at solid and semiconductor electrodes than at liquid mercury electrodes, it is not surprising that the majority of photoemission double-layer studies have been made at the latter. A detailed and clear account of the double layer itself has been given by Mohilner.[53] The photoemission influences broadly may be classified as primary if the electron emission step is affected directly, or secondary if subsequent reactions of the solvated electron with homogeneous acceptors in solution and/or the electrode are modified.

When the double-layer thickness does not exceed the de Broglie wavelength of the emitted electron and in the absence of specifically adsorbed ions and organic adsorbates, the interfacial structure has little influence on photoemission directly. In other words, the double layer in concentrated electrolytes is transparent and the electrons can tunnel through the inner and outer (diffuse) layers. Experimentally, variation of the electrolyte concentration using an uncharged scavenger (N_2O) permits two effects to be resolved. At high concentrations, the photoemission currents at mercury decrease as the concentration is increased due to a decrease in the mean solvation length,[54] whereas in dilute systems, the influence of the outer Helmholtz layer potential, Ψ_1, is evidenced. Under the latter conditions, the diffuse layer thickness is larger than the electron wavelength and the primary effect is to modify the migration of solvated electrons and any charged reaction products. At negative Ψ_1 potentials, for example, the rate of photoemission also is influenced in a manner similar to the Frumkin correction for the rate of electron transfer reactions,

$$I = \text{const.} \left[h\nu - h\nu_0 - e(\phi - \Psi_1) \right]^{5/2} \tag{37}$$

and the emission current is smaller than in concentrated electrolyte because an additional energy is required to overcome the diffuse layer potential barrier. This effect is compensated by an increase of the Debye screening length, κ^{-1}, with dilution. Experimental results and theoretical predictions as to which effects predominate are provided in Ref. 10.

The second, direct way that photoemission quantum yields can be modified is through scattering of electrons by adsorbed molecules or ions. In general, the behavior described above pertains; however, additional variations of the quantum yield can occur due to partial or total blocking of the surface. Specific anion adsorption of halides induces a negative shift of the Ψ_1 potential, which in turn modifies the migration of hydrated electrons in the vicinity of the point of zero charge.[55] Tetrabutylammonium cations, which provide monolayer coverages at relatively low bulk concentrations, affect the photocurrent both by altering the Ψ_1 potential and by a screening effect.[56] Adsorption of R_4N^+ ions decreases the photocurrents in N_2O-saturated 0.1 M KF, primarily due to blocking, but in dilute electrolytes the decrease is thought due largely to Ψ_1 potential effects again. Large organic molecules, which form thick surface layers, can cause appreciable diminution of the photocurrents. If the layer is sufficiently thick, an exponential rather than a $\frac{5}{2}$-power law results. Combination of photoemission and differential capacitance measurements permits adsorbed layer thicknesses and the dielectric constant of the film to be established.[10]

Grider[20] has studied photoemission from copper with adsorbed Br^-, I^-, R_4N^+ ion, thiourea, and pyridine. He proposed that the major effect attributable to adsorption of halide ions and thiourea was partial relaxation of the conservation of electron momentum parallel to the surface. In the case of R_4N^+ ion adsorption, the structure of the yield ratio of photocurrents produced by p- and s-polarized light shifted with photon energy, and this was explained by an increase in the local density of electrons near the surface due to the adsorption, which can cause scattering effects. Pyridine, upon adsorption, similarly affected the yield ratio structure at bulk concentrations $>10^{-6}$ M, but a clear understanding of this result remains to be attained. Grider et al.,[57] also have provided evidence that partial relaxation of parallel momentum conservation occurs for *in situ* photoemission from a single-crystal copper cylinder in 0.5 M H_2SO_4. A complete breakdown of parallel energy conservation resulted with thiourea adsorption in the same system. Apparently, photoemission is often dependent on surface conditions, and a general quantitative understanding of adsorption phenomena in terms of the parameters in Eq. (36) remains to be formulated.

Applications of Photoemission to Double-Layer Studies

At any particular potential, the thermodynamic work needed to transfer an electron from a metal surface into solution is determined only by the magnitude of the potential, regardless of the nature of the metal (the actual photoemissive work function may be slightly larger, depending on the internal momentum distribution,[10] but in what follows we shall equate the thermodynamic and photoemissive work functions). This independence of the threshold potential has been confirmed for liquid (e.g., Ref. 36), and solid metals (e.g., Ref. 40). At the electrified interface, it is well known that the electrochemical potential of species i, $\bar{\mu}_i$, is given by

$$\bar{\mu}_i = \mu_i + z_i F \Phi \tag{38}$$

in which μ_i is the chemical potential, and the inner (or Galvani) potential difference (Φ) is given by

$$\Phi = (\Psi + \chi) \tag{39}$$

where Ψ is the outer (or Volta) potential which is measurable, and χ is the surface dipole contribution resulting from the chemical interphasial interactions. The work function (which can be measured also) in these terms is

$$w_{ms} = \bar{\mu}_i - z_i F \chi \tag{40}$$

for the metal/solution transfer. From photoemission data, we may write approximately

$$w_{ms} = h\nu - e\Psi_0 \tag{41}$$

where ν is the light frequency and Ψ_0 is the threshold potential, which can be obtained from the photocurrent-voltage plot, $I^{0,4}$ versus E. Considering these relations for a mercury electrode held at the potential of zero charge, $E_{pzc} = -0.19$ V versus SHE, the following equation holds from a consideration of the thermodynamic cycle shown in Figure 6:

$$w_{ms} = w_{mv} + eE' + \Delta G_{sv} \tag{42}$$

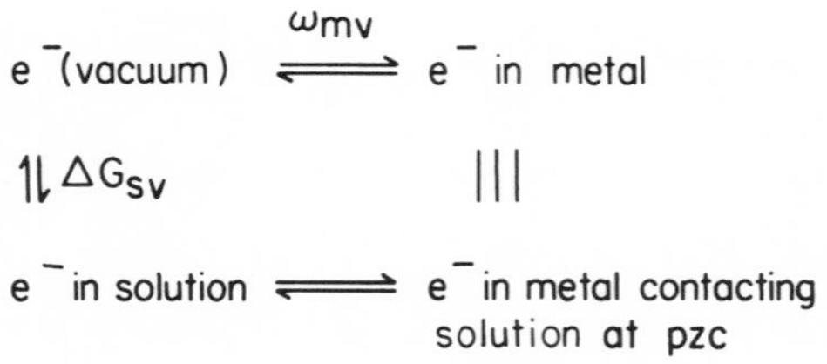

$$e^-(vacuum) \xrightleftharpoons{\omega_{mv}} e^- \text{ in metal}$$

$$\Big\updownarrow \Delta G_{sv} \qquad\qquad |||$$

$$e^- \text{ in solution} \rightleftharpoons e^- \text{ in metal contacting solution at pzc}$$

FIGURE 6. Thermodynamic cycle for photoemission at potential of zero charge.[10]

where the primes refer to measurements at the potential of zero charge. The term ΔG_{sv}, or the difference in work to transfer an electron from vacuum into the solution, represents the free energy of the interaction of an epithermal electron with a polar phase. Using the values chosen by Gurevich *et al.*,[10] $w'_{ms} = 2.97$ eV, $w_{mv} = 4.51$ eV, and taking the mercury/aqueous solution Volta potential at the pzc, $E' = -0.26$ eV, $\Delta G_{sv} = -1.28$ eV, which is larger (less negative) than the free energy of hydration of the electron. Experimentally, it is not the solvation energy of the hydrated electron that is measured in photoemission but that of the "dry" electron because no account is made of the subsequent thermalization and hydration, nor of the solution/vacuum surface potential. The quantity U_{sv} corresponds to the bottom of the conduction band of the liquid and comprises the energies of electron polarization, nonelectrostatic interactions between the dry electron and solvent molecules, and the work to transfer the electron through the solution/vacuum interface. This free energy is not always negative as in the case of water (more strictly aqueous electrolytes) but varies considerably with the solvent[10,51,58] (Table III).

A second use of *in situ* photoemission is the measurement of the potential of zero charge. At low concentrations and in the absence of adsorbates, it is to be expected that the diffuse layer effect as well as effects due to charged acceptors will be absent at the pzc and thus the photocurrent curves of different electrolyte concentrations should intercept. This is shown in Figure 7 for a Cd electrode. Values of the pzc have been obtained for a variety of metal

TABLE III. Examples of Conduction Band Energies[a] for Various Liquids versus the Vacuum Level

Liquid	Conduction band energy (eV)
H_2O	-1.25[b,c]
Formamide	-1.1[c]
CH_3OH	-0.9[b]; -0.19[c]
C_2H_5OH	-0.3[b]; -0.15[c]
Acetonitrile	-0.16[c]
CH_4	-0.1[d]
DMSO	$+0.07$[b]; -0.05[c]
C_2H_6	$+0.2$[d]
HMPA	$+0.22$[c]
Liq. H_2	$> +2.0$[d]

[a] Lower-edge energy of a dry electron in the solvent, comprising energies of transfer through solution–vapor interface, electron polarization, and (nonelectrostatic) interaction with solvent molecules.
[b] Ref. 10.
[c] Ref. 58.
[d] Ref. 51.

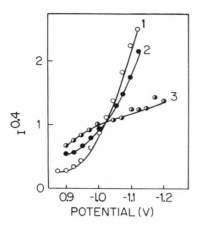

FIGURE 7. Determination of pzc at Cd in 0.001 M HCl with (1) zero, (2) 0.01 M KCl, and (3) 0.1 M KCl additions.[10]

electrodes, e.g., Hg, Pb, Bi, Cd, In, Sb, and the method may be less susceptible to surface conditions than the conventional differential capacitance technique. Photoemission also may be used to measure the potential at the outer Helmholtz plane, Ψ_1, and the adsorption characteristics of thick organic films on electrodes. Details are provided in Ref. (10) and the original literature.

3. Theoretical: Semiconductors

The electronic properties of solids usually are described in terms of the band model. When the isolated atoms, which are characterized by filled and vacant orbitals, are assembled into a lattice, new molecular orbitals form. These orbitals are so closely spaced that they form essentially continuous bands; the filled bonding orbitals form the valence band and the vacant antibonding orbitals form the conduction band. These bands may be separated by a forbidden region or bandgap of energy E_g. Depending on the magnitude of the bandgap energy, different types of conductors are obtained. For the case when the valence band and the conduction band overlap, the best conductors are obtained, i.e., metals. For larger values of E_g, semiconductors and insulators are obtained.

For the semiconductor/vacuum interface, photoemission investigations are a well-established branch of solid-state physics (see, e.g. Ref. 51, Chapter 11; 52, and 59). In particular, the theoretical work of Kane[60] and Gobeli and Allen[61] should be mentioned. Referring to Figure 8, the vacuum level, E_{vac}, is the energy at which an electron would emerge from the semiconductor's surface and appear in the vacuum with practically zero kinetic energy. The energy separating the edge of the conduction band from the vacuum level is the electron affinity, E_A. The thermodynamic work function, ϕ_{th}, in this case is the energy difference between the Fermi level and the vacuum level. The

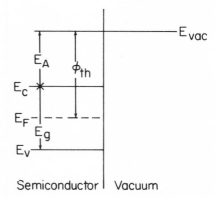

FIGURE 8. Energy levels at the intrinsic semi-
conductor/vacuum interface (no band bending
or surface states).

minimum conceivable energy necessary to achieve photoemission, E_T, is
given by

$$E_T = E_A + E_g \tag{43}$$

Actually, photoemission from semiconductors may proceed by volume (bulk)
or surface processes, but the most important of these is electronic transitions
from filled bands. This is because the conduction band normally is vacant and
the density of surface states is small relative to the crystal volume penetrated
by the radiation. To determine the actual value of E_T, the type of optical
interband transition must be taken into account—i.e., direct or indirect (Figure
9). Direct bulk transitions comprise the interaction of a single electron having
an initial energy E_i and a photon of energy $h\nu$. This results in the transition
of the electron to the second band. Since the photon momentum $(\sim h\nu/c)$ is
negligibly small, the quasi-momentum of the electron in the crystal does not
change; thus $\mathbf{P}_i = \mathbf{P}_f$, where $\mathbf{P}_f = \{\mathbf{P}_{f,x}, \mathbf{P}_{\parallel}\}$ is the momentum of the photoexcited electron in the conduction zone, and $\mathbf{P}_i = \{\mathbf{P}_{i,x}, \mathbf{P}_{\parallel}\}$ is the initial momentum of the electron in the valence band. In this case, every initial state at

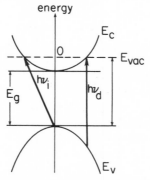

FIGURE 9. Direct and indirect transitions between parabolic semiconductor bands.

$E_i(\mathbf{P}_i)$ is associated with a final state at $E_f(\mathbf{P}_f)$ such that

$$E_f(\mathbf{P}_f) - E_i(\mathbf{P}_i) = h\nu \qquad (44)$$

Indirect transitions involve interactions with one or several phonons, as well as with trace impurities, vacancies, etc. In this instance, P_i differs from P_f. The momentum is conserved via phonon interactions. A phonon is a quantum lattice vibration; consequently, indirect transitions are less probable than the direct ones in high-purity crystals. For direct optical transitions and volume excitation, the photocurrent varies linearly with the frequency of the incident radiation,[61]

$$I \propto (\nu - \nu_0) \qquad (45)$$

As in the case of metal electrodes, photoemission yield measurements can be expressed in empirical "Fowler plots."

For *in situ* photoemission at the semiconductor/electrolyte interface, the generally accepted theory is that due to Gurevich and Pleskov.[10,62-64] This predicts a photocurrent–energy relation known as the $\frac{3}{2}$ law:

$$I \propto (h\nu - h\nu_0 - e\phi)^{3/2} \qquad (46)$$

Here the electrode potential ϕ is uncorrected for the electrical double layer contribution. As in the case of metallic electrodes, development of the theories for the vacuum system has aided our understanding of the more complex semiconductor/electrolyte interface.

3.1. Kane's Theory for Semiconductor/Vacuum Interfaces

The theory of photoelectric emission from semiconductors was developed by Kane in 1962. The semiconductor/vacuum system is more complicated than the metal/vacuum case. This is mainly because there is pronounced charge separation in semiconductors. Additionally, surface states may play an important role, and in many cases where potential is applied, the space charge region within the electrode has to be considered. Kane's theory may be divided into two categories: volume (or bulk) processes and surface processes (Table IV). For a direct, volume transition the "unity power law" is predicted and has been found experimentally for clean surfaces.[61] It will be developed in some detail below. The remainder of the power laws can be studied from the original paper.[60]

In order to start the mathematical development of this theory, several assumptions and conventions are introduced. The yield versus energy dependences are based entirely on "density of states" considerations assuming that matrix elements do not vary rapidly near threshold. The optical absorption should vary slowly near threshold. Strict energy conservation is always assumed. Energy losses are treated on an "all or nothing" basis, and this assumption is certainly not always adequate. Taylor expansions to lowest nonvanishing order for functions of the energy bands always are made around

TABLE IV. Threshold Energies and Energy Dependencies of the Photoelectric Yield at the Semiconductor/Vacuum Interface[a]

Excitation from:	Transition	Scattering process	Threshold	exponent
Valence band	Indirect	Unscattered	$E_A + E_c - E_v$	$\frac{5}{2}$
		Scattered		
	Direct	Unscattered	$E_A + E_c - E_v$	1
		Scattered	$E_A + E_c - E_v$	2
		Surface processes		
Valence band		Diffuse surface scattering	$E_A + E_c - E_v$	$\frac{5}{2}$
		Specular surface scattering	$E_A + E_c - E_v$	$\frac{3}{2}$
Discrete localized surface states below E_F			$E_A + E_c - E_{ss}$	1
Continuous distribution of surface states at E_F			$E_A + E_c - E_F$	2
Surface state band below E_F	Indirect		$E_A + E_c - E_F$	2
	Direct		$E_A + E_c - E_F$	1
Surface state band at E_F	Indirect		$E_A + E_c - E_F$	$\frac{5}{2}$
	Direct		$E_A + E_c - E_F$	$\frac{3}{2}$

[a] Ref. 60.

the threshold point. This latter approximation may go bad quickly when bands are close together in energy at the relevant threshold point in the k space: therefore, this approximation only applies for energies small compared to the appropriate band separations.

It is assumed that a complete set of Bloch states exist inside the crystal and join smoothly through the image barrier to free-electron states of equal energy and equal tangential vector. All calculations are made for a filled valence band and empty conduction band at $T = 0$ with no "band bending." By using the "golden rule,"[65] the rate of photon absorption in a semiconductor in the one-electron approximation is given by

$$T_{c,v} = \frac{2\pi}{h} \frac{e^2}{(mc)^2} \frac{2V}{(2\pi)^3} \int_{B\cdot Z} |\tilde{A} \cdot \tilde{P}_{c,v}(k)|^2 \delta[E_c(k) - E_v(k) - E] \, dk \quad (47)$$

where $T_{c,v}$ is the transition probability between energy eigenstates; subscripts c and v refer, respectively, to the conduction and valence bands. It is assumed that all transitions are from valence band to conduction band. $\tilde{A}$ is the vector potential of the light and V is the volume. A factor of 2 for spin is included, the integration goes over the Brillouin zone (BZ), and transitions are considered between pairs of bands with energy $E_i(k)$ where subscripts $i = c$ or v. E is the photon energy. The condition for escape may be written

$$\frac{\partial E_c(k)}{\partial k_n} > 0 \quad \text{where} \quad E_c(k) = \frac{h(k_t + k_n)}{2m} \quad (48)$$

Both expressions are valid as long as scattering and the energy loss in the volume and at the surface are neglected. The term k_t is the tangential vector which is conserved in the emission process but the normal k vector, k_n and k' in solid and vacuum, respectively, is not. The energy zero is taken to be an electron at infinity in vacuum. The quantum yield, electrons per photon, denoted by Y, then is given by

$$Y = \frac{\int' \sum_{c,v} |\tilde{A} \cdot \tilde{P}_{c,v}(k)|^2 \delta[E_c(k) - E_v(k) - E] \, dk}{\int_{BZ} \sum_{c,v} |\tilde{A} \cdot \tilde{P}_{c,v}(k)|^2 \delta[E_c(k) - E_v(k) - E] \, dk} \tag{49}$$

where the prime denotes integration over those values of k satisfying condition (48). Since the main interest is in the energy dependence near threshold of the quantum yield, it is assumed that the range of integration of the numerator in (49) collapses to a single point, or to a set of points equivalent by symmetry, k_d.

It was stated above that the energy dependence of the yield is based on "density of states." However, the calculation of the density of states in a three-dimensional system is frequently a difficult task,[66] requiring sophisticated numerical methods.[67] In this brief overview, only the basic outline of the mathematical derivation is possible. The density of states formally is reduced to a surface integral,[66,68] the energy conservation and escape conditions are imposed, and by expansions and approximations it is possible to derive that the yield, Y, is given by

$$Y = \gamma(E - E_T) \tag{50}$$

in which γ can be considered constant if the absorption and $P_{c,v}(k)$ are varying slowly in the vincinity of k_d. Linear dependences of this sort have been observed by Gobeli and Allen[61] and Scheer and Van Laar.[69]

Experimental Verification

Gobeli and Allen[61,70–72] studied Kane's theory in depth using atomically clean Si and Ge crystal surfaces obtained by cleavage in high vacuum.[73] The observed spectra and their dependence on sample doping are interpreted as being due to the volume excitation process which is modified by space charge band bending effects. They also demonstrated the existence of at least two types of emission processes near threshold.[71,72] Just above threshold, the yield followed an approximate cube law which, within experimental accuracy, is in agreement with the predicted "$\frac{5}{2}$-power law" for the volume, indirect excitation process. At higher photon energies a linear dependence sets in (Figure 10). This behavior can be expressed in the form of a sum of two processes:

$$Y(h\nu) = \gamma_1(h\nu - E_T^i)^3 + \gamma_2(h\nu - E_T^d) \tag{51}$$

where γ_1 and γ_2 are constants containing the light intensity and the absorption efficiencies for production of excited electrons with final states lying above the vacuum level. In general, it is expected that the spectral yield should rise

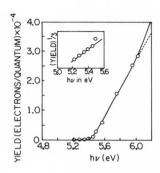

FIGURE 10. Photoyield/optical energy dependency for the p-Si/vacuum interface. Insert shows cube-root plot for foot of wave.[61]

from the indirect threshold (cubic law) and then show an abrupt increase as the direct photoelectric threshold is exceeded and this more efficient process dominates. To fit the data, the cubic term must drop out a few tenths of an electron volt above E_T^d (direct transition threshold energy). Proof of such direct and indirect processes is possible with a polarization experiment.[71] The direction of emission of electrons outside the surface was shown to depend on the direction of polarization of the incident light. Scattering would have destroyed the electron "memory" of the excitation process.

The low-efficiency component characterized by the threshold value E_T^i corresponds to the uppermost filled electronic states of the semiconductor. In principle, filled surface states lying between the valence band edge and the Fermi level at the surface can yield photoelectrons. This fact raises the problem that E_T^d cannot be identified with W, which is the energy difference between an electron in the highest valence band state at the surface and vacuum level because the E_T^d value exceeds W by the amount of kinetic energy, ΔE_v. On the other hand, E_d^i must correspond either to W or to a value between W and ϕ, where ϕ is the true work function which marks the position of the filled surface states. Nonetheless, E_d values have been considered by many researchers to be equal to W as long as surface effects (i.e., surface photoelectric effects) are not important. Fischer[74] also has concluded from his studies that surface state photoemission can occur in principle but only as a departure from the "cubic power law" in the photoemission tail.

Finally, it should be pointed out that the experimental conditions are very critical for semiconductor/vacuum photoemission. It has been demonstrated, for example, that nitrogen or oxygen interactions on the clean surfaces of cleaved crystals can produce drastic changes in the rate law and efficiencies of photoemission.[72] Therefore, a great deal of work has been done in developing techniques to produce "perfect" surfaces. As will be seen below, surface conditions undoubtedly play a strong role in the case of semiconductor/electrolyte photoemission.

Gurevich[63] has developed a theory for the case of electronic photoemission into concentrated electrolyte solutions. The approach is analogous to the metal/electrolyte case. The main feature of this theory is based upon the threshold theory of emission, and the law is well known as the "$\frac{3}{2}$-power law."

3.2. Gurevich's Quantum Mechanical $\frac{3}{2}$ Law for In Situ Photoemission

Electronic emission from semiconductor surfaces into electrolytes, as for photoemission from metal electrodes, is energetically feasible at frequencies roughly in the visible and near-ultraviolet parts of the spectrum. However, there is a whole series of qualitative features which differ from the metal case. The first difference arises from the fact that the chemical potential, μ, which determines the value of the thermodynamic work function, ϕ_{th}, is located at the forbidden band.[10] This means that electrons with initial energies E_i (where E_i is equal to μ) are absent within the semiconductor. Thus, in order to have a single photon emission, the required energy is $h\nu_0$, which is equal to the photoemission work function, ϕ_p. Energy $h\nu$ will therefore be equal to $E_g + E_A$, where E_g is the bandgap energy and E_A is the electron affinity. The values of ϕ_p and E_A depend not only on the bulk properties of the crystal, but also on the properties of the medium and the interface.

To develop the $\frac{3}{2}$ law, the first condition to be considered is that the band transition must be direct. Therefore,

$$P_i = P_f \tag{52}$$

and

$$E_f(P_f) - E_i(P_i) = E(\nu) \tag{53}$$

Expression (53) is the law of conservation of energy. Another condition is that the energy of the incident photon must be greater than $\phi_p [E(\nu) > \phi_p]$; hence, $E(\nu)$ is greater than $E_g + E_A$. The difference $E(\nu) - \phi_p$ depends on the scattering law in the bands. The next condition is based on strictly direct transitions:

$$P_{i,x} = P_{f,x} \tag{54}$$

This holds as long as the electrons are excited at sufficient depth in the crystal. In mathematical terms, $a_x/d \ll 1$, where a_x is the lattice constant in the x-direction and d is defined in the simplest case as the smaller of the following characteristic dimensions; the distance of attenuation of the field of the light wave in the semiconductor, d_ν, and the pathlengths of the photoexcited electron as regards inelastic and elastic interactions, d_{in} and d_{el}. Usually d_ν and d_{el} are much greater than a but d_{in} must be greater than a. If the condition $a/d \ll 1$ is not met, the concept of direct transition no longer applies and the picture set forth above must be modified.

A second important difference between photoemission from semiconductors and that from metals results from the law of change of density of states for the states which contribute to photoemission in the near-threshold region. For the case of metals, the initial energy states lie in the vicinity of the Fermi level and the density of states is a slowly changing function. However, for the semiconductor case, the density of states reverts to zero at energies corresponding to the photoelectric threshold, since the highest filled state of energy $E_v(P_v)$ coincides with the top of the valence band. This leads to a relatively large contribution by low-energy electrons to the emission current and affects the character of the dependence of I_p on the difference $E(\nu) - \phi_p$.

To begin the formal treatment of this theory, an analogous expression to Eq. (23) is used[63]

$$I_p = \int J[\nu, E_f(P_f), p_{\parallel}]F[E_i(P_i), P_{\parallel}]\rho[E_i(P_i), P_{\parallel}] dE_i(P_i) dP_{\parallel} \qquad (55)$$

where $J[\nu, E_f(P_f(P_f)), P_{\parallel}]$ is the photoemission current due to electrons with an intial energy $E_i(P_i)$; $E_f(P_f) = E_i(P_i) + E(\nu)$; the components $P_{\parallel} = \{P_y, P_z\}$ of the quasi-momentum $P_{\parallel}$ parallel to the surface; $F[E_i(P_i), P_{\parallel}]$ is the filling probability; and $\rho[E_i(P_i), P_{\parallel}]$ is the density of the corresponding initial states.

The law of conservation of the tangential quasi-momentum has to be taken into account as in the case of metals. The integration range in Eq. (55) is determined by the laws of energy and momentum conservation, as mentioned previously in the metal/electrolyte case:

$$2m[E_i(P_i) + E(\nu)] - P_{\parallel} > 0 \qquad (56)$$

Finally, as for that case, there is a restriction with respect to the investigated radiation frequency, namely, the region of frequencies of interest lies at the near-threshold.

The value of $J[E_f(P_f), \nu, P_{\parallel}]$ is obtained in the same manner as for the case of metals, by using the threshold model and the quantum mechanical operator J_x. Hence, as a result, Eqs. (57) are obtained:

$$J_x = \frac{P}{m} \frac{|\Lambda|^2}{|f_{(0,p)}|^2} = \frac{J_x\{f(x)\}}{|f_{(0,p)}|^2}|\Lambda|^2; \qquad J_x\{f(x)\} = \frac{P}{m} \qquad (57)$$

Here, $|\Lambda|^2$ is a constant independent of $P_{v,x}$ and $P_{\parallel}$, and $f_{(0,p)} = f_{(x,p)_{x=0}}$. Once again, making analogy with the metal/electrolyte case, the acquisition of a value for $J_x\{f(x)\}/f_{(0,p)}|^2$, which is determined by the limiting condition as $x \to \infty$, requires no knowledge of the solution for the internal problem ($x < 0$). It is sufficient to know the solution outside the crystal when $x > \delta$, where δ is of the order of angstroms (δ is the width of the transition zone). Moreover, the passage through the transition zone is assured as long as the condition $|d \ln f(x)/dx| < 1$ at $x = 0$ is fulfilled. The condition pertains in the development of the "$\frac{5}{2}$-power law." Returning to $|\Lambda|^2$, it should be noted that this constant quantity retains here its approximate nondependence on $E_f(P_f)$, which is the final energy of the emitted electron, if the width of the energy range $E_f(P_f)$ is smaller than the energy of the electron's motion in the region $x < 0$ and the resonant levels are absent. Therefore, for $|\Lambda|^2$ to keep its feature of having a constant value, the following condition must hold[9,63]; $E_f(P_f)/x \ll$ 1. Notice that a similar inequality is required for $|\Lambda|^2$ to be a constant for the metal/electrolyte derivation, except that there the inequality is $E_f/E_f \ll 1$ and obviously E_f must be replaced by χ for the semiconductor case.

We now study the frequency dependence of the photoemission current. As a first consideration, surface photoemission will be developed. Similar to the case of metal/electrolyte junctions, image forces are neglected in the mathematical treatment by assuming that the dense part of the double layer can be included in the region $0 < x < \delta$, so that the potential $u(x)$ can be considered negligible in the $x > \delta$ region. This assumption is valid also for volume photoemission processes into an electrolyte. Consider firstly $E_p(\nu) < E(\nu) < E'(\nu)$, where $E_p(\nu) = \phi_p$ and $E'(\nu) > E_p(\nu)$. Here, photoexcitation occurs in the vicinity of the surface (surface photoemission), and the emission of initial electrons found in the vicinity of the upper edge of the valence band is possible. Under these conditions, direct transitions in the bulk which lead to photoemission are energetically forbidden. Photoemission occurs only by way of indirect transition where $P_i = P_f$, due to the presence of the surface or to participation in the interaction of phonons, impurities, defects, etc. (bulk photoeffect).[64] If for this case the effective mass approximation, m^*, can be used, $E_i(P_i) = E_v - P_i/2m^*$, where E_v is the top of the valence band and is less than zero and m^* is the effective mass of the electron in the valence band before excitation, $m^* > 0$. With the appropriate escape conditions and simplifications, it has been shown that the general result predicts

$$I_p = \beta[E(\nu) - E_p(\nu)]^2 \tag{58}$$

where β is independent of the term $[E(\nu) - E_p(\nu)]$.

Now, if the incident energy is increased to a point where it is much greater than the photoemission work function or threshold energy, ϕ_p, and even greater than $E'(\nu)$, the condition $E(\nu) > E'(\nu)$ holds. At this level of energy, electrons can be photogenerated in the bulk with sufficient energy to escape from the semiconductor. The phototransition taking place will be of the direct mode, namely, there is momentum conservation, $P_i = P_f$. Assuming that the resulting excitation occurs between two parabolic bands with an extremum at $k = 0$, the parabolic dispersion law is applicable to this case. This law is basically the so-called "effective mass theory,"[68,75,76] wherein the initial and final energies of the electron are defined as follows:

$$E_i(\mathbf{P}_i) = E_v - (P_{i,x}^2 + P_{\parallel}^2)/2m_v^* \tag{59}$$

$$E_f(\mathbf{P}_f) = E_c + (P_{f,x}^2 + P_{\parallel}^2)/2m_c^* \tag{60}$$

By now applying the necessary restrictions, and performing the appropriate integration, the following expression results:

$$I_p = C[E(\nu) - E'(\nu)]^{3/2} \tag{61}$$

that is, the "$\frac{3}{2}$ law." Here C is a constant independent of the energy difference $E(\nu) - E'(\nu)$.

This quantum mechanical treatment can be used also for describing the semiconductor/vacuum interface. As for the metal/vacuum system, image forces play an important role. In fact, if image forces are considered in Gurevich's theory,[63] a similar expression to Kane's result is obtained, namely, the "unity power law." The theory predicts that the $\frac{3}{2}$-power law may be obeyed for emission into an electrolyte in all cases where the "unity power law" holds for emission into vacuum. Kane dervied a $\frac{3}{2}$ law for the vacuum case which pertains to emission from surface states in which the transition is direct and the threshold point lies at the Fermi level, E_F (Table IV).

To summarize the semiconductor/electrolyte theory of photoemission,

$$I_p = \begin{cases} 0 & \text{for } E(\nu) < E_g + E_A \\ B[E(\nu) - E_p(\nu)]^2 & \text{for } E_p < E(\nu) < E'(\nu) \\ C[E(\nu) - E'(\nu)]^{3/2} & \text{for } E(\nu) > E'(\nu) \end{cases} \tag{62}$$

where $E_p(\nu) = \phi_p = E_g + E_A$, $E'(\nu) = E_p + (m_c^*/m_v^*)E_A$, and $E(\nu)$ is the energy of the incident photon. B and C are constants which are independent of $[E(\nu) - E_p(\nu)]$ and $[E(\nu) - E'(\nu)]$, respectively.

Finally, it is necessary to examine the effect of an applied external potential, ϕ. If a semiconductor electrode is polarized, the potential is distributed in a complicated manner between the space charge region in the semiconductor and the Helmholtz double layer.[10,64,77] Knowledge of electrical double layers at semiconductor electrodes is extremely scant,[78] so any effects that arise are not well quantified (Section 2.5) and the dependence of I_p on ϕ may be complex. It is usual to work with concentrated electrolytes to minimize Helmholtz double layer effects. Unlike the case of photoemission at metal electrodes, however, it is considered that an external potential does not appreciably change the photoemission process at the semiconductor/electrolyte interface. This is because generally the distance over which the space charge potential drop occurs is far larger than the de Broglie wavelength of the electrons. Accordingly, the influence of an applied external potential (as a first approximation) is to introduce only an excess potential drop in the Helmholtz layer. With no band bending (flat band, fb),

$$E_A(\phi) = E_A(\text{fb}) + e(\Psi_1) \tag{63}$$

and for the most important condition of Eqs. (62), the direct transition case,

$$I_p = C[E(\nu) - E_0'(\nu) - e(1 + m_c^*/m_v^*)\Psi_1]^{3/2} \tag{64}$$

For highly doped samples, tunneling contributions may be present as a consequence of extreme bending (Section 3.4), in addition to effects on the threshold energy.

Photoemission from Surface States

The role of surface states has been discussed theoretically by Gurevich et al.[10] but no experimental results are available at this time. The simplest case is when the surface states are assumed to form a band. If the upper energy of this band is less than that of the Fermi level, the band will be completely filled. Using a parabolic dispersion law, Gurevich et al. have calculated that the photocurrent from a filled band would obey a relation of the form

$$I_{ss} \propto (h\nu - h\nu_0^{ss})^{3/2} \qquad (65)$$

This has the same exponent dependency as direct volume transitions. Such currents are not easily distinguishable therefore, especially if their quantum efficiencies are low relative to that of the bulk semiconductor or if indirect transitions occur at similar energies. If the surface state band is partially filled, the photocurrent relation with this theory becomes exponential:

$$I_{ss} \propto \exp\left[(E_F - E_l)/kT\right] \qquad (66)$$

where E_l, the lower-energy edge of the surface state level, is larger than the Fermi energy. In this case, the dependence of the surface state photocurrent on potential would be exponential.

Experimental Verification

According to Gurevich's theory, the dependence of I_p on λ is given by the $\frac{3}{2}$ law, while the dependence of I_p on Ψ has a more complicated character if double-layer effects are present. Experimentally, Gurevich's law has not been widely tested; in fact, very few studies of semiconductor/electrolyte photoemission have been made. It is usual to test the theory with plots of (photocurrent)$^{0.66}$ versus optical energy at constant potential or of (photocurrent)$^{0.66}$ versus potential at constant $h\nu$. The latter experiments are easier to perform. One of the difficulties in studying photoemission from semiconductors is that photocurrents are possible due to electron transfer reactions at the valence and conduction band edges, as discussed in Chapter 9 of Ref. 10. Hence, these photofaradaic currents (and possibly tunneling contributions) have to be distinguished from photoemission currents. A thorough study of the dark background of the electrolyte in the presence and absence of the scavenger is advised. Severeyn and Gale[79] have reported "anomalous" dark catalytic currents using N_2O in oxygen-free 0.1 M LiCl at freshly etched p-GaAs electrodes. Such currents occurred at about 300 mV negative of the threshold for photoemission at $\lambda = 365$ nm. A second problem is the surface preparation and condition. These topics are discussed more fully in Section 4.2.

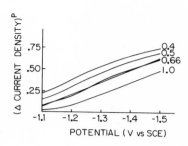

FIGURE 11. Difference (N_2O – Ar) photocurrents for the p-GaAs/0.1 M KCl electrolyte (pH 11.4) raised to various exponents as a function of electrode potential.[84]

Krotova and Pleskov[80] were first to appraise the $\frac{3}{2}$-power law using p-Ge in 0.2 M KCl with N_2O scavenger. This work, together with subsequent studies at p-Ge,[81,82] p-GaAs,[21,83,84] and p-InP[21] do show agreement with a $\frac{3}{2}$-power law. For example, in the recent study by Severeyn and Gale,[84] the photoemission current differences in proper form, $(i_{N_2O} - i_{Ar})^{0.66}$, followed a linear relation with voltage to the very foot of the onset threshold (Figure 11). However, this raises a further important problem; namely, the theoretical behavior of the volume process has a similar form to that of surface states. Unfortunately, too, the above semiconductors irreversibly form surface oxide films in aqueous electrolytes. Figure 12 illustrates the decrease in the photocurrent as a function of time of contact between freshly etched GaAs and aqueous 0.1 M LiCl. In this example, the rate dependency changes from a $\frac{2}{3}$ exponent to exponential form as a film grows on the electrode. So, although the quantum efficiency for direct, volume transitions may be far larger than that for a surface process, the extent of electron scattering by interfacial compound formation might be a modulation factor at III-V semiconductor/aqueous interfaces. Additionally, bulk photoemission has to be discriminated from that arising from surface regions and films.

There have been some recent investigations of photoelectron injection into electrolytes which examine the fate and energetics of the solvated electron in liquid ammonia, rather than the process rate laws.[85-88] Vondrak and coworkers[89] have proposed a procedure to estimate the effective threshold energy from the integral current produced by white light versus $1/T$. The method was tested with oxidized W, Na_xWO_3, and TiO electrodes.

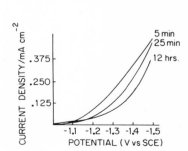

FIGURE 12. Time dependence of difference currents (conditions as for Figure 11) (by permission).[84]

3.3. Bockris and Uosaki Treatment

A general theory has been proposed by Bockris and Uosaki[27,90-93] in which it is considered that the charge transfer from a semiconductor to an acceptor in the electrolyte is the rate-determining step. Their approach is quite different from Gurevich's treatment; in fact, they take into account many features that are neglected by Gurevich, such as the effect of the interfacial properties, penetration through the barrier, and the probability of the acceptor states in solution. Basically, the model contains three regions wherein photocurrents are produced (Figure 13). These regions define the potential energy barrier that the electrons must tunnel through or pass over. The potential barrier for electron transfer from the semiconductor to an acceptor in the outer Helmholtz plane (OHP) is constructed considering the following interactions:

1. Interaction of an electron with the dipole potential of adsorbed solvent, U_{H_2O}.
2. Interaction with ions in the OHP and their images, U_{im}. This is a major difference from Gurevich's approach.
3. Optical Born energy of the electron. This contribution is due to the interaction of the electron with the second layer of solvent molecules. Apparently, this type of interaction is quite complicated to evaluate; hence this second layer is considered to be a continuum. This means that the electron derives energy from the continuum due to optical Born charging according to

$$U_{OpBorn} = -e_0^2(1 - 1/E_{op})/2r_c)\tag{67}$$

where r_c is the radius of the electron cavity and E_{op} is the optical dielectric.

The barrier maximum, U, was calculated with respect to the bottom of the

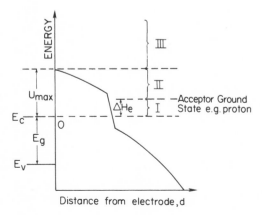

FIGURE 13. Bockris–Uosaki model of the potential energy barrier for electron transfer at p-type semiconductor/solution interface. U_{max} is the maximum barrier height with respect to the bottom of the conduction band.[27]

conduction band and is given by

$$U = E_A + U_{H_2O} + U_{im} + U_{OpBorn} \tag{68}$$

As mentioned above, the three regions in Figure 13 contribute to a net photocurrent. At energies corresponding to region I, the electrons pass through the barrier and are scavenged by the acceptor, e.g., proton. The current for this region can be expressed by

$$I_{P_I} = e_0 \frac{C_A}{C_T} \int_0^{\Delta H_e} N_e(E) G_A(E) P_T(E)\, dE \tag{69}$$

in which C_A and C_T are the total number of acceptors and of sites, respectively, per unit area of the OHP, and ΔH_e is the enthalpy for the electron transfer from semiconductor to acceptor. The first term in the integral, $N_e(E)$, is the number of electrons with energy E which strike the semiconductor surface per unit time and area; $G_A(E)$ is the distribution function for the vibrational-rotational states of an acceptor; and $P_T(E)$ is the WKB tunneling probability of an electron across the potential barrier of height U_m and width l (Section 2.2). The reader is referred to Ref. 27 and the original articles[90-93] for details of the mathematical derivations involved.

In the second region, the photoelectrons do not find an acceptor state at the level of the conduction band but are trapped by states in the solvent to become solvated electrons. The current density for this region may be represented by

$$I_{P_{II}} = e_0 \frac{C_A}{C_T} \int_{\Delta H_e}^{U_{max}} N_e(E) P_T(E)\, dE \tag{70}$$

Finally, in the third energy region, the photoelectrons may pass over the barrier and into the solvent. The expression for the current is now given by

$$I_{P_{III}} = e_0 \frac{C_A}{C_T} \int_{U_{max}}^{\infty} N_e(E)\, dE \tag{71}$$

and the net photocurrent is given by

$$I_{net} = I_{P_I} + I_{P_{II}} + I_{P_{III}} \tag{72}$$

Note that in regions II and III, when the electron is accepted by the solvent, solvated electrons are formed which means that Eqs. (70) and (71) are applicable to photoemission of electrons. Calculations by Bockris and Uosaki have given fair agreement with experiment, except the quantum efficiencies were lower than the experimental ones. This is in no way surprising because there is a cumulative set of uncertainties from all parameters about which assumptions had to be made. From a qualitative viewpoint, this theory is the most comprehensive in dealing with the fate of the electron in the solvent. Unfortunately, photocurrent rate expressions are not derived for experimental testing.

3.4. Hot Carrier Effects: The Nozik–Williams Model

To conclude this section on semiconductor theories, hot carrier injection of electrons from semiconductors into electrolytes should be discussed. In photoelectrochemistry, it generally has been assumed that the energy of injected photogenerated carriers is given by the position of the minority carrier band edge, i.e., the Gerischer postulate.[94] The occupational probabilities of the electronic states are described in terms of position-dependent quasi-Fermi levels for electrons and positive holes which are assumed to be separately equilibrated amongst their respective bands and defect states. This model, or view, has been modified by Nozik and coworkers,[95-104] who approximated the region of the semiconductor/electrolyte interface as a semiconductor heterojunction. This focused attention on the irreversibilities associated with minority charge carrier injection. The irreversibilities refer to a model in which the minority carrier does not undergo thermalization due to carrier–phonon collision prior to injection into the electrolyte. In other words, the minority carrier is injected without full intraband relaxation or being photoemitted in the usual sense and the process is termed "hot carrier injection."

In temporal terms, this mechanism can occur if the thermalization time, τ_{th}, of the photogenerated carriers in the semiconductor space charge layers is greater than both the charge transfer (or tunneling) time, τ_t, of the carriers into the electrolyte and the effective relaxation time of the injected carriers in the electrolyte, τ_r (Figure 14). One can summarize that the major criterion for this process to happen is that $\tau_{th} > \tau_t, \tau_r$. Values for the various characteristic times have been estimated using classical[102] and quantum mechanical approaches.[97,99] For the quantum mechanical approach, the electrolyte is considered as a large-bandgap semiconductor, and the photoexcited carriers

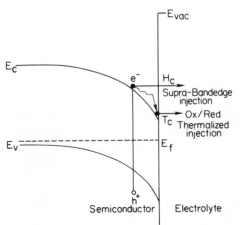

FIGURE 14. Hot carrier injection for p-type semiconductor.

find themselves in the potential well created by the position-dependent potential in the semiconductor/electrolyte barrier. This well has characteristic quantum levels,[97] and carriers can be injected from these levels into the electrolyte. The mathematical outline of this quantum mechanical treatment is well developed in Ref. 97, and the results obtained are comparable to those based on classical approaches.[102] Hot carrier injection is favored in semiconductor electrodes which have a low effective mass for the minority carrier. This is because this low effective mass will produce more widely spaced quantized levels in the depletion layer, which then results in long intraband thermalization. Two types of hot electrons can be distinguished. Type I are those injected at energies between the conduction band edge in the bulk and that at the surface and require an electronic field in the depletion region for their formation. Type II hot electrons depart the semiconductor at energies above the conduction band edge in the bulk. It is thought that the occurrence of hot carrier injection in photoelectrochemical reactions might significantly improve practical cell performances.

Some experimental evidence is available which supports the occurrence of hot carrier injection processes.[96,101,103]

4. Experimental Techniques

The first consideration is the choice of model system, and this may involve an aqueous electrolyte or nonaqueous media such as an organic electrolyte or molten salt. The selection of electrolyte and scavenger will depend on the energetics of the electrolyte reduction (good stability to reduction is needed) and upon the acceptor reactivity. Apparatus and electrochemical cells for metal/electrolyte and semiconductor/electrolyte systems are similar generally. Some guidelines are presented below to assist with experimental practice (see also Chapter 3 in Ref. 10). Theses, also, are a good source of practical information, e.g., Refs. 20, 21, and 81.

4.1. Choice of Scavenger and Electrolyte

Often, the scavenger should react rapidly with solvated electrons but be nonelectroactive at the electrode in dark to large negative potentials. The review of the physical properties and chemistry of solvated electrons by Matheson[105] is helpful in the selection of suitable scavengers. It is helpful, too, to choose an acceptor whose kinetic parameters and follow-up chemical reactions are known, unless the interest is in the reaction chemistry of the scavenger *per se*. Nitrous oxide has proven to be an excellent acceptor for aqueous systems, both at metallic and semiconducting electrodes. This is because it is not readily reduced at most electrodes (-1.8 V versus SCE at Hg), it is a neutral molecule, and its homogeneous chemistry and reaction products are well characterized.[79] Additionally, it is available as a high-purity product for medical use. There have been very few rate studies of semiconductor photoemission in nonaqueous systems to date[21]; the organic acceptors used in metal electrode studies may be appropriate.

It is always advisable to prepare solutions freshly from purified stock reagents. Preparation of ultrapure electrolyte is needed to ensure a minimum of extraneous impurities that can act as electron acceptors. Even dissolved CO_2 and (especially) traces of oxygen or protons may be a problem. To prepare high-quality distilled water, the authors have found the Gilmont continuous still to be satisfactory (model MOD VI, Gilmont Instruments Inc., Great Neck, NY). This still is designed to eliminate CO_2 and volatile contaminating gases.[106] The water is collected in polyethylene and used when fresh. Salts for electrolytes should be recrystallized several times from purified solvent, using glass microfiber filters to remove dust and particulates. Organic solvent purification is more involved, and the electrochemical and organic literature provides specific procedures, e.g., Refs. 107 and 108. Lithium and fluoride salts should be recrystallized from plasticware. Solvents and electrolytes should not absorb the UV radiation strongly and this, too, limits the choice of acceptors.

4.2. Cell Design and Electrode Preparation

Good cell design is an art, and for photoemission studies an extra emphasis is on airtightness. Glassware and cells should be leached in 20% (v/v) HCl and HNO_3 and thoroughly rinsed (NO_3^- ion is an electron scavenger). Ground glass seals are preferable, fitted with Teflon liners or moistened with electrolyte, rather than lubricants which can seep into the electrolyte. Argon and nitrogen inert gas must be scrupulously scrubbed of traces of oxygen. Degassing takes longer for organic than for aqueous systems. Freeze-thaw degassing is an option if the cell is vacuum tight. Because the optical radiation is usually near-UV, cell windows, or even the cell itself, are quartz or a suitably transparent material. If reagents are expensive and/or difficult to prepare, it is often better to design a cell with a small electrolyte volume. Designs used for spectroelectrochemistry may be adaptable, e.g., Ref. 109.

The preparation of the electrode surface should be such as to permit reproducible experiments, and, hopefully, provide a well-defined interphasial system. Mercury does not require surface treatment, of course, but solid electrodes may require mechanical polishing and chemical and/or electrochemical polishing. The mounting of semiconductor electrodes may present even greater difficulties because many of the etchants used are extremely corrosive. This dictates the choice of encapsulating insulators. Bozhkov *et al.*[110] have proposed cleavage of semiconductor crystals in the liquid dielectric as a means to obtain atomically pure surfaces. This procedure is ideal but may prove expensive for general purpose work. Ohmic contact to semiconductors has to be ensured. Procedures for this can be found in the solid-state physics and photoelectrochemical cell literature.

4.3. Optics, Apparatus, and Methods

Because the quantum yield in photoemission is low (10^{-2} to 10^{-5} electron/photon), it is necessary to have a powerful light source. A 250 W (or, better, 1000-W) Hg or Xe high-pressure lamp with quartz optics and stabilized power supply, or an ultraviolet laser source, is required to ensure adequate intensities. Consequently, many *in situ* experiments have been made with a fairly broad bandpass as the monochromator must permit adequate light throughput. Mechanical modulation with a synchronous beam chopper permits selective amplification of the response when used in combination with a lock-in amplifier. Alternatively, lamp modulation, short-period flash lamps, or pulsed lasers may be employed for excitation. Introductory tests and articles on the use of electrochemical instrumentation, lock-in amplifiers, and correlation methods are available, e.g., Refs. 111-114. Modern potentiostats may have bandpasses 200 kHz or larger, so for many experiments the apparatus response is not a limiting factor. The better the sensitivity of the current detection is, the better the optical resolution achievable. Quantum efficiency measurements may be at selected steady-state wavelengths (e.g., the peaks in the Hg lamp, for example) or obtained by slowly sweeping the wavelength with a monochromator driving motor. The latter may be computer controlled in an active interactive mode. Figure 15 is a block diagram of the apparatus for quantum efficiency measurements by potential modulation. The PAR 174A, in hold mode, produces ~50-ms square-wave pulses to a voltage at which photoemission can occur, per 0.5 s for example. The wavelength is scanned slowly and the photocurrent accessed with a sample-and-hold amplifier, digitized, and stored. The advantage of this technique is that the acceptor is not depleted at the electrode because its concentration is restored by convection during the 0.5-s rest period.[21] Rapid-sweep monochromators are available commercially, too, but have not been used for measurements of photoemission into condensed phases.

A schematic of an arrangement for voltage sweep measurements is shown in Figure 16. This is, perhaps, one of the simplest experiments as it entails constant illumination at fixed wavelength while ramping the voltage of the working electrode. It is customary in this type of experiment to record the current response in the dark, with the electrode illuminated in the absence of scavenger, and with the electrode illuminated in the presence of scavenger. Passive or active microcomputer use is highly advantagous since rate laws are

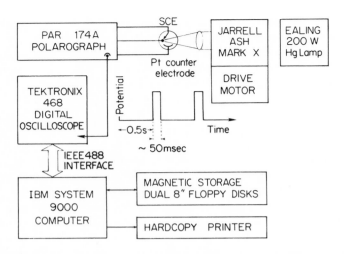

FIGURE 15. Schematic of apparatus for potential modulation quantum efficiency measurements. Central inset illustrates potential waveform.

difficult to ascertain and the microcomputer provides fast data processing and better control over experimental parameters, and more data points can be routinely processed than would be conveniently computed by hand.[115] Statistical methods of data processing are indispensable; it has been usual to linearize data and use linear correlation methods (see, e.g., Ref. 79). The authors favor data acquisition using instruments fitted with the IEEE-488 interface (e.g., digital voltmeters, oscilloscopes, lock-in amplifiers) because of their versatility in configuration, multiplex capability, and speed. When real-time data processing is not required, software drivers may be written in a high-level language. Gale and coworkers[21] have developed an optical pulse FFT method for

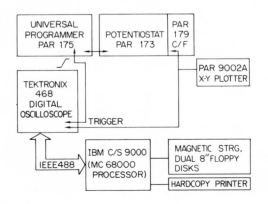

FIGURE 16. Schematic of voltage sweep apparatus for photoemission studies (by permission).[84]

evaluation of photoemission currents. Computer-based data processing of this type may prove useful for kinetics analysis (see Section 5.2).

5. Conclusions

There are two broad aspects of studies of photoemission from electrodes into condensed media—the description of the physical mechanisms *per se* and the investigation of the subsequent interactions of the photoemitted electron in the solution phase. In this chapter, the theory presented has dealt almost exclusively with the first of these topics, i.e., the various attempts at an exact, quantitative description of the microscopic interactions occurring in photoemission. The two concluding sections below contain brief summaries of the current status of each of these two fields of research.

5.1. Physical Mechanistic Studies

One immediate conclusion is that the experimental validations of theories for both metal and semiconductor photoemission into vacuum certainly are more substantial than those for the respective condensed media cases. This is not surprising in view of the relative simplicity afforded by the lack of a second material phase, in addition to the ability to control the surface compositions and structures better under vacuum conditions. Further, the metal case is simplified insomuch as the internally excited electrons are considered to be free within the electrode phase, whereas for the more complex semiconductor formulism, the criteria permit direct and indirect transitions. Nonetheless, vacuum photoemission studies of semiconductors provide a fairly detailed picture which, importantly, can be separated by mechanism. The experimental objective in most studies has been to identify the source of the photoelectrons through application of the appropriate theoretical rate law. At each interface, an additional complication which applies under certain circumstances is the prospect of electron tunneling currents. In general, the influence of tunneling of the rate expressions has not been well quantified, and theoretical as well as experimental advances are needed. One further complication might be the presence of an intrinsic tail of localized states extending below the solvent conduction band edge.[116] Photoinjection to these states should occur at energies similar to those needed for tunneling. These two effects may have to be distinguished by an understanding of their mechanistic origins and knowledge of the density of states of the solvent.

The initial thrust in studying photoemission from metal electrodes into solutions was to confirm the form of the photoemission rate laws at amorphous electrodes. Mercury was the natural choice of substrate because of its wide range of polarizability and its surface integrity. Although the rate measurements agreed well with the quantum mechanical theory of Gurevich *et al.* (Sections

2.3 and 3.2), the question of the role of image potential is unresolved, with the Russian workers assuming complete charge screening of the photoemitted electron. This question is a thorny one which pertains in the semiconductor case too. If cumulative electronic screening is possible by a concerted, internal mechanism, major differences might be expected between the metal and semiconductor systems because free charge is far less mobile in the latter. However, the experimental results available at semiconductor electrodes similarly seem to support Gurevich's quantum mechanical theory with no image effects included. This question has to be resolved by short-timescale rate analyses. Steady-state photocurrent measurements of the type presently made reflect a rate law which pertains to a quasi-equilibrium condition. The emerging picture for a direct transition at a semiconductor electrode under potentiostatic control may be stated as follows. Firstly, the electrons are emitted on a picosecond timescale, modifying the band bending instantaneously by leaving positive holes in the valence band (the vacuum model). Presumably, even before current can flow through the electrode (limited by the respective carrier mobilities), internal recombination via thermal electrons should begin to occur and relax this band bending effect. Next, current flow to the electrode occurs in response to the potential control on the timescale of the space charge time constant (RC). Thus, in the illuminated steady state, the bands will be partially modified from the dark condition at the same potential, and a small, steady-state concentration of holes will influence the photocurrent rate. The magnitude of this resultant image effect, therefore, depends on the timescale of the photocurrent measurement and the rate of internal recombination. Obviously, optical detection of products is necessary to evaluate photoemission rate laws on timescales faster than the internal carrier mobilities. An appropriate theoretical analysis would include wave mechanical and thermal activation relaxation. The possibility remains, too, that for these condensed media cases the experimental results may be unsatisfactory due to surface complications.

More recently, photoemission studies have been extended to polarized light excitation of single-crystal substrates of known epitaxy. This approach is definitely rewarding because it provides an additional parameter to delineate the roles of volume and surface contributions, as well as clarify the basic internal excitation mechanism. We are not aware of any semiconductor/electrolyte studies using these techniques, but their application will prove extremely helpful in resolving the various contributions to the net photocurrents and scattering effects. There is, clearly, a need for better experimentation and theoretical development for the photoemission phenomena into condensed media from each class of electrode.

5.2. Solvated Electron Chemistry

Early work on the chemical interactions of solvated electrons arose due to an interest in effects caused by ionizing radiation. A further interest is in characterizing the thermodynamic properties of liquid interactions with electrons. Photoemission from electrodes affords the ability to produce solvated electrons relatively easily and with good quantitative control of follow-up reactions, provided the subsequent electrochemical stoichiometries are known. It is surprising, perhaps, that the ability to generate solvated electrons heterogeneously by photoemission under fairly well controlled conditions has not been exploited more from the standpoint of providing a useful kinetics probe for mechanistic studies. For example, photoemission–ESR studies are one viable means for research of free radical reactions.

Most of the developments in this area, since the first studies by Barker with N_2O, have been made by the Russian schools of researchers, and, again, only rate studies at metal electrodes are available. (Severeyn[21] achieved photoemission from p-GaAs and p-InP into organic electrolytes but no rate studies were conducted.) Chapter 7 in the book by Gurevich et al.[10] reviews the electrochemically important case of the formation of atomic hydrogen. Chapter 8 of that book surveys the use of photoemission as a method of studying homogeneous reactions that involve free radicals. The scavengers discussed include NO_3^-, CO_2, alcohol reactions with H· and OH·, and SF_6. Most of the studies have been made at mercury electrodes (Table V). For example, rate constants for the reactions of isopropyl, butyl, and amyl alcohols, phenol, and aniline with OH· radical have been determined.[118] Samvelyan et al.[119] report that the rates of electron capture by chloro- and bromobenzene are far smaller than the dissociation rate of the radical anion intermediate. Alternatively, acetone as a scavenger provides a relatively long-lived $(CH_3)_2COH$· radical intermediate which surface absorbs with a life of about 0.1 s. There has been also a fairly recent study of the kinetics of CO_2 reduction at mercury.[20] Many basic reactions still need to be studied. For example, the interaction of the photogenerated electron in acetonitrile (a versatile electrochemical solvent) is not fully understood.[58,124,125] For fast kinetics studies, there is a further need to develop methodologies which interface photoemission with other spectroelectrochemical techniques.

TABLE V. Kinetics Studies of Scavengers Using Photoemission at the Hg/Electrolyte Interface

System	Reference
N_2O product reactions with aqueous CH_3OH, C_2H_5OH	18
Disintegration rate of NO_3^{2-}	117
OH· radical reaction rates with isopropyl, butyl and amyl alcohols, phenol, aniline	118
C_6H_5Cl and C_6H_5Br scavengers in DMF; acetone radical anion formation	119
Reduction and oxidation of CO_2 reduction intermediates	120, cf. 121
Hydroxyalkyl radical kinetics	122
Reduction and oxidation of $(CH_3)_2COH$· radicals	123

Acknowledgments

Part of this chapter was completed when R. J. G. was a summer fellow of the Associated Western Universities at the Solar Energy Research Institute, Golden, Colorado (1985). We are grateful to Professor Neil Kestner, of the LSU Chemistry Department, for reading this chapter and recommending some improvements.

References

1. H. Hertz, *Ann. Phys.* **31**, 421, 983 (1887).
2. A. Einstein, *Ann. Phys.* **17**, 132 (1905).
3. O. W. Richardson, *Phil. Mag.* **23**, 615 (1912); **24**, 570 (1912).
4. J. J. Thomson, *Phil. Mag.* **2**, 674 (1926).
5. W. Uspensky, *Z. Physik* **40**, 456 (1926); A. L. Hughes and L. A. DuBridge, *Photoelectric Phenomena*, McGraw-Hill, New York (1932), p. 202.
6. G. Wentzel, *Probleme der Modernen Physik*, Sommerfeld Festschrift, Leipzig (1928), p. 79
7. R. H. Fowler, *Phys. Rev.* **38**, 45 (1931).
8. A. Modinos, *Field, Thermionic and Secondary Electron Emission Spectroscopy*, Plenum Press, New York (1984).
9. E. C. Becquerel, *C. R. Acad. Sci.* **9**, 143 (1839).
10. Yu. Ya. Gurevich, Yu. V. Pleskov, and Z. A. Rotenberg, *Photoelectrochemistry*, Consultants Bureau, New York (1980).
11. H. Berg, *Naturwissenschaften* **47**, 320 (1960).
12. H. Berg and P. Reissmann, *J. Electroanal. Chem.* **24**, 427 (1970).
13. M. Heyrovsky, *Nature* **200**, 1356 (1965).
14. M. Heyrovsky, *Croat. Chem. Acta.* **45**, 247 (1973).
15. L. I. Korshunov, Ya. M. Zolotovitskii, and V. A. Benderskii, *Russ. Chem. Rev.* **40**(8), 69 (1971).
16. G. C. Barker and A. W. Gardner, *Osnovnye voprosy sovremennoi teoreticheskvi Electrokhimii* p. 118, MIR, Moscow (1965).
17. G. C. Barker, A. W. Gardner, and G. Bottura, *J. Electroanal. Chem.* **45**, 21 (1973).
18. G. C. Barker, *Electrochim. Acta* **13**, 1221 (1968).
19. A. M. Brodskii and Yu. Ya. Gurevich, *Sov. Phys. JETP* **27**, 114 (1968).
20. D. E. Grider, Ph.D. thesis, Iowa State University (1980).
21. R. Borjas Severeyn, Ph.D. thesis, Louisiana State University (1984).
22. R. H. Fowler, *Phys. Rev.* **38**, 45 (1931).
23. R. H. Fowler, *Statistical Mechanics*, Cambridge Press, Cambridge, Chapter XI (1955).
24. L. A. DuBridge and W. W. Roehr, *Phys. Rev.* **39**, 99 (1932).
25. L. A. DuBridge, *Phys. Rev.* **39**, 108 (1932).
26. D. B. Matthews and S. U. M. Khan, *Aust. J. Chem.* **28**, 253 (1975).
27. J. O'M. Bockris and S. U. M. Khan, *Quantum Electrochemistry*, Plenum Press, New York (1979), Chapter 12.
28. A. I. Baz, Ya. B. Zeldovich, and A. M. Perelomov, *Scattering, Reactions, and Disintegration in Nonrelativistic Quantum Mechanics*, Nauka, Moscow (1966).
29. R. G. Newton, *Scattering Theory of Waves and Particles*, McGraw-Hill, New York (1967).
30. A. B. Brodskii and Yu. Ya. Gurevich, *Dokl. Akad. Nauk SSSR* **178**(4), 868 (1968).
31. Yu. Ya. Gurevich, A. M. Brodskii, and V. G. Levich, *Electrokhimiya* **3**, 1302 (1967).
32. Yu. V. Pleskov and Z. A. Rotenberg, *J. Electroanal. Chem.* **20**, 1 (1969).

33. A. M. Brodskii, Yu. Ya. Gurevich, and V. G. Levich, *Phys. Status Solidi* **40**, 139 (1970).
34. H. Gerischer, D. M. Kolb, and J. K. Sass, *Adv. Phys.* **27**, 437 (1978).
35. L. I. Korshunov, Ya. M. Zolotovitskii, and V. A. Benderskii, *Electrokhimiya* **4**, 499 (1968).
36. Z. A. Rotenberg and Yu. V. Pleskov, *Electrokhimiya* **4**, 826 (1968).
37. R. DeLevie and J. C. Kreuser, *J. Electroanal. Chem.* **21**, 221 (1969).
38. Yu. V. Pleskov, Z. A. Rotenberg, and V. I. Lakomov, *Electrokhimiya* **6**, 1787 (1970).
39. Yu. V. Pleskov and Z. A. Rotenberg, *Russ. Chem. Rev.* **41**, 21 (1972).
40. Ya. M. Zolotovitskii, L. I. Korshunov, and V. A. Benderskii, *Izv. Akad. Nauk. SSSR* **4**, 802 (1972).
41. Z. A. Rotenberg, *Electrokhimiya* **10**, 1031 (1974).
42. V. V. Yakushev, A. M. Skundin, and V. S. Bagotskii, *Elektrokhimiya* **20**, 99 (1984).
43. G. D. Mahan, in: *Electron and Ion Scattering of Solids* (L. Fiermans, J. Vennik, and W. Dekeyser, eds.), NATO Adv. Study Inst. Ser., Ser. B. Physics, Vol. 32, pp. 1–53, Plenum Press, New York.
44. G. D. Mahan, *Phys. Rev. B* **2**, 4334 (1970).
45. P. J. Feibelman and D. E. Eastman, *Phys. Rev. B* **10**, 4932 (1974).
46. I. Adawi, *Phys. Rev.* **134**, A788 (1964).
47. J. K. Sass and H. J. Lewerenz, *J. Phys. C* **5**, 277 (1972).
48. A. M. Brodskii and Yu. Ya. Gurevich, *Theory of Electronic Emission from Metals*, Nauka, Moscow (1973).
49. W. N. Hansen, in: *Advances in Electrochemistry and Electrochemical Engineering* (P. Delahay and C. W. Tobias, eds.), Vol. 9, pp. 1–60, John Wiley and Sons (1973).
50. L. I. Korshunov, Ya. M. Zolotovitskii, and V. A. Benderskii, *Elektrokhimiya* **5**, 716 (1969).
51. J. K. Sass and H. Gerischer, in: *Photoemission and the Electronic Properties of Surfaces*, (B. Feuerbacher *et al.*, eds.), Chapter 16, Wiley-Interscience, New York (1978).
52. B. Feuerbacher and R. F. Willis, *J. Phys. C* **9**, 169 (1976).
53. D. Mohilner, *Electronanalytical Chemistry* (A. J. Bard, ed.), Vol. 1, p. 241, Marcel Dekker, New York (1966).
54. Z. A. Rotenberg, V. I. Lakomov, and Yu. V. Pleskov, *Elektrokhimiya* **9**, 11 (1973).
55. Z. A. Rottenberg, V. I. Lakomov, and Yu. V. Pleskov, *Elektrokhimiya* **9**, 152 (1973).
56. Z. A. Rotenberg and Yu. V. Pleskov, *Electrokhimiya* **5**, 982 (1969).
57. D. Grider, P. Lange, H. Neff, and K. Ho, *Vacuum* **31**, 563 (1981).
58. R. G. Kokilashvili, V. V. Eletskii, and Yu. V. Pleskov, *Elektrokhimiya* **20**, 1075 (1984).
59. W. E. Spicer and R. C. Eden, *Int. Conf. on the Physics of Semiconductors, Proceedings*, July 23–29, 1968, Vol. 1, Nauka, Leningrad (1968).
60. E. O. Kane, *Phys. Rev.* **127**, 131 (1962).
61. G. W. Gobeli and F. G. Allen, *Phys. Rev.* **127**, 141 (1962).
62. M. D. Krotova and Yu. V. Pleskov, *Sov. Phys. Solid State* **15**, 1871 (1974).
63. Yu. Ya. Gurevich, *Elektrokhimiya* **8**, 1564 (1972).
64. Yu. Ya. Gurevich, M. D. Krotova, and Yu. V. Pleskov, *J. Electroanal. Chem.* **75**, 339 (1977).
65. K. Gottfied, *Quantum Mechanics, Vol. 1: Fundamentals*, The Benjamin Cumming Co., London (1966), pp. 443–444.
66. J. Callaway, *Quantum Theory of the Solid State*, Academic Press, New York (1974), pp. 24–28.
67. G. Gilat, *J. Comp. Phys.* **10**, 432 (1972).
68. Ref. 66, pp. 521–525.
69. J. J. Scheer and J. Van Laar, *Phillips Research Reports* **16**, 323 (1961).
70. G. W. Gobeli and F. G. Allen, in: *Solid Surfaces* (H. C. Gattor, ed.), p. 402, North-Holland Publishing Co., Amsterdam (1964).
71. G. W. Gobeli, F. G. Allen, and E. O. Kane, *Phys. Rev. Lett.* **21**, 94 (1964).
72. G. W. Gobeli and F. G. Allen, in: *Semiconductors and Semimetals* (R. K. Willardson and A. C. Beer, eds.), Vol. 2, pp. 253–280, Academic Press, New York (1966).
73. G. W. Gobeli and F. G. Allen, *J. Chem. Phys. Solids* **14**, 23 (1960).
74. T. E. Fischer, *Surface Sci.* **13**, 30 (1969).

84   Ricardo Borjas Severeyn and Robert J. Gale

75. W. A. Harrison, *Solid State Theory*, Dover Publications, New York (1979), pp. 140–142.
76. C. Kittle, *Introduction to Solid State Theory*, 5th Edition, John Wiley, New York (1976), pp. 218–219.
77. Yu. Ya. Gurevich and Yu. V. Pleskov, in: *Semiconductors and Semimetals* (R. K. Willardson and A. C. Beer, eds.), Vol. 19, Chapter 4, Academic Press, New York (1983).
78. Yu. V. Pleskov, in: *Comprehensive Treatise of Electrochemistry* (J. O'M. Bockris, B. E. Conway, and E. Yeager, eds), Vol. 1, Chapter 6, Plenum Press, New York (1980).
79. R. Borjas Severeyn and R. J. Gale, *J. Electroanal. Chem.* **150**, 619 (1983).
80. M. D. Krotova and Yu. V. Pleskov, *Fiz. Tverd. Tela.* **15**, 2806 (1973).
81. E. Meyer, Dissertation, Tech. Univ., Munich (1973).
82. G. V. Boikova, M. D. Krotova, and Yu. V. Pleskov, *Elektrokhimiya* **12**, 922 (1976).
83. Yu. Ya. Gurevich, M. D. Krotova, and Yu. Ya. Pleskov, *J. Electroanal. Chem.* **75**, 339 (1977).
84. R. Borjas Severeyn and R. J. Gale, *Mol. Cryst. Liq. Cryst.* **107**, 227 (1984).
85. C. E. Krohn and J. S. Thompson, *Chem. Phys. Lett.* **65**, 132 (1979).
86. R. E. Malpas, K. Itaya, and A. J. Bard, *J. Am. Chem. Soc.* **101**, 2535 (1979).
87. G. V. Amerongen, D. Guyomard, R. Heindl, M. Herlem, and J.-L. Sculfort, *J. Electrochem. Soc.* **129**, 1998 (1982).
88. M. Herlem, D. Guyomard, C. Mathieu, J. Belloni, and J.-L. Sculfort, *J. Phys. Chem.* **88**, 3826 (1984).
89. J. Vondrak, J. Buldska, I. Jakubec, and J. Velek, *Electrochim. Acta* **29**, 315 (1984).
90. J. O'M. Bockris and K. Uosaki, *J. Electrochem. Soc.* **125**, 223 (1977).
91. J. O'M. Bockris, K. Uosaki, and H. Kita, *J. Appl. Phys.* **52**, 808 (1981).
92. J. O'M. Bockris and K. Uosaki, in: (J. B. Goodenough and M. Stanley Whittingham, eds.) *Adv. Chem. Ser. No. 163, Solid State Chemistry of Energy Conversion and Storage*, pp. 33–70, American Chemical Society, Washington, D.C. (1977).
93. J. O'M. Bockris and K. Uosaki, *Int. J. Hydrogen Energy* **3**, 157 (1977).
94. H. Gerischer, in: *Advances in Electrochemistry and Electrochemical Engineering* (P. Delahay and C. W. Tobias, eds.), Vol. 1, p. 139, John Wiley and Sons, New York (1961).
95. A. J. Nozik, *Ann. Rev. Phys. Chem.* **29**, 189 (1978).
96. A. J. Nozik, in: *Photochemical Conversion and Storage of Solar Energy* (J. S. Connolly, ed.), Chapter 10, Academic Press, New York (1981).
97. D. S. Boudreaux, F. Williams, and A. J. Nozik, *J. Appl. Phys.* **51**, 2158 (1980).
98. G. Cooper, J. A. Turner, B. A. Parkinson, and A. J. Nozik, *J. Appl. Phys.* **54**, 6463 (1983).
99. F. Williams, and A. J. Nozik, *Nature* **271**, 137 (1978).
100. J. A. Turner and A. J. Nozik, *Appl. Phys. Lett.* **41**, 101 (1982).
101. A. J. Nozik, in: *Photovoltaic and Photoelectrochemical Solar Energy Conversion* (F. Cardon, W. P. Gomes, and W. Dekeyser, eds.), pp. 263–312, Plenum Press, London, published in cooperation with NATO Scientific Affairs Division (1981).
102. A. J. Nozik, D. S. Boudreaux, R. R. Chance, and F. Williams, in: *Interfacial Photoprocesses: Energy Conversion and Synthesis* (M. S. Wrighton, ed.), *Adv. Chem. Ser. No. 184*, pp. 155–171, American Chemical Society, Washington, D. C. (1980).
103. J. A. Turner, J. Manassen, and A. J. Nozik, *Appl. Phys. Lett.* **37**, 488 (1980).
104. F. Williams and A. J. Nozik, *Nature* **312**, 21 (1984).
105. M. S. Matheson, in: *Physical Chemistry: An Advanced Treatise* (H. Eyring, ed.), Vol. 7, p. 533, Academic Press, New York (1975).
106. R. Gilmont and S. J. Silvis, *Am. Lab.*, December, 46 (1974).
107. D. T. Sawyer and J. L. Roberts, Jr., *Experimental Electrochemistry for Chemists*, John Wiley and Sons, New York (1974).
108. D. D. Perrin, W. L. F. Armarego, and D. R. Perrin, *Purification of Laboratory Chemicals*, 2nd Edition, Pergamon Press, New York (1981).
109. T. E. Furtak and D. W. Lynch, *J. Electroanal. Chem.* **79**, 1 (1977).
110. V. G. Bozhkov, G. A. Kataev, G. F. Kovtuneko, and K. V. Saldatenko, *Elektrokhimiya* **7**, 549 (1971).

111. R. Greef, *Instrumental Methods in Electrochemisty*, Ellis Horwood Series in Physical Chemistry, Halsted, New York (1985).
112. M. C. H. McKubre and D. D. Macdonald, in: *Comprehensive Treatise of Electrochemistry* (R. E. White, J. O'M. Bockris, B. E. Conway, and E. Yeager, eds.), Vol. 8, Chapter 1, Plenum Press, New York (1984).
113. E. H. Fisher, *Laser Focus*, November (1977).
114. G. M. Hieftje and G. Horlick, *Am. Lab.*, March, 76 (1981).
115. T. H. Ridgway and H. B. Mark, Jr., in: *Comprehensive Treatise of Electochemistry* (R. E. White, J. O'M. Bockris, B. E. Conway, and E. Yeager, eds.), Vol. 8, Chapter 2, Plenum Press, New York (1984).
116. G. T. Bennet and J. C. Thompson, *J. Chem. Phys.* **84**, 1901 (1986).
117. Z. A. Rotenburg, V. I. Lakomov, and Yu. V. Pleskov, *J. Electroanal. Chem.* **27**, 403 (1970).
118. V. V. Eletskii and Yu. V. Pleskov, *Elektrokhimiya* **10**, 179 (1974).
119. S. Kh. Samvelyan, Z. A. Rotenberg, and V. G. Mairanovskii, *Elektrokhimiya* **19**, 1319 (1983).
120. S. D. Babenko, V. A. Benderskii, A. G. Krivenko, and V. A. Kurmaz, *J. Electroanal. Chem.* **159**, 163 (1983).
121. D. J. Schiffrin, *Faraday Discuss Chem. Soc.* **56**, 75–95 (1973).
122. Z. A. Rotenberg and N. M. Rufman, *Elektrokhimiya* **19**, 1349 (1983).
123. Z. A. Rotenberg and N. M. Rufman, *J. Electroanal. Chem.* **175**, 153 (1984).
124. D. Bradley and J. Wilkinson, *J. Chem. Soc.* (*A*), 531 (1967).
125. A. Singh, H. D. Gesser, and A. R. Scott, *Chem. Phys. Lett.* **2**, 271 (1968).

4

UV–Visible Reflectance Spectroscopy

Dieter M. Kolb

1. Introduction

Within the last two decades, interfacial electrochemistry has become very much an integral part of modern surface science.[1-6] In the search for a microscopic description of the electrochemical double layer, which obviously is necessary for a detailed understanding of electrode processes, electrochemists have increasingly recognized the potential power of surface science concepts and techniques for obtaining information which would complement the thermodynamic description derived from classical electrochemical methods. As a consequence, (1) new, nonelectrochemical techniques have been introduced and combined with the traditional electrochemical methods, (2) atomically well-defined single-crystal surfaces have been increasingly used as electrodes rather than polycrystalline material, and (3) structure-sensitive methods such as low-energy electron diffraction (LEED), developed for surface characterization under ultrahigh-vacuum conditions, have been employed to study electrode surfaces before and after electrochemical experiments.[4,5] In addition, atomistic concepts for interactions at surfaces have been introduced from surface science into electrochemistry to modify or replace the macroscopic and highly phenomenological models of the electrochemical double layer, which were developed from the analysis of thermodynamic data.

Reflectance spectroscopy was among the first nontraditional techniques used to study electrode surfaces.[7,8] This method can provide information on an atomic or molecular level, is nondestructive, and, quite importantly for many electrochemists, is applicable *in situ*, i.e., in the presence of the bulk electrolyte. Although the bulk electrolyte imposes limitations on the spectral range due to its transmission properties, which allow an easy access of the 1-to-6-eV range only, these limits have recently been extended considerably into the infrared and the X-ray region by the development of optical thin-layer cells (see Chapters 5 and 6).

Dieter M. Kolb • Fritz Haber Institute of Max Planck Gesellschaft, D-1000 Berlin 33, Federal Republic of Germany.

When reflectance spectroscopy was first used for the *in situ* study of electrode surfaces, the main emphasis was placed on monitoring the formation of adsorbates and thin films, e.g., the underpotential deposition of metals or the formation of oxides, because a rather strict correlation between coverage and adsorbate- or film-induced reflectance change was generally observed.[9,10] In the beginning of the seventies, the evaluation of adsorbate optical constants from measurements of $\Delta R/R$ with the help of stratified multilayer models seemed possible,[11-13] which stimulated the interest in reflectance studies of electrode surfaces. However, the evaluation of film optical constants for monolayer and submonolayer thicknesses by the classical continuum theory was not very satisfactory, and microscopic theories, which could have been handled by an experimentalist, were not available. At that stage, reflectance spectroscopy ultimately did not meet the expectations and the general interest faded. Several years later, optical studies with single-crystal electrodes were started,[14] which led to the discovery of a number of effects highly relevant to questions of the physical and chemical properties of the metal-electrolyte interface. Ever since, reflectance spectroscopy seems well established in interfacial electrochemistry, and with its variety of experimental techniques, such as electroreflectance, differential reflectance, surface plasmon excitation, etc., an even wider variety of problems could be addressed: the optical properties of electrode surfaces and their change with potential, the study of surface states and surface reconstruction, the optical properties of adsorbates and the influence of the electric field in the double layer on them, just to mention a few. Even structural information on adsorbates could be obtained besides the information on their electronic properties. All this, however, emerged from the combination of optical spectroscopy and single-crystal surfaces.

The aim of this chapter is to provide some background on the theory of the optical properties of solids and to outline what we can learn from reflectance measurements with a brief survey of the various techniques. With respect to the first point, the review by McIntyre[15] is still highly recommended. The main emphasis of this article (and this may be the main difference to McIntyre's) is on the presentation and discussion of experimental results which are now available and which demonstrate that, indeed, new information can be obtained by this technique—information which is not accessible by purely electrochemical methods and yet is important for solving the puzzle about the microscopic structure of the electrochemical interface.

2. Physical Optics

2.1. Optical Constants†

The response of a medium to an incoming electromagnetic wave is in most cases sufficiently characterized by *one* frequency-dependent quantity, the

† For a more thorough introduction, see Refs. 16–18.

complex refractive index $\hat{n}(\omega)$:

$$\hat{n}(\omega) = n(\omega) - ik(\omega) \tag{1}$$

Describing the electromagnetic plane wave by its time-dependent value of the electric vector,

$$\mathbf{E} = \mathbf{E}_0 \exp[i(\omega t - \mathbf{k} \cdot \mathbf{r})] \tag{2a}$$

where $\mathbf{E}_0$ is the wave amplitude and $\mathbf{k}$ the wave vector ($|\mathbf{k}| = \hat{n}(\omega/c) = 2\pi\hat{n}/\lambda$ with λ = vacuum wavelength), it becomes evident that the real and imaginary parts of the complex refractive index represent, respectively, refraction and attenuation of the wave by the solid.

For an electromagnetic wave propagating in the z-direction, Eq. (2a) can be rewritten as:

$$\mathbf{E} = \mathbf{E}_0 \exp(-2\pi kz/\lambda) \exp[i(\omega t - 2\pi nz/\lambda)] \tag{2b}$$

The first exp-term represents the exponential decay of the wave amplitude with penetration into the (absorbing) medium, while the second exponential term describes the time and phase dependence of the electromagnetic wave, traveling with phase velocity $v = c/n$. The two quantities n and k are called the refractive index and extinction coefficient, respectively, and are commonly referred to as the *optical constants*.

Since the intensity of the light is directly proportional to the mean square electric field strength, $\langle \mathbf{E}^2 \rangle$, we can derive from Beer's law [$I = I_0 \exp(-\alpha z)$] the following expression for the absorption coefficient α:

$$\alpha = 4\pi k/\lambda \tag{3}$$

Maxwell's relation, $\hat{\varepsilon} = \hat{n}^2$, leads us from the optical constants (n, k) to a yet more fundamental parameter, the complex dielectric function, $\hat{\varepsilon}$. The real and imaginary parts of $\hat{\varepsilon} = \varepsilon' - i\varepsilon''$ can be related to the optical constants:

$$\varepsilon' = n^2 - k^2 \tag{4a}$$

and

$$\varepsilon'' = 2nk \tag{4b}$$

We mention in passing that the real and imaginary parts of the complex dielectric function, $\hat{\varepsilon}$, can also be related to, respectively, the polarizability $\tilde{\alpha}(\omega)$ and the optical conductivity $\sigma(\omega)$ (i.e., the electrical conductivity at optical frequencies) of the medium:

$$\varepsilon' = 1 + 4\pi\tilde{\alpha} \tag{5a}$$

and

$$\varepsilon'' = 4\pi\sigma/\omega \tag{5b}$$

The quantities ε'' and σ describe directly the energy dissipation in the medium. The rate of energy loss, $-\dot{w}$, from an electromagnetic wave per unit volume of the medium is given by:

$$-\dot{w} = \sigma\langle E^2 \rangle = \frac{\omega}{4\pi}\varepsilon''\langle E^2 \rangle \tag{6}$$

It is important to note that ε' and ε'', which describe dispersion and absorption in a medium, are not independent variables but are interconnected by the Kramers-Kronig relation (causality relation), according to [16,19]:

$$\varepsilon'(\omega_0) - 1 = \frac{2}{\pi}\int_0^\infty \frac{\omega\varepsilon''(\omega)}{\omega^2 - \omega_0^2}\,d\omega \tag{7a}$$

and

$$\varepsilon''(\omega_0) = -\frac{2\omega_0}{\pi}\int_0^\infty \frac{\varepsilon'(\omega) - 1}{\omega^2 - \omega_0^2}\,d\omega \tag{7b}$$

Any optical excitation (absorption) process can, in principle, be described by a classical Lorentz oscillator or a sum of oscillators, yielding†:

$$\hat{\varepsilon} = 1 + \frac{(4\pi Ne^2/m)}{(\omega_0^2 - \omega^2) + i\gamma\omega} \tag{8}$$

where ω_0 is the resonance frequency of the oscillator, γ represents the damping (energy loss), and N is the number of identical oscillators per unit volume. All other symbols have their usual meaning. The overall shapes of the real and imaginary part of $\hat{\varepsilon}$ for a single oscillator are sketched in Figure 1. The physical origin of such an oscillator can be sought in bound electrons, free electrons (of a metal), lattice vibrations (phonons), or impurities. In the UV-visible range, free and bound electrons are the main contributors to the optical response of an atom, molecule, or solid. Hence, the dielectric function in that photon energy range can be correlated with elementary electronic excitation processes and consequently with the electronic properties of the medium under investigation (e.g., with the electronic band structure of a solid).

† Note that the sign of the damping term $i\gamma\omega$ depends on the definition for the plane wave (Eq. 2). A time variation $e^{i\omega t}$ leads to $\hat{n} = n - ik$ and to $+i\gamma\omega$, while $e^{-i\omega t}$ yields $\hat{n} = n + ik$ and $-i\gamma\omega$.

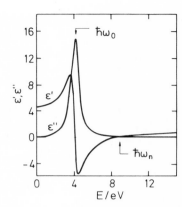

FIGURE 1. Frequency dependence of ε' and ε'' for a Lorentz oscillator with resonance at $\hbar\omega_0 = 4$ eV. After Ref. 16.

In the following, the contributions of free (f) and bound (b) electrons to the dielectric properties of a solid (preferably a metal) will be briefly discussed.

The Free-Electron Contribution to $\hat{\varepsilon}$

The free electrons in a metal are treated as a collective of zero-frequency oscillators ($\omega_0 = 0$), i.e., the restoring force is zero. This leads to the well-known Drude relation:

$$\hat{\varepsilon}_f = 1 - \frac{\omega_n^2}{\omega^2 - i\omega\tau^{-1}} \tag{9}$$

where $\omega_n = (4\pi Ne^2/m^*)^{1/2}$ is the so-called plasma frequency of a free-electron metal (see also Section 2.4), N is the number of free electrons per unit volume and with an effective mass, m^*, and $\tau = \gamma^{-1}$ is the time between two collisions of the free electron with the crystal lattice (the reciprocal of which is directly related to the energy absorption). Separating the real and imaginary parts in Eq. (9) yields the dielectric constants of a free-electron metal:

$$\varepsilon'_f = 1 - \frac{\omega_n^2}{\omega^2 + \tau^{-2}} \tag{10a}$$

and

$$\varepsilon''_f = \frac{\omega_n^2}{\omega\tau(\omega^2 + \tau^{-2})} \tag{10b}$$

For most metals, we find $\omega \gg \tau^{-1}$ in the UV–visible range (e.g., for silver $\tau = 0.38 \times 10^{-13}$ s,[20] while at 2 eV, $\omega = 3 \times 10^{15}$ s^{-1}), and hence Eqs. (10) can be approximated by:

$$\varepsilon'_f \approx 1 - \omega_n^2/\omega^2 \tag{11a}$$

and

$$\varepsilon''_f \approx 0 \tag{11b}$$

The free-electron behavior of metals is often termed an *intraband* transition, because, in an energy band diagram, it is thought to arise from the optical excitation of an electron within the same band. It should be kept in mind, however, that the free-electron behavior is not connected with energy absorption but merely results from enforced oscillations of the electrons (hence the high reflectivity). For true intraband transitions, momentum conservation has to be assured (vertical transitions within the reduced energy band diagram) and therefore scattering of the free electrons by the crystal lattice has to be invoked. The reflectivity R of a free-electron metal (e.g., alkali, Al, or Hg) is sketched in Figure 2. It is characterized by a high reflectivity of the metal up to a photon energy $\hbar\omega_n$ and a sharp drop in R for $\hbar\omega > \hbar\omega_n$. Note that for $\omega = \omega_n$, ε'_f passes through zero (Eq. 11a).

The Bound-Electron Contribution to $\hat{\varepsilon}$ (Interband Transitions)

Excitation of (bound) electrons in atoms or molecules is characterized by a rather well-defined excitation energy which peaks at $\hbar\omega_0 = E_f - E_i$, the energy difference between the initial and the final state of the electron. In a solid, the electronic states are described by energy bands rather than discrete energy levels and hence interband transitions have a threshold energy.

Within the electronic band structure description of a solid, the absorption as represented by $\varepsilon''(\omega)$ is the sum of all possible transitions between occupied and empty electronic states of the same **k**-vector within the reduced energy band scheme (vertical or direct transitions). Note that **k** in this context represents the wave vector of the electron in the crystal, the electron being described by a Bloch wave. Let the final and initial states be separated by the

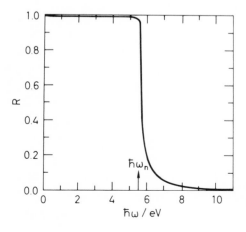

FIGURE 2. Reflectivity R of a free-electron metal with plasma frequency ω_n. After Ref. 16.

energy difference $\hbar\omega = E_f - E_i$; then the transition probability, W_{if} $[= -\dot{w}/\hbar\omega$; see Eq. (6)], for an electron from state φ_i to state φ_f is given by[16]:

$$W_{if} = \frac{e^2 4\pi^3 \hbar^2}{\pi m^2 \omega \varepsilon'} \cdot |\mathbf{eM}_{if}|^2 \cdot \delta(E_f - E_i - \hbar\omega) \tag{12}$$

$\mathbf{eM}_{if}$ is the matrix element of the perturbation, which is **k**-dependent and describes the overlap of the wave functions for the electron in the final and the initial state:

$$\mathbf{eM}_{if} = \int \varphi_f^* \mathbf{e}\nabla\varphi_i \, d\mathbf{r} \tag{13}$$

with **e** a unit vector in the direction of the electric field. Because of the correlation between ε_{if}'' and W_{if}

$$\varepsilon_{if}'' = 4\pi\sigma/\omega = \varepsilon' W_{if}/\omega \tag{14}$$

we can derive the following expression for ε_{if}'' [16]:

$$\varepsilon_{if}''(\omega) = \frac{e^2 \hbar^2}{\pi m^2 \omega^2} \int_{BZ} |\mathbf{eM}_{if}|^2 \cdot \delta(E_f - E_i - \hbar\omega) \, d\mathbf{k} \tag{15}$$

The matrix element is generally assumed to vary slowly with energy and to be the same for all isoenergetic transition in the Brillouin zone (BZ). We can then define the so-called joint density-of-state (JDOS) function, $J(\hbar\omega)$, as:

$$J(\hbar\omega) = \frac{1}{8\pi^3} \int_{BZ} \delta(E_f - E_i - \hbar\omega) \, d\mathbf{k} \tag{16}$$

Transforming the volume integral in Eq. (16) into an integral over the energy surface, $E_f - E_i =$ constant, in the BZ, we obtain:

$$\varepsilon_{if}''(\omega) \sim J(\hbar\omega) = \frac{1}{8\pi^3} \int_{E_f - E_i = \hbar\omega} \frac{d\mathbf{s}}{|\nabla_k E_f - \nabla_k E_i|} \tag{17}$$

Hence, the dielectric function can yield information on the JDOS and especially on "critical points" in the BZ, where the energy bands involved in the transition run parallel, giving rise to a singularity in the integrand in Eq. (17). Such critical points introduce slope discontinuities in $\varepsilon''(\omega)$ and therefore markedly determine the prominent structure in the spectral shape of the optical response of a solid.[21]

Plasmon Excitation in a Solid

Plasma oscillations or plasmons arise from collective oscillations of an electron gas as the response of the electrons to a transverse electromagnetic wave at the plasma frequency. As will be shown in Sections 2.4 and 4.1, plasmon excitation can play an important role in metal optics in general and in the electroreflectance studies of electrode surfaces in particular. The necessary condition for plasma oscillations is[22,23]:

$$\hat{\varepsilon} = 0 \tag{18}$$

which means that $\varepsilon' = 0$ and $\varepsilon'' \ll 1$ define the energy of plasmon excitation. The plasma frequency of a free-electron metal is $\omega_n = (4\pi Ne^2/m^*)^{1/2}$, where the necessary condition $\varepsilon_f' = 1 - \omega_n^2/\omega^2 = 0$ is fulfilled (damping neglected). For so-called nearly-free-electron metals like Cu, Ag, and Au, however, where bound electrons (interband transitions) contribute significantly to the overall dielectric response $\hat{\varepsilon}$, the requirement $\hat{\varepsilon} = \hat{\varepsilon}_b - \omega_n^2/\omega^2 = 0$ (see below) leads now to a plasma frequency $\omega_p = \omega_n/\sqrt{\varepsilon_b'}$, which is different from that for the free-electron metal, ω_n.

We briefly note that the so-called energy–loss function $\mathrm{Im}(1/\hat{\varepsilon})$ which characterizes the response of a solid to incident electrons in, for example, transmission electron energy loss spectroscopy (EELS) has a pronounced maximum for $\varepsilon' = 0$ and $\varepsilon'' \ll 1$,[23]

$$\mathrm{Im}\,(\hat{\varepsilon}^{-1}) = \frac{\varepsilon''}{\varepsilon'^2 + \varepsilon''^2} \tag{19}$$

The Optical Properties of Real Metals: Au, Ag, and Pt

Real and imaginary parts of the complex dielectric function for Au and Ag, as derived from a Kramers–Kronig analysis of the reflectivity data, are shown in Figures 3 and 4. In order to demonstrate the various contributions to the overall optical response of the metal, we separate $\hat{\varepsilon}$ into free- and bound-electron parts,†

$$\hat{\varepsilon} = \hat{\varepsilon}_f + \hat{\varepsilon}_b - 1 = \hat{\varepsilon}_b - \frac{\omega_n^2}{(\omega^2 - i\omega\tau^{-1})} \tag{20}$$

by calculating $\hat{\varepsilon}_f$ via Eq. (9) and subtracting it from the experimentally determined $\hat{\varepsilon}$ to obtain $\hat{\varepsilon}_b$. For photon energies below 2.5 eV, the optical properties of Au are dominated by the free-electron behavior. Around 2.5 eV,

---

† In the past, this separation was done as $\hat{\varepsilon} = \hat{\varepsilon}_f + \hat{\varepsilon}_b$ (see, e.g., Ref. 15), where in the absence of any bound-electron contribution, the term $\hat{\varepsilon}_b$ was dropped ($\hat{\varepsilon}_b = 0$). However, it seems physically more realistic to sum over the various susceptibilities rather than the dielectric functions ($\chi_i = \varepsilon_i - 1$), because in the absence of a polarizable medium, χ becomes zero whereas ε becomes unity (e.g., for vacuum, $\varepsilon = 1$ and not zero). For the separation procedure, we therefore use the Ansatz:

$$\hat{\chi} = \hat{\chi}_f + \hat{\chi}_b,$$

which leads to Eq. (20): $\hat{\varepsilon} = \hat{\varepsilon}_f + \delta\hat{\varepsilon}_b$, with $\delta\hat{\varepsilon}_b = \hat{\varepsilon}_b - 1$.[24]

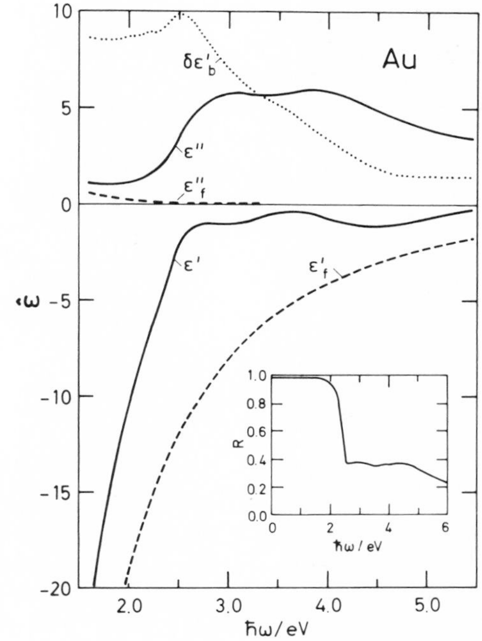

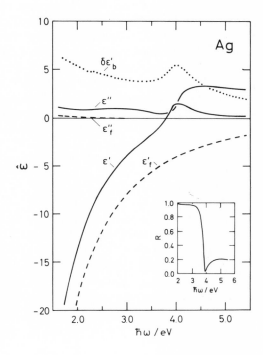

FIGURE 3. Frequency dependence of real and imaginary parts of the dielectric constant of Au, $\hat{\varepsilon} = \varepsilon' - i\varepsilon''$, with decomposition into free-electron (f) and bound-electron (b) contributions. The normal-incidence reflectivity of Au (in air) is shown in the inset. After Ref. 15.

FIGURE 4. Frequency dependence of real and imaginary parts of the dielectric constant of Ag with decomposition into free- and bound-electron contributions. The normal-incidence reflectivity spectrum of Ag is shown in the inset.

however, an interband transition from the d-band to the sp-band at the Fermi energy, E_F, sets in, which causes $\varepsilon_b'' \approx \varepsilon''$ to rise due to the absorption process. As a consequence of this interband transition, the reflectivity drops around 2.5 eV, causing the yellowish color of gold.

A similar behavior is observed for Ag, the only difference being that the d-band of Ag is about 4 eV below E_F rather than 2.5 eV as in Au. Hence, ε_b'' starts to rise at 4.0 eV due to the onset of the $4d \rightarrow 5s(E_F)$ interband transition. The corresponding variation of ε_b' now causes ε' to pass through zero at rather low photon energies, and thus the necessary condition for plasmon excitation is fulfilled at $\hbar\omega_p = 3.8$ eV. The sudden drop in the reflectivity of Ag at 3.8 eV (see inset of Figure 4) is characteristic of plasmon excitation, while the increase in R shortly after the plasma frequency, which creates the pronounced dip at 3.9 eV in the reflectivity spectrum of Ag, is due to the onset of the interband transition (as seen in the rise of ε_b'' at that energy). The interband transition in Ag is thus responsible for the plasma frequency being found at 3.8 eV rather than at 9.0 eV as calculated from the free-electron model, Eq. (9). The influence of the d-electrons on the plasma oscillations can be visualized in terms of a screening effect, which reduces the force constant for the oscillations.[16]

By comparing Figures 3 and 4, it becomes obvious that the interband transition in Au occurs at too low an energy (where $|\varepsilon_f'|$ is still large) for ε' to be pushed clear through zero, although the value for $|\varepsilon'|$ becomes quite small around 2.5 eV. Therefore, plasmon excitation is not clearly observed for gold, since the conditions $\varepsilon' = 0$ and $\varepsilon'' \ll 1$ are not strictly fulfilled, although one may speak of a pseudo-plasmon excitation in this energy range (e.g., the energy loss function shows a broad and asymmetric peak around 2.5 eV).

Finally, the optical properties of platinum, another important metal in electrochemistry, are shown in Figure 5. For this metal the d-band is located right at and below the Fermi level, and hence the onset of the interband transitions lies in the infrared region. Consequently, the reflectivity of Pt drops below unity already at very low photon energies and decreases continuously throughout the UV-visible region. A table with optical constants of some selected materials is given in Appendix II.

2.2. The Reflectivity of an Interface

The reflectivity of an interface is calculated from Fresnel's equations, which are derived from the boundary conditions that the tangential (t) component of the electric field vector, E_t, and the normal (n) component of the electric displacement vector, $D_n = \varepsilon \cdot E_n$, are continuous across the interface.[16-18] We define the so-called Fresnel reflection coefficient r as the ratio of the complex amplitudes of the electric field vectors of the reflected and the incident waves:

$$\hat{r} = \hat{E}_r / \hat{E}_i \qquad (21a)$$

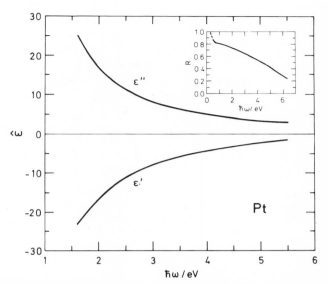

FIGURE 5. Frequency dependence of real and imaginary parts of the dielectric constant of Pt. The normal-incidence reflectivity of Pt (in air) is shown in the inset. After Ref 15.

The reflectivity R of the interface is then given by:

$$R = \hat{r} \cdot \hat{r}^* = |\hat{r}|^2 \tag{21b}$$

where $\hat{r}^*$ is the complex conjugate of $\hat{r}$. Similarly, a Fresnel transmission coefficient $\hat{t}$ can be defined from the complex amplitudes of the electric field vectors of the transmitted and the incident beam.[18]

The Fresnel coefficients are a function of the polarization and the angle of incidence of the incident beam. We commonly work with linearly polarized light, where the electric field vector is either perpendicular (s, $\perp$) or parallel (p, $\|$) to the plane of incidence. For s-polarization, the electric field vector $\mathbf{E}$ is parallel to the surface at all angles of incidence, while for p-polarized light at oblique angles of incidence, the electric field vector has a component normal to the surface. At normal incidence, the physical difference between s- and p-polarization vanishes.

For oblique angles of incidence, it is useful to define for each phase j an angle-dependent quantity, ξ_j, as:

$$\xi_j = \hat{n}_j \cos \varphi_j = (\hat{n}_j^2 - n_1^2 \sin^2 \varphi_1)^{1/2} \tag{22}$$

where ξ_j represents in essence the normal component of the wave vector of the light, and n_1 and φ_1 are, respectively, the real refractive index and the real angle of incidence for the transparent ambient phase (1), while $\hat{n}_j$ is the complex refractive index of phase j. If phase j is absorbing (complex n_j), the angle φ_j becomes a complex quantity and hence loses its meaning as the angle for the refracted beam in medium j. A complex angle of incidence relates to

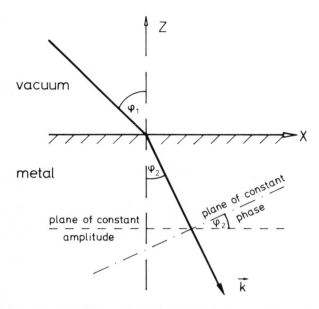

FIGURE 6. Schematic of an obliquely damped electromagnetic wave. Note that the planes of constant amplitude and constant phase do not coincide.

the fact that in an absorbing medium the light wave is "obliquely damped," i.e., the planes of constant phase and of constant amplitude no longer coincide.[17] This is shown schematically in Figure 6. Note that the amplitude (hence the intensity) decreases with the distance to the surface of the absorbing medium and therefore the plane of constant amplitude is in that case always parallel to the surface.

The Fresnel reflection coefficients for the interface between two phases j and k, for perpendicular and parallel polarized light, are given by[18]:

$$\hat{r}_{\perp jk} = \frac{\xi_j - \xi_k}{\xi_j + \xi_k} \tag{23a}$$

and

$$\hat{r}_{\parallel jk} = \frac{\hat{\varepsilon}_k \xi_j - \hat{\varepsilon}_j \xi_k}{\hat{\varepsilon}_k \xi_j + \hat{\varepsilon}_j \xi_k} \tag{23b}$$

For normal incidence ($\varphi_1 = 0$),

$$\hat{r}_{\perp jk} = -\hat{r}_{\parallel jk} = (\hat{n}_j - \hat{n}_k)/(\hat{n}_j + \hat{n}_k) \tag{24}$$

In the following, a few useful expressions for the reflectivity $R = \hat{r} \cdot \hat{r}^*$ (Eq. 21b) of an interface between a transparent and an absorbing medium (e.g., metal/air or metal/electrolyte interface) are given.

At normal incidence, the reflectivity of a metal ($\hat{n} = n - ik$) adjacent to an ambient medium of refractive index, n_1, is:

$$R = \frac{(n - n_1)^2 + k^2}{(n + n_1)^2 + k^2} \qquad (25)$$

For oblique angle of incidence, we find the following expressions[25]:

$$R_\perp = \frac{(a - \cos \varphi_1)^2 + b^2}{(a + \cos \varphi_1)^2 + b^2} \qquad (26a)$$

and

$$\left(\frac{R_\parallel}{R_\perp}\right) = \frac{(a - \sin \varphi_1 \tan \varphi_1)^2 + b^2}{(a + \sin \varphi_1 \tan \varphi_1)^2 + b^2} \qquad (26b)$$

where a and b are defined by the equation:

$$(a - ib)^2 = \left(\frac{n - ik}{n_1}\right)^2 - \sin^2 \varphi_1 \qquad (26c)$$

From Eqs. (26a) and (26b) it is readily seen that for $\varphi_1 = 45°$, $R_\parallel = R_\perp^2$. The angular dependences of $R_\perp$ and $R_\parallel$ for two different cases are shown in Figure 7. The right-hand side of the figure represents the reflectivity of an interface between two nonabsorbing media (e.g., quartz/air), while on the left-hand side the situation for a metal/air interface is depicted. In both cases, $R_\perp$ increases monotonically from its minimum value at $\varphi_1 = 0°$ to unity at $\varphi_1 = 90°$, whereas $R_\parallel$ exhibits a minimum at the so-called pseudo-Brewster angle. For a nonabsorbing medium ($k = 0$), $R_\parallel$ becomes zero at the Brewster angle, φ_B, which is defined by:

$$\tan \varphi_B = n/n_1 \qquad (27)$$

i.e., at φ_B reflected and transmitted beams are orthogonal to each other.

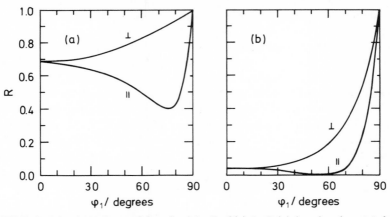

FIGURE 7. Angular dependence of the reflectivity R of (a) the Pt/air interface ($n = 1.6$; $k = 3.7$; 2.5 eV) and (b) the quartz/air interface ($n = 1.5$; $k = 0.0$).

Since later in this chapter we shall be interested in the study of atoms, ions, or molecules at or very close to the electrode surface, it will be important to have some ideas about the (local) electric field strength right at the surface. This should yield further insight into the absorption processes occurring at the interface.

When light is reflected from a surface, the electric fields of the coherent incident and reflected plane waves sum up vectorially and produce a standing-wave electric field in the incident medium. As we mentioned earlier, the boundary conditions for electromagnetic waves require that the tangential components of the electric field vector be continuous across the interface, and so must be the normal component of the electric displacement vector. These conditions are the basis of the relations for the mean square electric field strengths (i.e., the local intensity of the light) at the interface. Such equations are derived and discussed in Ref. 26. The angular dependence of various components of the mean square electric field strength, $\langle E^2 \rangle$, normalized to that of the incident beam, $\langle E_0^2 \rangle$, are shown in Figure 8 for two different cases.[15] The coordinate system is chosen such that perpendicular polarized light would have its electric field in the y-direction, while parallel polarized radiation has components of the electric field vector in the x- and the z-direction, with the z-axis being normal to the surface and the plane of incidence being defined by x and z. Note that $\langle E_y^2 \rangle$ is not shown in Fig. 8, since it is similar to $\langle E_{\parallel,x}^2 \rangle$. For a perfect metal (i.e., high optical conductivity and R close to unity), the components of the electric field strength parallel to the surface have nodes at the surface because of a phase change upon reflection by 180°, which causes cancellation of the field components for the incoming and reflected beams (see inset of Figure 8a). Only the z-component, $\langle E_{\parallel,z}^2 \rangle$, has an antinode at the surface. The conditions for a perfect metal are realized for most metals in the infrared region, where $k \gg n$ and $R \approx 1$, and hence $\langle E_{\parallel,x}^2 \rangle$ and $\langle E_y^2 \rangle$ are practically zero at all angles of incidence. Only for p-polarized light at high angles of incidence does $\langle E_{\parallel,z}^2 \rangle$ acquire significant values. Consequently, an atom, molecule, or thin layer at the surface of a highly reflecting substrate cannot be detected by optical spectroscopy with s-polarized light, nor at normal incidence as the electric field strength of the radiation is virtually zero at the surface. Therefore, infrared reflectance spectroscopy is always performed with p-polarized light and employing grazing angles of incidence (note also Chapter 5 of this book). In the UV–visible range, on the other hand, where the onset of interband transitions causes a marked reduction of the reflectivity, the tangential field components gain sufficient magnitude, so that light can now interact strongly with adsorbed species, even at normal incidence (Figure 8b).

2.3. Three-Phase System and Linear Approximation

In optical studies of electrode surfaces, we have to consider three phases: (1) the bulk electrolyte, (2) the interfacial region, and (3) the bulk electrode

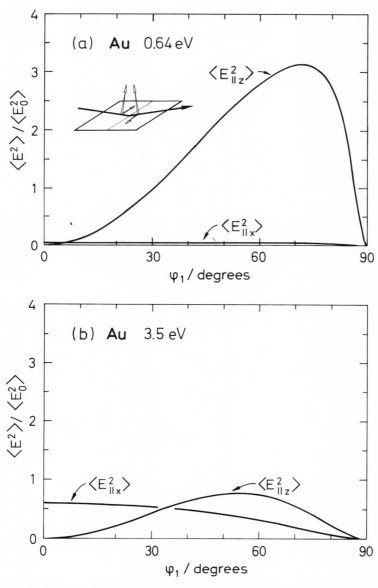

FIGURE 8. Angular dependence of the electric field strength for p-polarized light at a gold surface under (a) highly reflecting conditions ($n_1 = 1.33$; $n_3 = 1.09$; $k_3 = 13.43$; $R_{\varphi_1=0} = 0.97$) and (b) moderately reflecting conditions ($n_1 = 1.33$; $n_3 = 1.64$; $k_3 = 1.75$; $R_{\varphi_1=0} = 0.27$).

102   *Dieter M. Kolb*

material. We define the interfacial region as having optical properties different
from those of the two adjacent semi-infinite bulk phases. This region can
represent the Helmholtz layer proper, an adsorbate or film deposit on the
electrode surface, the electrode surface region itself, or any combination of
the three. For the sake of simplicity, we describe the complete electrode/elec-
trolyte interface by a three-phase system with sharp boundaries as depicted
in Figure 9.[11] The bulk electrolyte as a transparent ambient (phase 1) is
characterized by a real dielectric constant, whereas film and substrate phases
(2 and 3, respectively) are described by complex dielectric constants. In
addition, phase 2 is assumed for the moment to be a homogeneous, isotropic
thin film of uniform thickness d. The reflectivity of the three-phase system,
$R_{123} = f(\varepsilon_1, \hat{\varepsilon}_2, \hat{\varepsilon}_3, d, \lambda, \varphi_1)$, hence contains information on the optical properties $\hat{\varepsilon}_2$ and the thickness d of the interphase.[2]

The accurate determination of absolute reflectance values for surfaces
immersed in an electrolyte and enclosed in an electrochemical cell is practically
impossible, because of undefined intensity losses due to reflection and scattering at cell windows and in the electrolyte. Attempts to overcome these difficulties by using a light detector which works in the electrolyte were not very
convincing.[27] Therefore, it is much more convenient to measure relative
changes $\Delta R/R$, where ΔR is a reflectance change of the electrode surface
caused by some electrochemical manipulation (usually an electrode potential
change). Because all intensity losses of the light beam not associated with the
electrochemical interface cancel, $\Delta R/R$ is directly given by the corresponding
intensity change, $\Delta I/I$, measured with a photomultiplier outside the cell.

In the study of adsorbates, we define $\Delta R/R$ as the normalized reflectance
change for a bare surface due to the formation of an adlayer[11]:

$$\Delta R/R = R_{123}/R_{13} - 1 = [R(d) - R(0)]/R(0) \qquad (28)$$

With the help of Fresnel's equations, the exact expressions for $\Delta R/R$ can be
obtained by calculating the reflection coefficients for the three-phase system
$[R_{123} \equiv R(d)]$ and for the two-phase system $[R_{13} \equiv R(0)]$. Such calculations,

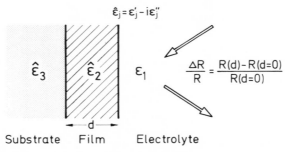

FIGURE 9. Three-phase model with sharp boundaries for the electrode/electrolyte interface.
After Ref. 11.

although quite cumbersome when done by hand, are easily performed by computers. For very thin films $(d \ll \lambda)$, a linear approximation in d/λ for the Fresnel equations has been derived[11,15] which leads to rather simple expressions in $\Delta R/R$. These equations are very valuable since they allow a more direct insight into the correlations between optical constants and measured quantities than the exact treatment. The expressions derived for $d \ll \lambda$ are[11]:

$$\left(\frac{\Delta R}{R}\right)_{\perp} = \frac{8\pi \, d\sqrt{\varepsilon_1}\,\cos\varphi_1}{\lambda} \cdot \mathrm{Im}\left(\frac{\hat{\varepsilon}_2 - \hat{\varepsilon}_3}{\varepsilon_1 - \hat{\varepsilon}_3}\right) \tag{29a}$$

and

$$\left(\frac{\Delta R}{R}\right)_{\parallel} = \frac{8\pi \, d\sqrt{\varepsilon_1}\,\cos\varphi_1}{\lambda} \cdot \mathrm{Im}\left\{\left(\frac{\hat{\varepsilon}_2 - \hat{\varepsilon}_3}{\varepsilon_1 - \hat{\varepsilon}_3}\right)\left[\frac{1 - (\varepsilon_1/\hat{\varepsilon}_2\hat{\varepsilon}_3)(\hat{\varepsilon}_2 + \hat{\varepsilon}_3)\sin^2\varphi_1}{1 - (1/\hat{\varepsilon}_3)(\varepsilon_1 + \hat{\varepsilon}_3)\sin^2\varphi_1}\right]\right\} \tag{29b}$$

$\mathrm{Im}\,(x)$ means that the imaginary part of the function x has to be taken.

For normal incidence $(\varphi_1 = 0)$, Eqs. (29a) and (29b) reduce to[15]:

$$\left(\frac{\Delta R}{R}\right)_{\varphi_1=0} = 8\pi\sqrt{\varepsilon_1}\,\frac{d}{\lambda}\left[\frac{(\varepsilon_1 - \varepsilon_2')\varepsilon_3'' - (\varepsilon_1 - \varepsilon_3')\varepsilon_2''}{(\varepsilon_1 - \varepsilon_3')^2 + \varepsilon_3''^2}\right] \tag{30}$$

Equation (30) turns out to be very useful in obtaining an estimate for the magnitude of $\Delta R/R$. For example, it is easily seen from Eq. (30) that reflectance spectroscopy is not very sensitive in detecting dielectric (nonabsorbing) films $(\varepsilon_2'' = 0)$ while strongly absorbing films are readily discernible in $\Delta R/R$. In the first case, ellipsometry is far superior.[28]

We briefly illustrate the above statement by calculating $\Delta R/R$ for two different adlayers on a silver substrate $(\varepsilon_3' = -24, \varepsilon_3'' = 1$ for $\lambda = 800$ nm) in contact with an electrolyte for which $\varepsilon_1 = 1.77$. The first example is a Pb monolayer formed at underpotentials. For convenience, we assume bulk optical properties for the Pb monolayer $(\varepsilon_2' = -15, \varepsilon_2'' = 20, d = 0.4$ nm), which is certainly not correct but may serve the purpose. Equation (30) yields a relative reflectance decrease of about 1% for the Pb monolayer on silver, which is in good agreement with experimental findings.[29] The second case deals with the Helmholtz layer as a purely nonabsorbing film in the visible–near-UV range. We assume that the optical properties of the interfacial water change from those of bulk water to, say, $\varepsilon_2' = 1.87$ and $\varepsilon_2'' = 0$ when stepping the potential of the Ag electrode from the potential of zero charge (pzc) to some extreme value. This yields a $\Delta R/R$ value for $d = 0.4$ nm of -2.5×10^{-6} ! The effect in this case is about three to four orders of magnitude smaller, which demonstrates the enormous difficulties encountered in the study of the Helmholtz layer proper in the wavelength region of optical transparency. We shall address this point again in Section 4.4.

Figure 10 illustrates the angular dependence of $\Delta R/R$ for s- and p-polarized light for two different cases. In the first one, optical constants are

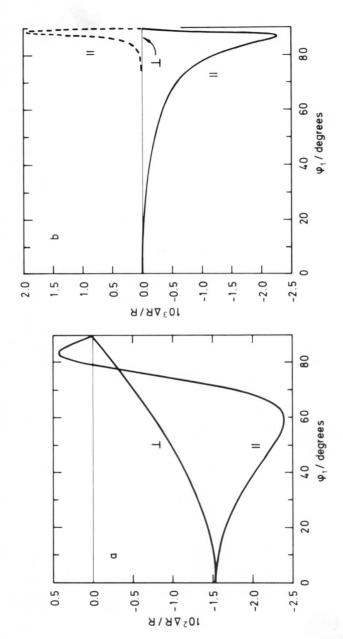

FIGURE 10. Angular dependence of the normalized change in reflectivity, $\Delta R/R$, caused by the deposition of a thin film on a moderately reflecting (a) and a highly reflecting (b) metal substrate. The optical parameters are: (a) $n_1 = 1.33$; $n_2 = 3.0$; $k_2 = 1.5$; $n_3 = 2.0$; $k_3 = 4.0$; $d/\lambda = 1 \times 10^{-3}$; (b) $n_1 = 1.0$; $n_2 = 1.3$; $k_2 = 0.1$; $n_3 = 3.0$; $k_3 = 30.0$; $d/\lambda = 1 \times 10^{-4}$. s- and p-polarization. After Ref. 15.

chosen which are typical for a monomolecular oxide layer on platinum in an aqueous solution and at wavelengths in the visible–near-UV range. The second case (Figure 10b) deals with a film on a highly reflecting surface in the near-IR range. Both examples demonstrate that measurements with p-polarized light at medium to high angles of incidence yield the largest values of $\Delta R/R$. Nevertheless, as will be pointed out on various occasions throughout this chapter, experiments at normal incidence are often preferable in the visible–near-UV range because of complications in metal optics with electric field components normal to the surface (see, e.g., Section 2.4 on nonlocal optics).

In some cases, one may wish to describe the interfacial region between bulk substrate and bulk electrolyte by more than just one (homogeneous) phase 2. For example, an adsorbate layer will change to some extent the optical properties of the substrate surface. As a consequence, the three-phase system of Figure 9 should be replaced by a four-phase system, with two thin films sandwiched between the bulk phases.[30,31] In order to get an idea about the contributions to $\Delta R/R$ of the various films in a multilayer system, the linear approximation theory has been expanded.[31] It can be shown that the total $\Delta R/R$ of such a system is given by the sum over all $\Delta R/R$ values due to each thin layer. For $d = \sum_j d_j \ll \lambda$, Eq. (29a) becomes[31]:

$$\left(\frac{\Delta R}{R}\right)_\perp = \frac{8\pi\sqrt{\varepsilon_1}\cos\varphi_1}{\lambda}\sum_j d_j \operatorname{Im}\left(\frac{\hat{\varepsilon}_{2j} - \hat{\varepsilon}_3}{\varepsilon_1 - \hat{\varepsilon}_3}\right) \tag{31}$$

where the index j refers to j different layers of thickness d_j and dielectric function $\hat{\varepsilon}_{2j}$ between bulk phases 1 and 3. Equation (29b) for p-polarized light changes accordingly.[31] Clearly, because of the linear approximation, $\Delta R/R$ is independent of the sequence of the j thin layers.

It is apparent from Eqs. (29) that the $\Delta R/R$ spectra of a surface film on an absorbing substrate do not always allow for a straightforward interpretation of the film optical constants. The spectra are by no means transmission-like but are markedly influenced by the optical properties of the substrate. Therefore, it appears to be desirable to evaluate directly the film dielectric function, $\hat{\varepsilon}_2 = \varepsilon_2' - i\varepsilon_2''$, from the experimentally determined quantity $\Delta R/R$ rather than attempting a detailed interpretation of the $\Delta R/R$ spectra alone. In the following, we briefly discuss three different methods for evaluating film optical constants.

Inversion of the Linear Approximation Equations

By measuring $(\Delta R/R)_\perp$ and $(\Delta R/R)_\parallel$ at one angle of incidence, the film optical properties can be obtained from Eqs. (29a) and (29b) by inverting and solving them for the two unknown parameters ε_2' and ε_2''.[13] The third unknown parameter, the film thickness d, has to be determined independently, e.g., from coulometric measurements, or estimated. However, since in Eqs. (29), d appears only in the sensitivity factor $(8\pi d\sqrt{\varepsilon_1}\cos\varphi_1/\lambda)$ and not in the shape factor $\text{Im}\,(\hat{\varepsilon}_i, \varphi_1)$, the uncertainty in d should only affect the absolute magnitude of ε_2' and ε_2'', but not so much the shape of the spectral dependence of these quantities, the spectral shape containing the most relevant information. The final equation to solve for the film optical constants is to the third power. When the number of real roots is three instead of one, it may be difficult to decide which is the physically meaningful solution, and hence it may not be possible to make an assignment at one wavelength with certainty. However, when a wide spectral range in $\Delta R/R$ is evaluated, only one set of solutions is inherently continuous in most cases, indicating that this is the true physical solution. The other two solutions, which become identical just before turning complex, can then be discarded.[13] With this method, we have to bear in mind that the initial equations [Eqs. (29a) and (29b)] have been derived for isotropic thin films. Especially in the case of monolayer deposits, however, anisotropy of the optical constants will most likely arise. We shall return to this problem later.

Iteration Methods

When $(\Delta R/R)_{\parallel}$ is measured over a wide range of angles of incidence (say, $10° \le \varphi_1 \le 80°$), the film optical constants can be found by computational iteration to the best fit of the measured $\Delta R/R(\varphi_1)$ curve. This is not possible for s-polarization, since $(\Delta R/R)_{\perp}$ is simply proportional to $\cos \varphi_1$ (Eq. 29a). Although such an angle dependence study of $(\Delta R/R)_{\parallel}$ is quite cumbersome when done over the whole wavelength region, the advantage is that anisotropy effects of thin films can be investigated. As will be shown below, the optical response of a uniaxial absorbing film is described by two complex dielectric functions, $\hat{\varepsilon}_{2n}$ and $\hat{\varepsilon}_{2t}$, where the indices n and t refer to normal and tangential. This yields four unknown parameters to be determined by the fitting procedure. Usually, various minima with sufficiently small mean square deviations are found, depending on the starting point for the iteration, which often makes any decision on the true optical constants very difficult or speculative. The following procedure turned out to be quite useful[32]: Since for $\varphi_1 < 45°$ the $(\Delta R/R)_{\parallel}$ values are mostly determined by the tangential component of the film optical properties, this region is fitted first with $\hat{\varepsilon}_{2t}$ only, by using the isotropic model. Secondly, the $\Delta R/R$ values for $\varphi_1 > 45°$ are then fitted by variation of $\hat{\varepsilon}_{2n}$ but with those $\hat{\varepsilon}_{2t}$ values obtained at the smaller angles of incidence. Finally, the iteration procedure is repeated for the whole range of φ_1 with a variation again of all four optical parameters.

Kramers–Kronig Analysis

Kramers–Kronig analysis (KKA) has been frequently used to determine the optical constants of solids from normal-incidence reflectance spectra. This is well documented in the literature.[33] The causality relation can also be employed to evaluate the film optical constants from spectra of $\Delta R/R$, as was first demonstrated by Plieth and Naegele.[34,35]

In a two-phase system, the complex Fresnel reflection coefficient $\hat{r}_{13}$ may be expressed in terms of its argument, the square root of the reflectivity, and its phase angle, which then can be related to the optical constants. For example, at normal incidence,

$$\hat{r}_{13} = R_{13}^{1/2} \cdot \exp{(i\delta_{13}^r)} = \frac{n - ik - n_1}{n - ik + n_1} \tag{32}$$

Therefore, by determining the phase change δ_{13}^r upon reflection, the optical constants can be calculated. Since phase angle and reflectivity are related to each other by the Kramers–Kronig dispersion relation

$$\delta_{13}^r(\omega_0) = \frac{\omega_0}{\pi} \int_0^\infty \frac{\ln{[R(\omega)/R(\omega_0)]}}{\omega^2 - \omega_0^2} \, d\omega \tag{33}$$

measurement of the reflectance $R(\omega)$ over the entire frequency range yields δ_{13}^r and therefore ultimately allows for an evaluation of the optical constants.[33] Difficulties arise, however, from the fact that $R(\omega)$ usually is determined only within a limited frequency range and hence assumptions have to be made about the high-energy tail of $R(\omega)$ (see, e.g., p. 116 of Ref. 15). Alternatively, additional experiments with other methods have to be performed for a proper tail fit, e.g., transmission experiments or the determination of the optical constants at one wavelength by measuring R_p at two angles of incidence. These problems have been discussed in the literature.[36]

For a three-phase system (Figure 9), the ratio of the Fresnel reflection coefficients, $\hat{r}_{123}$ and $\hat{r}_{13}$, are related to the difference $\Delta\delta^r = \delta_{123}^r - \delta_{13}^r$ in the phase angles of the three-phase and the two-phase system[34,35]:

$$\frac{\hat{r}_{123}}{\hat{r}_{13}} = \left(\frac{R_{123}}{R_{13}}\right)^{1/2} \cdot \exp{(i\Delta\delta^r)} = f(\varepsilon_1, \hat{\varepsilon}_2, \hat{\varepsilon}_3, d, \lambda, \varphi_1) \tag{34}$$

For $\Delta\delta^r$ the following expression was derived by Plieth and Naegele for s-polarized light[34]:

$$\Delta\delta_\perp^r(\omega_0) = \frac{1}{\pi} \int_0^\infty \frac{\omega \ln{[R_{123}(\omega)/R_{13}(\omega)]}}{\omega^2 - \omega_0^2} \, d\omega \tag{35}$$

An explicit evaluation of ε_2' and ε_2'' calls for the linear approximation instead of the exact expression for $\hat{r}_{123}/\hat{r}_{13}$:

$$\left(\frac{\hat{r}_{123}}{\hat{r}_{13}}\right)_\perp = 1 - i\frac{4\pi d\sqrt{\varepsilon_1}\cos{\varphi_1}}{\lambda} \cdot \frac{\hat{\varepsilon}_2 - \hat{\varepsilon}_3}{\varepsilon_1 - \hat{\varepsilon}_3} \tag{36}$$

Substituting $(\hat{r}_{123}/\hat{r}_{13})$ and $\Delta\delta^r$ of Eq. (34) by Eqs. (36) and (35), respectively, yields an expression for the real and the imaginary part of the film dielectric constant, $\hat{\varepsilon}_2$. Because $\Delta R/R \ll 1$ for almost all cases, we can use the approximation: $\ln(R_{123}/R_{13}) \approx \Delta R/R$.

The big advantages of the KKA, besides the relatively small experimental work required for obtaining the necessary data, are twofold. Firstly, it is possible to evaluate the two optical constants of a (assumed isotropic) thin film by measurements with s-polarized light only. Hence, problems arising from nonlocal effects (Section 2.4) are avoided. Secondly, if we are forced to assume an anisotropic film dielectric function, $\hat{\varepsilon}_{2t}$ and $\hat{\varepsilon}_{2n}$ can, in principle, be obtained from a KKA of the $(\Delta R/R)_{\perp}$ and $(\Delta R/R)_{\parallel}$ spectra.[37]

A difficulty of this method (as is generally true for KKA) lies in the very limited wavelength region for which the $\Delta R/R$ spectra can be determined. Extrapolation to high energies is not as simple as in the case of the two-phase system ($R \to 0$ for $E \to \infty$), because ΔR and R (or R_{123} and R_{13}) tend to go to zero with $E \to \infty$, leaving the extrapolation of $\Delta R/R$ an open question. Another problem with KKA is that it requires some knowledge of where exactly in the three-phase system the phase change $\Delta \delta^r$ occurs. This is briefly demonstrated in Figure 11. Let us assume that $\Delta R/R$ is zero over the entire frequency range. This yields, according to Eqs. (34)–(36), a film dielectric constant, $\hat{\varepsilon}_2$, identical to that of the substrate, $\hat{\varepsilon}_3$. We know, however, that film optical properties $\hat{\varepsilon}_2 = \varepsilon_1$ lead to the same result. As is seen from Eq. (36), to obtain the second solution a phase change of $\Delta \delta^r = 4\pi d \sqrt{\varepsilon_1} \cos \varphi_1/\lambda$ is required. If we define the position of phase change to be at the phase boundary (1–2), then the above equations are correct for the case shown in Figure 11b (e.g., oxide growth). If the film grows into the electrolyte (e.g., metal deposition, Figure 11c), then we have to add to Eq. (36) the correction term $-i4\pi d \sqrt{\varepsilon_1} \cos \varphi_1/\lambda$.[38] The determination of the absolute phase change at a single wavelength by, for example, interferometry would remove this ambiguity.[39]

Anisotropic Film Optical Constants

Monomolecular films and submonolayer adsorbates are expected to exhibit an appreciable amount of anisotropy in their optical properties, especially when there is a strong interaction either with the substrate or within

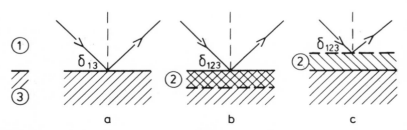

FIGURE 11. Schematic illustration for the different positions of the phase jump δ_{123} in a three-phase system, depending on the film growth.

the adlayer. The complex dielectric constant of the film phase has then to be written as a uniaxial tensor

$$\hat{\varepsilon}_2 = \begin{vmatrix} \hat{\varepsilon}_{2t} & 0 & 0 \\ 0 & \hat{\varepsilon}_{2t} & 0 \\ 0 & 0 & \hat{\varepsilon}_{2n} \end{vmatrix}$$

where the subscripts t and n refer to tangential and normal components, respectively. Dignam *et al.*[40] have derived quantitative expressions, using the linear approximation, of $\Delta R/R$ for a uniaxial absorbing film on an isotropic substrate. These are:

$$\left(\frac{\Delta R}{R}\right)_{\perp} = \frac{8\pi d\sqrt{\varepsilon_1}\cos\varphi_1}{\lambda}\, \mathrm{Im}\left(\frac{\hat{\varepsilon}_{2t} - \hat{\varepsilon}_3}{\varepsilon_1 - \hat{\varepsilon}_3}\right) \tag{37a}$$

and

$$\left(\frac{\Delta R}{R}\right)_{\|} = \frac{8\pi d\sqrt{\varepsilon_1}\cos\varphi_1}{\lambda}\, \mathrm{Im}\left\{\left(\frac{\hat{\varepsilon}_{2t} - \hat{\varepsilon}_3}{\varepsilon_1 - \hat{\varepsilon}_3}\right)\right.$$
$$\left.\times \left(\frac{1 - (\varepsilon_1/\hat{\varepsilon}_{2n}\hat{\varepsilon}_3)[(\hat{\varepsilon}_3^2 - \hat{\varepsilon}_{2t}\hat{\varepsilon}_{2n})/(\hat{\varepsilon}_3 - \hat{\varepsilon}_{2t})]\sin^2\varphi_1}{1 - (1/\hat{\varepsilon}_3)(\varepsilon_1 + \hat{\varepsilon}_3)\sin^2\varphi_1}\right)\right\} \tag{37b}$$

More recently, the problem of film anisotropy has been treated again in a more thorough fashion by Plieth.[41] He pointed out that the more fundamental parameter for describing the film optical response normal to the surface is $\hat{\varepsilon}_n^{-1}$ rather than $\hat{\varepsilon}_n$ proper, because of the close correlation between the normal component of the electric field vector and the so-called loss function, $\mathrm{Im}\,\hat{\varepsilon}^{-1}$.

Since in the case of a uniaxial absorbing film, four unknown parameters instead of two have to be determined, the evaluation procedure for the film optical constants becomes quite complicated. Besides, it may not be obvious in the first place whether the more simple "isotropic" case is sufficient or not. It has been pointed out that, for example, a thin, transparent but anisotropic film on a metallic substrate can yield $\Delta R/R$ values which, when used to calculate the optical constants for an isotropic film, may falsely indicate absorbing properties.[40]

We briefly demonstrate in the following the influence of film anisotropy on $\Delta R/R$ spectra and on evaluation procedures.[32] We assume for the film optical properties that $\hat{\varepsilon}_{2t}$ is described by two Lorentzian oscillators with resonance energies at 2.1 and 4.1 eV and $\hat{\varepsilon}_{2n}$ by three Lorentzian oscillators at 2.1, 2.8, and 4.1 eV. The resulting imaginary parts of $\hat{\varepsilon}_{2t}$ and $\hat{\varepsilon}_{2n}$ for our model film are shown in Figure 12a. With the help of Eqs. (37a) and (37b), we then calculate the corresponding $\Delta R/R$ spectra for a 0.3 nm thick model film on platinum in contact with an electrolyte ($\varepsilon_1 = \varepsilon_{H_2O}$). The result for $\varphi_1 = 45°$ is shown in Figure 13. If we now evaluate the film optical properties from the angular dependence of $(\Delta R/R)_{\|}$ as described in Section 2.3 (Iteration Methods), the data shown in Figure 12b are obtained for the imaginary parts. The starting point at 1.6 eV for the numerical iteration was arbitrarily chosen

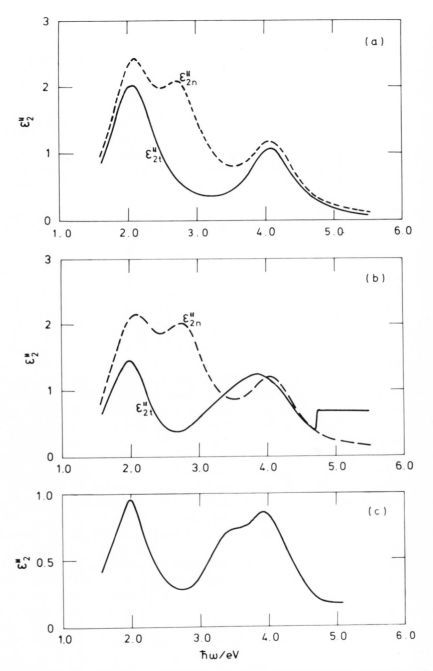

FIGURE 12. (a) Tangential (——) and normal (- - -) components of the imaginary part of the dielectric function of an anisotropic model film. (b) Evaluation of the film dielectric function from the angular dependence of $\Delta R/R$. (c) Evaluation of the film dielectric function from $\Delta R/R$, ignoring film anisotropy. See text.

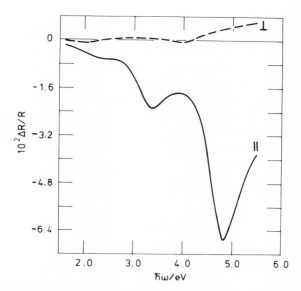

FIGURE 13. Calculated $\Delta R/R$-spectrum for the model film (Figure 12a) on Pt. $\varphi_1 = 45°$; $d = 0.3$ nm; $\varepsilon_1 = \varepsilon_{H_2O}$; s- and p-polarization. After Ref. 32.

as $\varepsilon'_{2t} = 1$ and $\varepsilon''_{2t} = 1$. Comparison of Figures 12a and 12b reveals a satisfactory agreement. The discontinuity of ε''_{2t} at 4.7 eV indicates a physically unrealistic solution for $\hbar\omega \geq 4.7$ eV, which is also reflected in a relatively large mean square deviation of about 10^{-4} as opposed to 10^{-9} for $\hbar\omega < 4.7$ eV. It may be interesting to see the result of an "isotropic" evaluation of the $\Delta R/R$ spectra in Figure 13 by inverting Eqs. (29a) and (29b) and solving them for an isotropic $\hat{\varepsilon}_2$, as described in Section 2.3 (see pp. 105–106). The result is shown in Figure 12c. We note that a d value of 0.4 nm had to be chosen in order to obtain one continuous set of physically realistic values. Except for this, the data look not too bad. At least, the absorption bands at 2.1 and 4.1 eV are reproduced reasonably well; the third (anisotropic) band at 2.8 eV may be the cause of the shoulder at 3.4 eV.

Before closing this section on the three-phase system and the evaluation of film optical properties, a word of caution seems appropriate. Some 15 years ago, when we started to evaluate monolayer optical constants by the three-phase model with sharp boundaries,[12,13] I was much more optimistic than I am today. The model discussed above (which still seems to be the only one around) has a number of severe drawbacks. For example, the validity of the concept of a macroscopic (continuum) theory for monolayer and submonolayer adsorbates was questioned quite early, although it was hoped that the model would yield an *effective* film dielectric constant which represents at least average values of the absorptive and refractive properties of the thin film. Application of the continuum theory is especially critical for the direction normal to the surface, where the local variations cannot be smeared out by averaging over a large distance as in the bulk. Microscopic theories of the optical and electronic properties of adatoms and adlayers have been described in the literature, but the connection between adlayer properties and observables, such as $\Delta R/R$, was again achieved via the simple three-phase model.[42]

As was mentioned before, we define the interfacial or thin-film region as the one whose optical constants differ from those of the adjacent bulk phases. However, the three-phase model with sharp boundaries, as sketched once more in Figure 14a, should be replaced by a model with a gradual change of $\hat{\varepsilon}$ from one phase to the next (Figure 14b). Calculation of the Fresnel reflection coefficients then requires integral boundary conditions.[38] The result is an integral value of the film dielectric constant, $\hat{\varepsilon}_2$, given by:

$$\hat{\varepsilon}_2 = \frac{1}{d} \int_{z_1}^{z_3} [\hat{\varepsilon}_2(z) - \hat{\varepsilon}_{ref}(z)] \, dz \tag{38}$$

where $\hat{\varepsilon}_{ref}(z)$ and $\hat{\varepsilon}_2(z)$ are the continuously changing dielectric constants in the transition region for the two-phase and the three-phase system, respectively. The integral expression, Eq. (38), and Figure 14 indicate that the three-phase model with sharp boundaries is meaningful only when the transition regions between phases 1 and 2 and 2 and 3 are small compared to the thickness of phase 2 proper. This is true for multilayer oxide films. For monolayers, however, an independent evaluation of $\hat{\varepsilon}_2$ and d rather than $\hat{\varepsilon}_2 \cdot d$ may be somewhat arbitrary.

A third problem arises specifically with the use of p-polarized light in the study of metal surfaces. The occurrence of so-called "nonlocal" effects, which are described in the next section, has usually been ignored in the past, and this led to serious misinterpretations of optical spectra. A proper inclusion of nonlocality in the metal optics is possible[43]; however, it complicates the situation to such an extent that a straightforward evaluation of film optical constants—as in the case of anisotropy—seems no longer feasible.

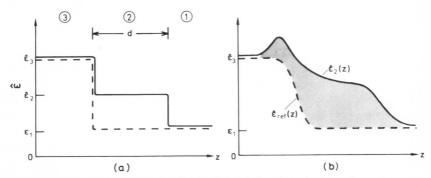

FIGURE 14. Schematic representation for the dielectric function of a three-phase system: (- - -) bare surface; (———) film-covered surface. (a) Sharp phase boundaries; (b) gradual variation of $\hat{\varepsilon}$ across the phase boundaries. After Ref. 41.

Therefore, as the reader will undoubtedly note, most case studies presented in this chapter refrain from an explicit evaluation of film optical constants. Instead, they demonstrate how physically relevant information is obtained directly from visual inspection of the experimental data.

2.4. Nonlocal Optics

Standard optics is based on the assumption that only transverse electromagnetic plane waves can propagate inside a solid, where the net charge is set to zero because of instantaneous screening. This is equivalent to div $\mathbf{E} = 0$ or $\mathbf{k} \cdot \mathbf{E} = 0$ (wave vector and wave amplitude are normal to each other). At high enough frequencies, however, the inertia of the conduction electrons in metals prevents such instantaneous screening, and as a consequence of this, charge density waves (so-called plasma waves) can be built up.[43] For many metals, this plasma excitation frequency is either in the vacuum–UV range and therefore in a range not easily accessible in normal *in situ* reflectance spectroscopy (e.g., Al, Ga, Hg, alkali metals) or plasma excitation occurs in the visible–near-UV range but is strongly damped by interband transitions (e.g., Cu). In both cases, the excitation of collective oscillations of the electrons (which plasma waves are) can be neglected. However, for metals like Ag and Au, the plasma frequency is around 3.8 and 2.5 eV (because of interband transitions, see Section 2.1), i.e., in a range where optical studies of electrode surfaces are routinely performed. In this frequency regime, the (longitudinal) plasma waves are eigenmodes of the metal ($\mathbf{k} \times \mathbf{E} = 0$), and they are optically excited with p-polarized light, with the electric field component normal to the surface inducing periodic charge density fluctuations.[43] The longitudinal waves propagate into the metal, in addition to the transverse waves (Figure 15), and therefore contribute to the reflectance and transmittance of the metal surface. A consequence of longitudinal waves is that the dielectric constant becomes a function of wave vector $\mathbf{k}$, i.e., $\hat{\varepsilon}(\omega, \mathbf{k})$ is now spatially dispersive. Or, in other words, the material equation

$$\mathbf{D}(\omega, \mathbf{r}) = \int \hat{\varepsilon}(\omega, \mathbf{r}, \mathbf{r}') \cdot \mathbf{E}(\omega, \mathbf{r}') \, d^3\mathbf{r}' \qquad (39)$$

is "nonlocal" in space, which means that the response $\mathbf{D}(\omega, \mathbf{r})$ is not only determined by the electric field strength $\mathbf{E}$ at position $\mathbf{r}$ but also by the $\mathbf{E}$ value in a certain neighborhood of $\mathbf{r}$. Fourier transformation brings Eq. (39) into the form:

$$\mathbf{D}(\omega, \mathbf{k}) = \hat{\varepsilon}(\omega, \mathbf{k}) \cdot \mathbf{E}(\omega, \mathbf{k}) \qquad (40)$$

Hence, nonlocality and spatial dispersion are two expressions for one and the same physical phenomenon.[43-45]

The simplest way to include spatial dispersion in the dielectric function of a free-electron metal is the so-called hydrodynamic approximation,[43] which

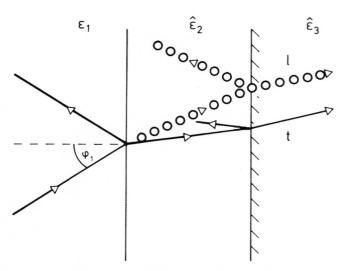

FIGURE 15. Schematic representation of transverse (——) and longitudinal (∘∘∘) waves, arising from reflection at an idealized three-phase system with two metallic phases (2 and 3). After Ref. 46.

yields

$$\hat{\varepsilon}_t(\omega) = 1 - \frac{\omega_n^2}{\omega^2 - i\omega\tau^{-1}} \tag{41a}$$

and

$$\hat{\varepsilon}_l(\omega, \mathbf{k}) = 1 - \frac{\omega_n^2}{\omega^2 - i\omega\tau^{-1} - \beta|\mathbf{k}|^2} \tag{41b}$$

where the subscripts t and l refer to transverse and longitudinal, respectively, and $\beta = \frac{3}{5}v_F^2$ with Fermi velocity v_F. Equation (41a) is identical to Eq. (9) and holds for s-polarized light. If damping is neglected ($\tau \gg \omega^{-1}$), then the condition for plasma excitation $\hat{\varepsilon}_l(\omega, \mathbf{k}) = 0$, yields the plasmon dispersion relation[23]:

$$\omega^2 = \omega_n^2 + \beta|\mathbf{k}|^2 \tag{42}$$

For a non-free-electron metal like Ag or Au, where $\hat{\varepsilon}$ contains contributions from free *and* bound electrons (see Section 2.1, p. 94)), the above expressions (41a) and (41b) have to be modified in the following way (neglecting again damping):

$$\hat{\varepsilon}_t(\omega) = \hat{\varepsilon}_b(\omega) - \frac{\omega_n^2}{\omega^2} \tag{43a}$$

and

$$\hat{\varepsilon}_l(\omega, \mathbf{k}) = \hat{\varepsilon}_b(\omega) - \frac{\omega_n^2}{\omega^2 - \beta|\mathbf{k}|^2} \tag{43b}$$

with $\hat{\varepsilon}_b(\omega)$ being the bound-electron contribution to the dielectric function [cf. Eq. (20)]. We remember that $\omega_n = (4\pi Ne^2/m^*)^{1/2}$ has now to be distinguished from ω_p, the plasma excitation frequency, which is shifted to much lower energies due to interband transitions (e.g., for Ag: $\omega_p \hat{=} 3.8$ eV and $\omega_n \hat{=} 9.0$ eV).

In order to calculate the reflectivity of a two- or three-phase system, including longitudinal waves (Figure 15), additional boundary conditions are required. This is still a matter of considerable controversy,[43,45] and we shall follow the treatment suggested by Forstmann[43] because of his relatively simple and straightforward approach and because of his success in interpreting optical spectra of electrode surfaces, which is already documented in the literature.[46-48]

To calculate the reflectivity of a two-phase system, a third boundary condition is necessary. The conditions for local optics, namely, continuity of $E_{\text{tangential}}$ and of D_{normal}, imply a discontinuity of E_{normal} and j_{normal} and lead to a delta-like surface charge. This infinite reservoir of current at the surface prevents any charge density fluctuations inside the metal. Such a singular surface charge, however, is unrealistic and has to be discarded on microscopic grounds. If we demand finite charge densities across the interface, besides finite fields, we obtain a third boundary condition: *the continuity of E_{normal}.* This leads to a spatial extension of the surface charge, and consequently an oscillating electric field normal to the surface (as for p-polarized light) can now induce charge density fluctuations which will be the origin of plasma waves.

With these three boundary conditions, the reflectivity coefficient for a two-phase system can be calculated, where the electric field inside the metal will be a superposition of transverse wave with amplitude $E_{0,t}$ and longitudinal wave with amplitude $E_{0,l}$ (Figure 15):

$$\mathbf{E} = \mathbf{E}_{0,t} \cdot \exp\left[i(k_x x + k_z z - \omega t)\right] + \mathbf{E}_{0,l} \cdot \exp\left[i(k_x x + \kappa z - \omega t)\right] \quad (44)$$

where k_z and κ are the z-components of the wave vector of the transverse and the longitudinal wave, respectively, while k_x is the corresponding x-component, which is the same for both waves because of k-parallel conservation. (Remember that k-parallel conservation leads, e.g., to Snellius' law.) The Fresnel reflection coefficient for p-polarization and for a metal/electrolyte interface is then derived as[43]:

$$\hat{r}_{\parallel 13} = \frac{\hat{\varepsilon}_3 \xi_1 - \varepsilon_1 \xi_3 - (k_x^2/\kappa)(\varepsilon_1 - \hat{\varepsilon}_3)}{\hat{\varepsilon}_3 \xi_1 + \varepsilon_1 \xi_3 + (k_x^2/\kappa)(\varepsilon_1 - \hat{\varepsilon}_3)} \quad (45)$$

with $\xi_j = \sqrt{\hat{\varepsilon}_j} \cdot \cos\varphi_j$ [compare with Eq. (23b) for local optics]. For s-polarized light, Eq. 23a still holds. In Eq. (45), $k_x = (2\pi/\lambda) \cdot n_1 \cdot \sin\varphi_1$, and $\kappa = (k^2 - k_x^2)^{1/2}$ has to be calculated either via Eq. (42) in the case of a free-electron metal or via $\beta k^2 = \omega^2 - \omega_n^2/\hat{\varepsilon}_b$ for a non-free-electron metal [set $\hat{\varepsilon}_l(\omega) = 0$ in Eq. (43b)].

The evaluation of a three-phase system where film and substrate phases are metals (e.g., deposition of a thin metal film on a metal substrate) requires yet another boundary condition for the metal/metal interface. Forstmann and Stenschke[49] postulated the continuity of the normal component of the energy current. This has been formulated as[43]

$$\frac{4\pi\beta}{\omega_n^2} \cdot \rho = \frac{\beta\hat{\varepsilon}_b}{\omega_n^2} \cdot \text{div } \mathbf{E} \text{ being continuous}$$

Unfortunately, the Fresnel reflection coefficient for the three-phase system with inclusion of nonlocal effects can no longer be written down explicitly, but has to be calculated by computer. In Appendix I, a guideline for such a calculation, taken from Forstmann and Gerhardts,[50] is given.

An important result of nonlocal optics is that for $\omega \approx \omega_p$, the electric fields at the surface but inside the metal are very small.[51] This has been rationalized in terms of destructive interference of transverse and longitudinal waves. As a consequence, the reflectance of the metal surface is insensitive in this energy range to changes of the surface properties because very little energy is absorbed in this region. Briefly, this is demonstrated in Figure 16, where the electroreflectance spectrum of an Ag film is shown for *p*- and *s*-polarized light at oblique angle of incidence. Ignoring for a moment the reasons for choosing such a layer structure of Ag on Cu (mainly to avoid the so-called 1/*R* effect; see later), the spectra of Figure 16 in essence represent the reflectance change induced by a potential modulation which causes the Ag surface optical properties to change within the first atomic layer. Clearly, the

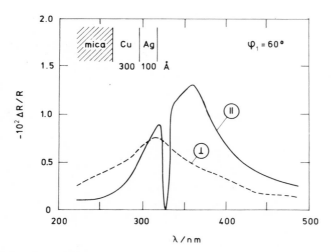

FIGURE 16. Electroreflectance spectra for a thin Ag overlayer on Cu in 0.5 *M* $NaClO_4$ for *s*- and *p*-polarized light. $\varphi_1 = 60°$. Potential step from -0.5 to 0.0 V versus SCE.

reflectance change for p-polarization drops to zero around ω_p (3.8 eV/330 nm) because of the above discussed insensitivity of the light to changes in the surface layer. Another experimental piece of evidence is found in photo-emission work on Al surfaces: the photoemission yield has a pronounced minimum at ω_p[52] which cannot be explained by standard optics but is well reproduced when nonlocal optics is applied.[53] Later on, we shall see several examples of electrode surface studies where nonlocal effects play an important role.

3. Experimental

3.1. Arrangements for Determining $\Delta R/R$

In most spectroelectrochemical surface studies, a relative reflectance change, $\Delta R/R$, caused by some defined electrochemical manipulation has to be determined to a high accuracy. As we pointed out earlier, $\Delta R/R$ is directly given by the corresponding intensity change, $\Delta I/I$, of the reflected beam, and therefore, the extremely difficult task of determining *in situ* absolute reflectance values is avoided, the latter being almost impossible because of unknown reflectance losses at cell windows and light scattering and absorption in the electrolyte. Typical values for $\Delta R/R$ range between 10^{-5} and 10^{-1}, which requires a high degree of noise reduction in experimentation. This calls either for modulation techniques and the use of phase-sensitive detection (lock-in amplifiers) or for high-speed signal averaging. Two different ways of carrying out electrochemical modulation spectroscopy are readily envisaged: by poten-tial modulation and by coverage modulation.

Potential modulation in the double-layer charging region of a metal is used in so-called electroreflectance studies, where the reflectance change induced by a potential variation (in the absence of any faradaic reaction) is measured. The optical apparatus for this type of measurement is quite simple, since a single-beam technique as shown in Figure 17 suffices.[12] Monochro-matic and (preferably) linearly polarized light is reflected from the electrode surface and focused onto a photomultiplier tube (PMT). The modulated PMT signal, which is proportional to ΔR, is amplified by a lock-in amplifier (LIA) and subsequently divided by the DC signal of the PMT (which is proportional to R) to yield the desired $\Delta R/R$. (Precisely speaking, this is only correct for $\Delta R \ll R$, because the DC signal is proportional to $R \pm \frac{1}{2}\Delta R_{pp}$, the sign depend-ing on the reference point. For very large $\Delta R/R$ values of 10% or more, a correction may be appropriate.) By this technique, signal-to-noise ratios of $\leq 10^{-5}$ in $\Delta R/R$ are routinely achieved with stabilized light sources and high-quality lock-in amplifiers.

While pure potential modulation is primarily used for the study of bare surfaces in the case of metal electrodes (Section 4) and of space charge layers

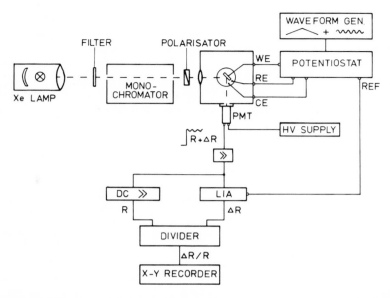

FIGURE 17. Single-beam arrangement for modulation spectroscopy (e.g., electroreflectance spectroscopy).

in the case of semiconductors,[21,54] coverage modulation is employed for investigating adsorbates. The coverage is, of course, also changed by a potential variation; however, the corresponding reflectance change is thought to be due to the change in coverage rather than in potential. In most cases one would like to achieve an "on–off" condition by square-wave potential modulation, where the resulting $\Delta R/R$ is caused by the adlayer formation, as schematically demonstrated in Figure 9. The above-described modulation method (Figure 17), however, will only work if the adsorption–desorption reaction is fast enough to allow frequencies of at least several hertz for the coverage modulation. Such an adsorption reaction, for example, is hydrogen on Pt.[55]

In many instances, the adsorption or film formation reaction will be too slow to be modulated at frequencies suitable for a lock-in amplifier. Besides, when single-crystal electrodes are used, surface restructuring or even complete loss of single crystallinity may occur when the surface reaction under study is performed many times. An example is oxygen adsorption on Pt(111) electrodes, which is known to create a stepped surface within only a few potential cycles.[56] For the study of adsorbates, the arrangement of Figure 18 has proven to be quite successful.[57,58] The rapid-scan spectrometer employed allows a single wavelength scan to be recorded within a few milliseconds. This is done by virtue of a small vibrating galvanometer mirror which reflects the light beam via a large spherical mirror onto the grating. Unlike conventional mono-

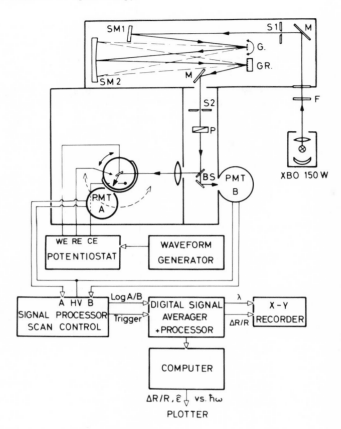

FIGURE 18. Rapid-scan spectrometer for *in situ* reflectance studies. S1, S2: entrance and exit slits; SM: spherical mirrors; G: vibrating galvanometer mirror; GR: grating; P: polarizing prism; BS: beam splitter; PMT A, B: photomultiplier tubes for sample and reference beams. After Ref. 57.

chromators, the grating is not rotated, but the angle of incidence for the incoming beam varies continuously at the grating with the vibrating mirror motion, yielding the desired wavelength scan at the exit slit. The fast scanning allows the use of signal averaging for noise reduction. By sampling about 1000 spectra at each potential (for the bare and the adsorbate-covered surface), which is done within a few seconds and which is usually sufficient for signal detection in the 10^{-3} range, the spectrum of $\Delta R/R$ can be obtained by one single potential step.

Although the above-described rapid-scan spectrometer (Figure 18) has been employed extensively and with great success in the author's laboratory, the instrumental design is now obsolete. Instead of using mechanically moving parts such as the vibrating mirror, diode arrays in combination with a poly-chromator seems a very promising way to go. In these so-called optical multichannel analyzers, which are already used in Raman spectroscopy,[59] the whole spectrum is spatially dispersed by a polychromator (which in the simplest case is a monochromator without an exit slit). The spectrum is then imaged onto the detector, which consists of an array of tiny photodiodes (e.g., 1024 diodes within an area of 25 mm × 2.5 mm). The repetition frequency for recording and sampling spectra is high (typically of the order of 64 Hz), although the spectral range at present is not yet as wide as one can obtain by using photomultipliers.

Figure 19 shows another dual-beam apparatus for recording at fixed wavelength and as a function of potential reflectance changes due to a surface reaction. When the electrode surface is bare, both light beams—the sample and the reference beam—are made equal in intensity with the help of an optical attenuator (e.g., a rotating comb for variable attenuation of the reference beam). When by a potential excursion an adsorbate is formed at the electrode, causing its reflectivity to change, the difference ΔR between bare and adsorbate-covered surface will show up directly as an AC signal because the reference beam still reflects the bare-surface condition.[60]

The wavelength range for *in situ* studies is limited to 220 nm $< \lambda <$ 1400 nm, mainly by the transmission properties of the aqueous electrolyte. The high-energy limit is usually determined by the onset of absorption in the

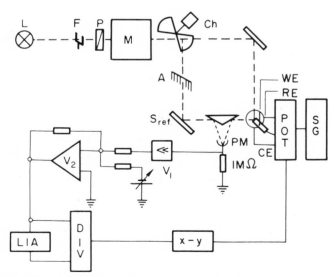

FIGURE 19. Dual-beam arrangement for determining adsorbate-induced reflectance changes. L: lamp; Ch: Chopper; A: variable light attenuator; S_{ref}: reference mirror; PM: photomultiplier; DIV: divider; LIA: lock-in amplifier. After Ref. 60.

electrolyte and in the polarizer (typically a Glan–Thompson prism), and by the rapidly decaying light intensity of the Xe arc lamp. The low-energy limit of 1400 nm is caused solely by the onset of water absorption. The use of thin-layer cells allows an expansion of the wavelength region well into the far-IR regime, as is demonstrated in Chapter 5.

In order to achieve an optimal signal-to-noise ratio, high-intensity light sources and sensitive detectors are required in addition to high-performance electronic equipment. For the 200 to 800-nm range, a Xe arc lamp (typically 150 W) and photomultiplier tubes with S20 characteristics and quartz windows (e.g., RCA C31034, EMI 9798 QB, Hamamatsu R 457) will be the best choice, while for the near-IR region (700–2000 nm), a 100-W tungsten–halogen lamp and a PbS photoresistor (possibly cooled by a Peltier device, e.g., Opto-Electronics OTC-21) seem to be the most obvious types of source and detector to choose. Edge filters have to be used with grating monochromators to eliminate the second order, when working in a wavelength range above, say, 600 nm. Measurements below 300 nm are often hampered by low signal-to-noise ratios, mainly because of low light intensity. Consequently, stray light from the monochromator becomes a problem which will pretend low $\Delta R/R$ values $(\Delta R/R = [(I_1 + I_{stray}) - (I_0 + I_{stray})]/(I_0 + I_{stray}) \approx \Delta I/(I_0 + I_{stray}))$ and obscures the wavelength dependence. It turned out that the situation can be markedly improved if for measurements in the near-UV range, the optics are aligned and optimized for UV radiation rather than for visible light as is usually done for convenience. Because of dispersion effects at lenses and condensers, which cause the focal lengths to change slightly with wavelength, the photon flux may be considerably smaller in the UV region, if the spectrometer was optimized in the visible region. Hence, after a first alignment by eye, a second adjustment via the photomultiplier should be performed in the UV region. Finally, we mention in passing that low-frequency noise due to vibrations of the building is often very disturbing and calls for a vibration damping of the table on which the spectrometer rests. A thick layer of rubber foam ($\sim$10–20 cm) underneath the spectrometer is cheap and serves the purpose.

3.2. Electrochemical Cells and Electrodes

The design of electrochemical cells has been described in the literature,[15,47,61] and several examples are reproduced in Figure 20. As has been pointed out in Ref. 15 (p. 103), multiple-reflection arrangements are not really advantageous for the visible–near-UV range because of the commonly low reflectivity of the electrode surfaces in this wavelength range. Besides, it is increasingly recognized that single crystals should be used in spectroelectrochemical studies, which also makes a single-reflection arrangement desirable because of limitations on the electrode surface area. The different aspects for the design of a spectroelectrochemical cell may be worth mentioning. In the

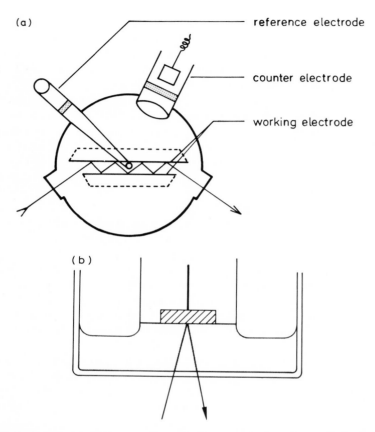

FIGURE 20. Optical cell arrangements for *in situ* reflectance experiments. (a) Multiple reflection cell. After Ref. 61. (b) Arrangement for the study of single-crystal surfaces by the so-called dipping technique. (c) Optical cell with cylindrical optics for easy variation of the angle of incidence. After Ref. 15. (d) Optical thin-layer cell for the study of metal deposits. After film formation the generator electrode (GE) is rotated out of the light path for optical measurements. After Ref. 47.

study of adsorbates, measurements at various angles of incidence are often useful in assessing structural information, such as orientation of molecules or anisotropy of adlayers. A cell, similar to that shown in Figure 20c, with a bent quartz window in combination with a φ-2φ goniometer table (note also Figure 18) has proven to be very convenient, as it allows for a continuous change of φ_1 from normal to grazing incidence. The second point relates to the use of single-crystal electrodes. In many electrochemical studies with single crystals, the so-called dipping technique[62] is employed, which enables one to have only the desired surface of the single crystal in contact with the electrolyte, without having to mask the other faces (Figure 20b). Such an arrangement, however, will limit φ_1 in most cases to near-normal incidence.

(c)

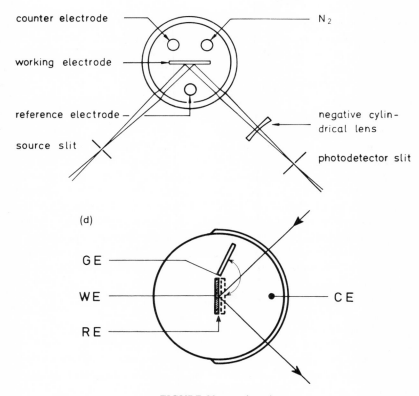

FIGURE 20—continued

For the sake of convenience, large-area electrodes (~ 1 cm^2) are commonly employed, such as polished metal sheets or films evaporated onto glass or quartz. These surfaces—despite their mirror finish—are usually polycrystalline, which means atomically rough and structurally ill-defined. Evaporation of Cu, Ag, and Au onto mica at elevated temperatures (e.g., 260°C for Ag and 360°C for Au) has proven to produce good-quality single-crystal surfaces of (111) orientation,[63] which are ideally suited for optical studies because of their large surface area.[64] Furthermore, it was shown that Au evaporated onto glass at about 400°C yields an atomically smooth film with mostly (111) surface orientation, but the crystallites are still randomly orientated around the [111] zone axis [such a film exhibits a LEED ring instead of LEED spots when studied by low-energy electron diffraction in UHV[65]]. Evaporation of Au onto glass at room temperature and subsequent annealing at 400°C leads to the same result,[65] whereas evaporation at room temperature without annealing leaves a microscopically rough surface.

4. The Metal/Electrolyte Interface

In the following, we deal with the study of electrode/electrolyte interfaces in the so-called double-layer charging region where no faradaic reaction takes place. In this region, the electrochemical interface behaves like a plate capacitor of molecular dimensions which is charged or discharged by appropriate potential variations. A textbook picture of the electrochemical double layer is reproduced in Figure 21.[66] The study of the structure and properties of the electrochemical interface constitutes one of the most important areas of research in today's physical electrochemistry. This interface is the site of extremely high electric fields, tremendous surface charges, and strong chemical forces, which all play an important role in governing electrode processes. Hence, when changing the electrode potential, the properties of the metal surface as well as the composition and structure of the electrolyte side of the double layer will be affected. As we shall see in a moment, the optical properties of an electrode surface are indeed markedly influenced by the electrode potential, an observation which allows us to extract important information, for example, about the surface band structure of a metal in contact with an electrolyte and the change of the surface properties with electrode potential.

4.1. Electroreflectance Studies of the Metal Surface

The reflectivity of a solid/electrolyte interface is often found to change with electrode potential, even in the absence of any faradaic reaction. This change in reflectance with potential, $\partial R/\partial U$, is commonly called electroreflectance (ER). In contrast to semiconductor ER,[21,54] where the externally applied static or low-frequency electric field penetrates several hundred nanometers into the bulk, thus probing the bulk band structure of the solid, for metals the perturbing electric field is screened within the first atomic layer because of the high charge carrier density; the so-called Thomas–Fermi screening length, $l_{TF} = v_F/(\sqrt{3} \cdot \omega_n)$, for a typical metal is about 0.05 nm (e.g., for silver $v_F = 1.4 \times 10^8$ cm $\cdot$ s^{-1} and $\omega_n = 1.37 \times 10^{16}$ s^{-1}). Therefore, ER for metals should be extremely surface sensitive. This is illustrated in Figure 22, where for a jellium metal, the surface charge density induced by a weak external electric field is shown as a function of distance normal to the surface.[67] The calculation clearly shows that the screening charge is located within a depth of about 0.2 to 0.3 nm at the surface, if we neglect the small density variations due to Friedel oscillations.

Soon after the first observation of an ER effect for a metal by Feinleib,[8] a systematic investigation of this effect was started at several places. The early results, mainly for polycrystalline Ag and Au surfaces were interpreted in terms of the "free-electron" model which was developed by Hansen[68] and later refined by McIntyre.[15,69] This model assumes that the change in the free-electron concentration at the surface, induced by the potential change, is

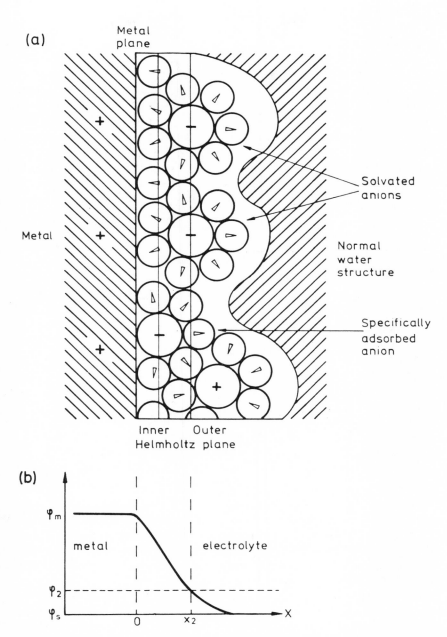

FIGURE 21. (a) Model of the electrochemical double layer. (b) Potential distribution across the double layer.

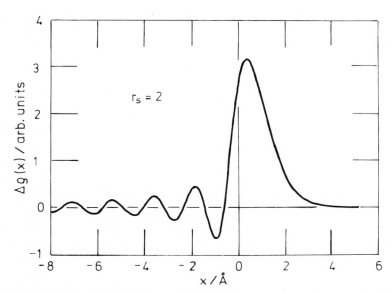

FIGURE 22. Surface charge density $\Delta\rho$ of a metal, induced by a weak external electric field. $x = 0$ denotes the edge of the positive-charge background. r_s = Wigner–Seitz radius. After Ref. 67.

mainly responsible for the observed effect. Along the lines given in Section 2, the metal surface dielectric constant, $\hat{\varepsilon}_2$, is split into contributions from free and bound electrons[29,69]

$$\hat{\varepsilon}_2 = \hat{\varepsilon}_{2f} + \hat{\varepsilon}_{2b} - 1 \tag{20a}$$

but bound electrons are taken to be perfectly screened by the free electrons; hence they do not respond to the applied electric field (and, therefore, $\hat{\varepsilon}_{2b} = \hat{\varepsilon}_{3b}$). Any change ΔN in the surface free-electron concentration will, however, change the surface dielectric function via $\hat{\varepsilon}_{2f}$, because of the well-known Drude relation:

$$\hat{\varepsilon}_{2f} = 1 - \frac{\omega_{n,2}^2}{(\omega^2 - i\omega\tau^{-1})} \tag{9a}$$

with $\omega_{n,2}^2 = 4\pi(N + \Delta N)e^2/m^*$ whereas $\omega_{n,3}^2 = 4\pi Ne^2/m^*$. Thus, charging the electrode surface causes a difference in surface and bulk dielectric constants which is given by:

$$\hat{\varepsilon}_2 - \hat{\varepsilon}_3 = (\hat{\varepsilon}_{3f} - 1)(\Delta N/N) \tag{46}$$

where $\Delta N/N$, the normalized potential-induced change in the free-electron concentration of the surface may be expressed as

$$\frac{\Delta N}{N} = \frac{C_{\mathrm{DL}}}{Ned} \cdot \Delta U \tag{46a}$$

with C_{DL} the integral double-layer capacity and d the penetration depth of the static electric field. Hence, the potential-induced reflectance change (electroreflectance) at normal incidence is given by:

$$\left(\frac{\Delta R}{R}\right)_{\varphi_1=0} = \frac{8\pi\sqrt{\varepsilon_1}\, d\cdot\Delta N}{\lambda N}\cdot \mathrm{Im}\left(\frac{\hat{\varepsilon}_{3f}-1}{\varepsilon_1-\hat{\varepsilon}_3}\right)$$

$$= \frac{8\pi\sqrt{\varepsilon_1}\, C_{DL}\cdot\Delta U}{\lambda Ne}\cdot \mathrm{Im}\left(\frac{\hat{\varepsilon}_{3f}-1}{\varepsilon_1-\hat{\varepsilon}_3}\right) \tag{47}$$

with $\Delta U = U_+ - U_-$. A calculation of the ER effect for Ag on this basis is reproduced in Figure 23a. It is only fair to say that the free-electron model has proven to reproduce the main features of an ER spectrum, especially for polycrystalline surfaces (e.g., the peak at 3.9 eV for Ag and at 2.5 eV for Au); however, we note that these gross features in $\Delta R/R$ arise mainly from the $1/R$ effect, which is always present for metals with rapidly varying reflectivity, such as Ag or Au. In recent years, emphasis was increasingly placed on investigating well-defined single-crystal surfaces rather than polycrystalline ones by ER spectroscopy. This led to many observations which could not be accounted for by this simple model (see, e.g., Figure 23b).[29,64] For example, the marked influence of the crystallographic orientation on the ER spectrum for different faces of the same metal strongly suggested that bound electrons do "feel" the modulating potential and hence the term $\partial\hat{\varepsilon}_{2b}/\partial U$ should not be neglected.[70,71] This implies that the ER spectra of single-crystal electrodes contain information about the band structure of the surface region and its dependence on the electrode potential.

The failure of the free-electron model to reproduce even pronounced effects in ER is demonstrated in Figure 24.[70] Here, the ER spectra for Cu(111) on mica are shown for s- and p-polarization at 45° angle of incidence (solid curves). The pronounced peaks in the experimental spectra at 2.2 eV, at the onset of the interband transition in Cu, cannot be reproduced by the free-electron model alone (dashed curves). A satisfactory agreement between theory and experiment—at least for the spectral shape—was, however, achieved when a shift of the interband transition energy with electrode potential of -0.2 eV/V was included in the calculation.[70] Although the absolute value of this potential-induced band structure change in Cu may be somewhat in question, the result certainly indicates that the electrochemical double layer can influence the electronic structure of metal surfaces.

Besides an actual change with potential in the electronic states at the surface due to the very high electric field, which leads to a modulation of the interband transition and hence to a contribution of bound electrons to ER, there can be a noticeable modulation of the interband transition proper, very much like the Franz–Keldysch effect in ER of semiconductors.[21,54] As was pointed out by Lynch,[72] this may occur via field-induced "indirect" interband transitions. Because of the high field strength attainable at metal/electrolyte

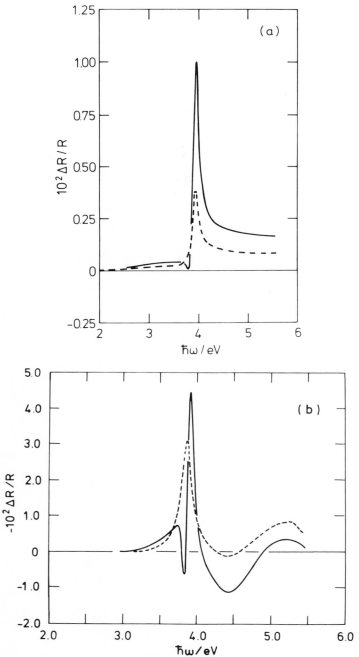

FIGURE 23. ER spectra of Ag. (a) Calculated by taking only free-electron effects into account ($\Delta N/N = +0.1$). After Ref. 32. (b) Experimental spectra for Ag(111) in 0.5 M NaClO$_4$ (pH 2). Potential step from -0.5 to 0.0 V versus SCE. After Refs. 29 and 64. ($- - -$) s-polarization, (———) p-polarization. $\varphi_1 = 45°$.

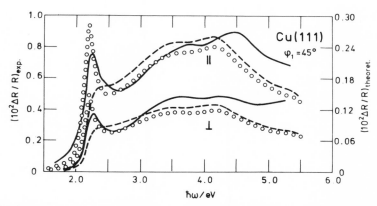

FIGURE 24. ER spectra of Cu(111) at 45°; *s*- and *p*-polarized light. (——) Experimental curves for $\Delta U = 500$ mV; (– – –) calculated by the free-electron model; (∘∘∘) adding to the free-electron model a 0.1-eV shift of the interband transition for this potential step. After Ref. 70.

interfaces, the uncertainty in the wave vector parallel to the applied field (normal to the surface) can be appreciably large (up to 0.1 Å^{-1}), hence allowing for k-conserving interband transitions over a larger variety of final states than in the absence of the field. This would also contribute to the ER effect.

Another interesting observation in ER spectroscopy of single-crystal metal electrodes is demonstrated in Figure 25, where the normal-incidence ER spectra of Cu(110) in 0.5 M H$_2$SO$_4$ are shown for two different crystallographic directions. This means that the electric vector of the linearly polarized light was rotated into different surface-crystallographic directions (remember that at near-normal incidence there is no difference between *s*- and *p*-polarization and the light electric vector is always parallel to the surface). This polarization anisotropy, which was first reported for Ag(110) by Furtak and Lynch,[14]

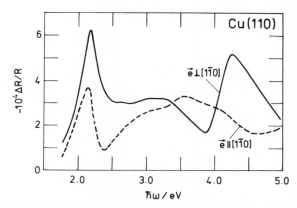

FIGURE 25. Normal-incidence ER spectra of Cu(110) in 0.5 M H$_2$SO$_4$ for two different directions of the electric vector. $U_0 = -0.3$ V versus SCE; $\Delta U_{pp} = 0.1$ V. After Ref. 70.

reflects the twofold symmetry of the surface and has been observed for the
(110) faces of Cu,[70] Ag,[14,73] and Au.[74,75] It demonstrates again that band
structure and crystal symmetry influence the ER effect. The dependence of the
ER effect on the surface crystallographic direction is shown in Figure 26 for
Ag(110) as the electric vector of the linearly polarized light is rotated by 360°.
We note that in most cases $\Delta R/R$ is largest for $\mathbf{e}\|[001]$ (i.e., across the atomic
rails) and smallest for $\mathbf{e}\|[110]$ (i.e., along the densely packed atomic rails).[71]
We briefly mention that the polarization anisotropy is found over a wide range
of photon energy and is not at all limited to the interband transition region.
As we shall see in Section 4.2, this is primarily due to surface states at the
metal/electrolyte interface, which often play a significant role in ER.

Finally, we discuss how the ER effects for the various crystallographic
surfaces of a particular metal compare.[71,76] This is demonstrated in the case
of Cu by the use of a single-crystal cylinder whose axis is parallel to [110].
The various high- and low-index faces can then be conveniently studied
optically by an appropriate rotation of the cylinder. The result of an ER study
at two different wavelengths is shown in Figure 27. While the anisotropy is
largest for the (110) and adjacent faces (see Figure 27c), the largest effect in
$\Delta R/R$ is clearly observed for the (113) face, which is the most open surface
(highest step density).

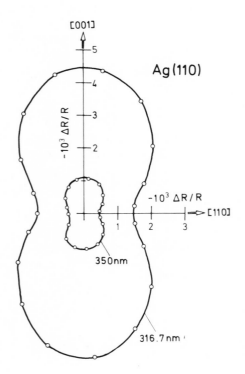

FIGURE 26. Normal-incidence ER for
Ag(110) as a function of surface crystallographic direction; two different
wavelengths. 0.1 M HClO$_4$. $U_0 = -0.4$ V
versus SCE. After Ref. 71.

The results of Figure 27 seem to indicate that the more open the surface structure is, the larger is the ER signal. In general, $\Delta R/R$ is found to be smallest for the most densely packed (111) surface.† This has been tentatively explained by a smoothing effect for the more open surfaces, which causes larger changes in the optical polarizability of the surface layer with potential.[70,71] The electron density contours for an open surface should reproduce the wavy structure of the positive background parallel to the surface. However, as was pointed out by Smoluchowski,[77] the smoothing effect tends to reduce any structure in the density profile parallel to the surface by accumulating negative charge in the valleys. Charging the metal electrode will affect the electron distribution not only perpendicular to the surface but also parallel to it. While for a densely packed surface or surface direction, the electron tail may be just pushed outward with negative charging without much change in shape, the density contours may change with charging at a more open surface (or surface direction) due to changes in the degree of smoothness. The increase of charge at negative potentials should smooth out the spatial variation of the electron density parallel to the surface, while at positive potentials the positive background contours should appear more strongly (Figure 28). Hence, the potential-induced *change* in the optical polarizability of a surface should be larger for a more open surface structure, causing larger $\Delta R/R$ values, as is found in the experiment.

The Role of Nonlocal Optics in ER

It has been predicted that standard optics (i.e., transverse electromagnetic waves only) should fail in describing the reflectivity of a metal near its volume plasma frequency, when p-polarized light is used.[43] In this frequency region (e.g., 3.8 eV for Ag and 2.5 eV for Au), longitudinal waves or plasma waves are eigenmodes of the metal, and these eigenmodes are optically excited with p-polarized light because the electric field component normal to the surface induces periodic charge density fluctuations. These longitudinal waves propagate into the metal, in addition to the transverse waves, and therefore should contribute to the reflectance and transmittance of a metal surface. We have discussed this already in Section 2.4.

One of the great virtues of ER at the metal/electrolyte interface is the very high sensitivity with which reflectance changes can be detected. Hence, theories on metal optics can be tested against experiment with high accuracy. In the following, the profound influence of nonlocal effects on the ER spectra of metals is demonstrated by two examples, gold and silver.

When the ER spectra of Au are calculated by standard optics, using the free-electron model, a pronounced peak in $-(\Delta R/R)$ below the volume plasma

---

† A comparison of Figures 24 and 27 reveals that the ER effect for Cu(111) on mica is much larger than for massive Cu(111). This is indicative of a poor-quality, defect-rich Cu(111) film on mica.

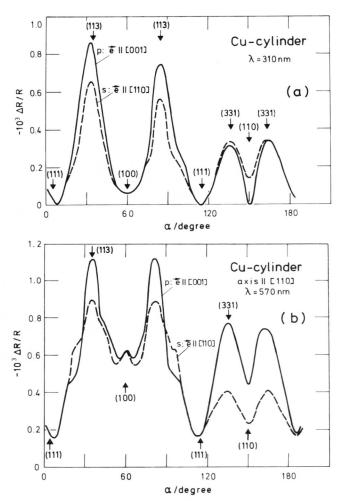

FIGURE 27. Normal-incidence ER for Cu as a function of crystallographic orientation (a, b) and polarization anisotropy (c). 0.5 M H_2SO_4. $U_0 = -0.3$ V versus SCE; $\Delta U_{pp} = 100$ mV. The various crystallographic orientations were obtained by rotating the Cu single-crystal cylinder about its [110] axis, as shown in the inset. After Ref. 71.

frequency of gold (2.5 eV) is predicted for *p*-polarization, especially for positive bias potentials (see Figure 29b, dotted line).[78] Such a feature has never been observed experimentally, and it is obviously an artifact arising from the use of improper metal optics. The experimental spectra in this wavelength region have about the same shape for *s*- and *p*-polarized light, with a difference in magnitude of about a factor of two for 45° angle of incidence.[15,69] When calculations with nonlocal optics are performed, the feature described above is no longer found in the theoretical spectrum (Figure 29, solid line), and a

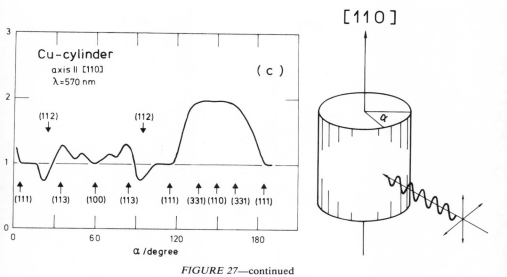

FIGURE 27—continued

spectral shape emerges which is in very good agreement with experiment.[48] In local optics, it is assumed that the surface layer has a plasma frequency of its own, which differs from the bulk plasmon frequency because of the field-induced change in the surface electron density. Charging the Au surface positively reduces the free-electron contribution to the real part of the Au dielectric constant ε' to such an extent that ε' of the surface layer becomes zero at lower photon energies.[78] The local treatment of the surface layer, for which now the plasmon resonance condition ($\varepsilon' = 0$) is fulfilled at energies below 2.5 eV, leads to the pronounced peak in $-\Delta R / R$ around 2.3 eV (Figure 29b). The correct, nonlocal optics treats the surface charge density not as a singularity but allows a spreading normal to the surface, which couples the plasmon waves of the surface layer to those of the bulk. This makes the $\Delta R / R$ values rather insensitive to the surface charge density, at least in the region of the plasma frequency, in agreement with experimental findings.[48]

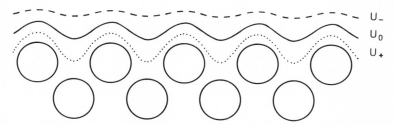

FIGURE 28. Schematic representation of the electron density profile at a metal surface for different charging.

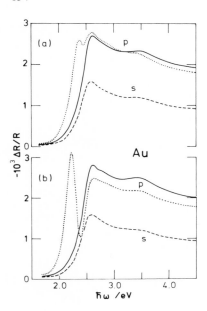

FIGURE 29. Calculated ER spectra for Au at 45° and for two different potential steps. (– – –) *s*-polarization; (——) *p*-polarization with plasma waves; (· · · ·) *p*-polarization with local optics, $d = 0.4$ nm. Free-electron density change from (a) 0.82 to 0.77 and (b) 0.70 to 0.65 of N_{bulk} (pzc: $N = 0.75$). Note that $\Delta N/N = 0.1$ corresponds roughly to $\Delta U = 1$ V. After Ref. 48.

There is a similar problem with the ER spectrum of Ag, especially when single-crystal surfaces are used. The most prominent features in the spectra for *p*-polarized light taken at oblique angles of incidence, say at 45°, are the peak in $-\Delta R/R$ at about 3.95 eV and the sharp dip around 3.85 eV, the volume plasmon frequency of Ag (Figure 30a). Model calculations with standard optics predict a pronounced derivative-like structure in $\Delta R/R$ at about 3.5 eV (dotted curve in Figure 30b) for a potential step of about 0.5 V in the positive direction [see Ref. 48 for details], which is not found in the experiment. When longitudinal waves are included in the calculation, a spectrum emerges (solid curve in Figure 30b) which is very similar to the measured one. Comparison of the two curves in Figure 30 demonstrates that standard optics fails once more to reproduce adequately the experimental ER spectrum of a metal in the vicinity of its plasma frequency, in this case for Ag especially for high positive charging. This is again because standard optics confines the charge density perturbation to a singular surface layer, which is not a realistic description of the interface. With spatial dispersion included, the charge perturbation at the surface associated with plasma waves spreads out and couples to the perturbation in the bulk. This coupling makes the resonance structure in $\Delta R/R$ rather insensitive to changes in the surface charge density, as found in the experiment. Such changes cause only a change in the coupling to the bulk plasmon. As already pointed out above in the case of Au (Figure 29), the newly created condition of $\varepsilon'_2 = 0$ at 3.5 eV arising from charging the surface layer positively does not lead to pronounced resonance structures as predicted by standard optics, because with the spatial extension of plasma waves, the

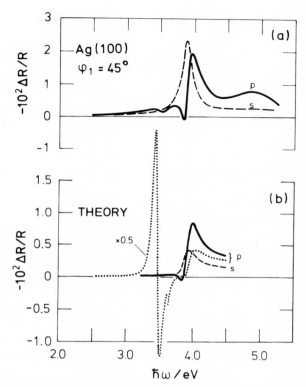

FIGURE 30. ER spectra of Ag at non-normal incidence. (a) Experimental result for Ag(100) in 0.5 M NaClO$_4$ (pzc: −0.9 V). Potential step from −0.5 to 0.0 V versus SCE. (b) Calculated ER spectra with (——) and without (· · · ·) inclusion of longitudinal waves. Free-electron density change from 0.71 to 0.66 of N_{bulk}. s- and p-polarized light. $\varphi_1 = 45°$. After Ref. 48.

surface layer can no longer play its own part but is coupled in its properties to the bulk.

Two more points may be worth mentioning in connection with nonlocal optics. First, as a result of a systematic study of the ER effect for Ag as a function of bias potential, it was concluded that at the pzc the metal surface has a conduction electron density which is clearly smaller than that of the bulk. In the corresponding three-phase model, the uncharged metal surface was successfully simulated by a surface layer of about 0.4 nm thickness and of an electron density $N_{surface} = 0.75\ N_{bulk}$.[46] The second point addresses the insensitivity of local optics to the form of surface layers, $\hat{\varepsilon}_2(z)$ (see Figure 14b), as only the integral information $\int \hat{\varepsilon}_2(z)\ dz$ is obtained because $\lambda \gg d$. Consequently, the ER effect can yield information only on the product $d \cdot \Delta N$ [see Eq. (47)]. It has recently been shown[79] that this need not be true when measurements are performed under conditions where volume plasmons are excited. Ellipsometric data on charged Au surfaces[80] could be explained by model calculations including nonlocal optics for d values of a given range only, namely, 0.25 to 0.30 nm. This finding could be an important contribution to optical spectroscopy of interfaces since it demonstrates that under certain conditions, $\Delta R/R$ can carry information on structural detail of monolayers and very thin films despite the very large wavelength of the probing light. The reason for this lies in the fact that longitudinal waves have wavelengths typically of the order of several nanometers (as opposed to several hundred nanometers for transverse waves) and are therefore much more sensitive to spatial variations of the optical properties normal to the surface than ordinary light. This shows that the study of nonlocal effects may well yield new and important information about the charge distribution at the electrochemical interface.

4.2. Surface States at the Metal/Electrolyte Interface

Surface states are electronic states which are strongly localized at the surface or interface of a solid. They arise as a consequence of lattice termination at the surface† and are energetically situated in so-called energy gaps.[81] They are mostly known from semiconductor surfaces where the existence of an absolute energy gap allows for easy detection of surface states.[82] Although the bulk band structure of metals, by definition, shows no energy gap, the *surface* band structure can reveal such gaps in certain crystallographic directions of the surface Brillouin zone. This is demonstrated in Figure 31 where the surface band structures (i.e., the projection of the bulk band structure onto a plane parallel to the surface) for Ag(hkl) are shown. In self-consistent pseudo-potential calculations for a number of metal surfaces, Ho and Liu as well as others have shown[83-88] that surface states do indeed exist in such energy gaps (note states A and B in Figure 31), which are split off the bulk states at the upper and lower edge of the gap. The energy of bulk as well as surface states depends on the local electrostatic potential the respective states feel. This fact actually leads to an interesting consequence for the electrochemical double layer with its strong potential gradient (i.e., high electric field) at the interface. Because surface states are so strongly localized at the surface, rapidly decaying along the surface normal in either direction, they should see a different portion of the externally applied electrostatic potential than bulk states do. As a consequence, the energetic position of the surface states would vary with the electrode potential in a way which is distinctly different from that of the bulk states. From this it follows that optical transitions from bulk states into surface states should be influenced in their energy by the electrode potential. As we shall see in the following, this dependency can be used to detect and identify surface states at the metal/electrolyte interface.[89]

† Such surface states are often called "intrinsic," in contrast to extrinsic surface states which arise from chemically different species at the surface.

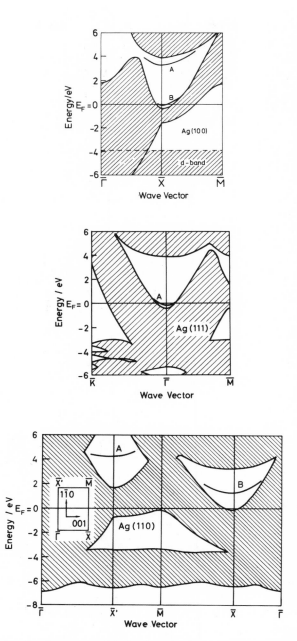

FIGURE 31. Surface band structure of Ag(100), Ag(111), and Ag(110), with surface states A and B. After Refs. 84 and 85.

When the electrode potential of a metal electrode is changed positively or negatively by, say, 1 V, then the whole bulk band structure (including the Fermi level) is shifted with respect to the vacuum level by exactly 1 eV downward (to higher work function) or upward (towards lower work function). Since the surface states penetrate somewhat into the double layer,[83] they experience only a certain fraction of the total potential drop across the Helmholtz layer. Hence, the surface states are shifted less in energy than the bulk states. As a result, the optical transition from an occupied bulk state into an empty surface state varies with potential, such that the transition energy increases as the potential moves in the positive direction. Such a behavior is demonstrated in Figure 32a for the ER spectra of Ag(100) in 0.5 M NaF[84,89] and in Figure 32b for the ER spectra of Au(100) in 0.02 M NaF[90] (for the latter, see also Figure 36). The two spectral features for Ag(100) around 3 and 1 eV (Figure 32a), which shift in energy with bias potential have been assigned to transitions from bulk states into the empty surface states A and B, respectively (see Figure 31a). A comparison of the experimentally observed transition energies and their dependence on the electrode potential with the calculations by Ho and Liu[84,85] gave an almost perfect agreement, which supported our assignment of the above-mentioned spectral features to optical transitions into surface states. Even the disappearance of the transition into B around -0.6 V (Figure 32a) is found to be in full agreement with theory, which predicts the surface state B to be shifted below the Fermi level at that potential.[84] This means that at the uncharged Ag(100) surface (the pzc is -0.9 V versus SCE), surface state B is occupied as confirmed by photoemission experiments *in vacuo*.[91,93] More recently, the unoccupied surface states of Cu, Ag, and Au single-crystal surfaces have been studied *in vacuo* by inverse photoelectron spectroscopy,[92–94] and the results are again in very good agreement with the electrochemical data for measurements at the pzc.

Optical transitions into empty surface states have also been observed for the (110) faces of Ag[89] and Au.[86] However, based on symmetry considerations, such transitions are allowed only for e∥[001], but not for e∥[110].[83] This is demonstrated for Ag(110) in Figure 33 in a slightly different manner. The potential dependence of the transitions into surface states can also be seen quite clearly, when $\Delta R/R$ is recorded as a function of bias potential for different wavelengths rather than as a function of wavelength for different bias potentials. Such curves are shown in Figure 33 for both polarization directions.[71] In Figure 33a, two surface state features are immediately recognized with e∥[001] around 0 V for 1000 nm and around -1.1 V for 800 nm and their shift with potential is marked. No such features are found with e∥[110], where the structures in $\Delta R/R$ do not shift with wavelength (Figure 33b). It is evident from a comparison of Figures 33a and 33b, that surface states contribute in a prominent way to the observed normal-incidence polarization anisotropy of (110) faces, e.g., of Ag(110), as discussed in the previous section.

As we have seen above, the energetic position of the surface states depends on the externally applied electric field, and therefore, information about the

(a)

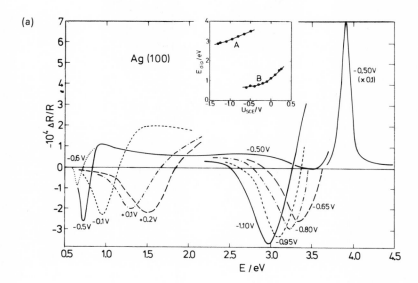

(b)

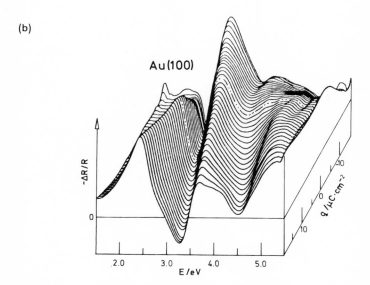

FIGURE 32. (a) Normal-incidence ER spectra of Ag(100) in 0.5 M NaF for various bias potentials, showing optical transitions into empty surface states. After Refs. 84 and 89. Inset: Transition energies as a function of electrode potential. (b) Normal-incidence ER spectra of Au(100) in 0.02 M NaF (pH 5) for various bias potentials (given in terms of surface charge q). After Ref. 90.

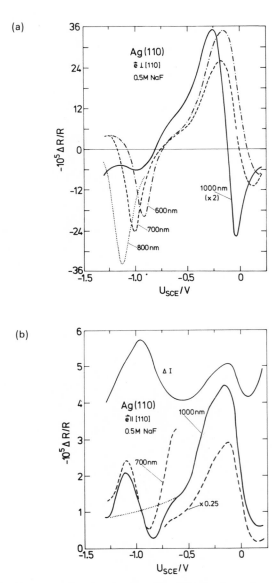

FIGURE 33. (a) ER signals of Ag(110) in 0.5 M NaF as a function of electrode potential for various wavelengths. Normal incidence; $\mathbf{e}\|[001]$; $\Delta U_{pp} = 100 \,\text{mV}$. After Ref. 71. (b) ER signals of Ag(110) in 0.5 M NaF as a function of electrode potential for two different wavelengths. Normal incidence; $\mathbf{e}\|[110]$; $\Delta U_{pp} = 100 \,\text{mV}$; ΔI = double-layer charging current; pzc: $-1.0 \,\text{V}$. After Ref. 71.

potential distribution near the metal surface hopefully is obtainable from such measurements. In Figure 34, a survey is given of the potential dependencies of various transitions into empty surface states.[89] In the simple picture of the electrochemical interface, i.e., a capacitor with about 0.3-nm plate distance,[5,66] a linear potential drop across the double layer is generally assumed. If we further assume that the surface states probe the region just outside the metal surface, say 0.1 nm away (see, e.g., Figure 2 in Ref. 85), then a shift of 0.3 eV/V would be expected for a bulk-to-surface state transition. From Figure 34 it is seen, however, that in general much higher slopes are found, certainly for Ag electrodes, which indicates that the potential difference between bulk and surface states is much larger than anticipated from the classical model with its linear potential drop across the Helmholtz layer. For example, the slope of unity which is found for Ag(100) implies that the potential difference between bulk and surface states, i.e., between metal surface and location of maximum density of surface states, is as large as the total potential drop across the Helmholtz layer. It is believed that these large shifts are indicative of a nonlinear, strongly varying potential distribution within the double layer, which could be caused by the microscopic structure of the interfacial water or specifically adsorbed ions. It has been speculated, for example, that the

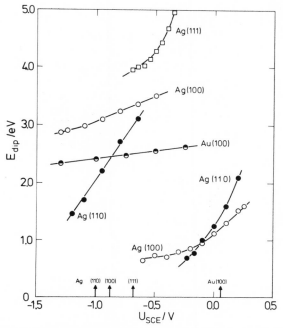

FIGURE 34. Shift of excitation energy for transitions into empty surface states (as evaluated from the dip in the corresponding ER spectra; see Figure 32) with electrode potential; 0.5 *M* NaF. The respective pzc's are marked by arrows. After Ref. 89.

finite size of the water dipoles may lead to regions of overscreening and underscreening, causing the electrostatic potential to oscillate along the surface normal.[84,85] The discreteness of charge in the double layer should also be taken into account as a possible source for strong (local) potential variations. It may be interesting to note that in Raman spectroscopy (surface enhanced) of adsorbed molecules, such as pyridine on Ag, similarly strong potential dependencies of the excitation wavelengths were reported.[95]

In Figure 35 we see the shift with electrode potential of the transition energy for excitation into surface state A of Ag(100) for two different electrolyte concentrations.[96] We note that the slope of the shift is clearly dependent on the potential drop across the inner Helmholtz layer, which is reduced in more dilute solutions as the diffuse part of the double layer comes into play. As one would expect, there is no concentration dependence of the surface state position at the potential of zero charge [−0.9 V versus SCE for Ag(100)], where the potential drop across the Helmholtz layer has vanished (at least for metals like Ag and Au).[97]

We then focus our attention on the lineshape of the absorption band for the optical excitation into surface states. As a consequence of the modulation technique applied for detecting electroreflectance signals, the optical transitions from bulk to surface states give rise to a derivative-like structure in the ER spectra, which shifts with bias potential. Integration of this structure leads to the actual absorption band. This absorption band for surface state B of Ag(100) has been shown to be surprisingly wide (about 1 eV) for the metal/electrolyte interface,[85] much wider than expected for the metal/vacuum interface (about 0.3 eV[98]). The small value in the latter case arises from the fact that the optical transition into the surface state does not start right from the top of the bulk band, as there bulk and surface states have the same p-like symmetry and hence the optical transition between them is forbidden.[84] With increasing photon energy, deeper-lying bulk states become involved in the transition, which have admixtures of allowed character, and hence absorption is observed.

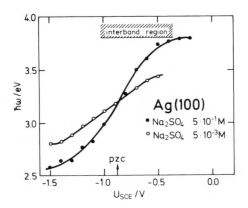

FIGURE 35. Shift in transition energy for excitation into surface state A of Ag(100) for two different electrolyte concentrations. After Ref. 96.

The broadening of the surface state level up to 1 eV could reflect the time fluctuations of the water dipoles in the double layer. It could also be caused by the discreteness of the charge, leading to an inhomogeneous potential distribution parallel to the surface. In any case, a lineshape analysis should throw some light onto the microscopic structure of the interfacial region and its potential dependence.

Surface band structure and surface states, which we have discussed in this section, are intimately correlated with the structure and the symmetry of a surface (note that the surface band structures of, for example, the (100) face of Cu, Ag, or Au are very similar, while those for different faces of one and the same metal are distinctly different).[85-88] Hence, any change in surface structure should have its impact on the surface states. Clean gold single-crystal surfaces are known to reconstruct when prepared under UHV conditions.[99] This means that the surface atoms occupy different sites than expected from a mere termination of the lattice. In general, the surface atoms have the tendency to form a densely packed layer in order to minimize their surface energy. A well-studied example is Au(100), which reconstructs into the so-called (5 × 20) structure where the surface atoms form a slightly buckled hexagonal close-packed (hcp) layer on top of the (100) bulk.[100] Adsorption usually removes the reconstruction and the surface retreats into the (1 × 1) structure. It has been shown that reconstructed gold surfaces are stable in the electrochemical cell, as long as no specific adsorption of anions occurs.[101,102] It has further been demonstrated that the reconstructed Au(100) surface shows no sign of the surface states A and B (shown in the inset of Figure 36) which are seen so prominently in the ER spectra of the unreconstructed (100) face. A comparison of the ER spectra of reconstructed and unreconstructed Au(100) as well as Au(111) (Figure 36) reveals the close similarity in the optical properties of Au(100)–(5 × 20) and Au(111), as one might expect, since both surfaces have a hcp structure. Furthermore, the complete absence of the surface state features [marked by A and B in the spectrum of the unreconstructed Au(100) surface] in the ER spectrum of Au(100)–(5 × 20) leads to a pronounced difference in the optical response for certain wavelengths of reconstructed (5 × 20) and unreconstructed (1 × 1) Au(100). This difference has been used to study *in situ* the kinetics of the adsorbate-induced transition from (5 × 20) to (1 × 1) (Figure 37).[101,102]

4.3. Surface Plasmon Studies

Surface plasmons (SPs) are collective propagating charge density waves, which are confined to the surface of a solid.[103-105] The charge perturbation is strongly decaying in the direction normal to the surface and is coupled to the electromagnetic fields inside and outside the solid. Since for $\omega \leq \omega_p/\sqrt{2}$, which is the relevant energy range for surface plasmon excitation with light, the charge density perturbation decays much faster than the transverse electric

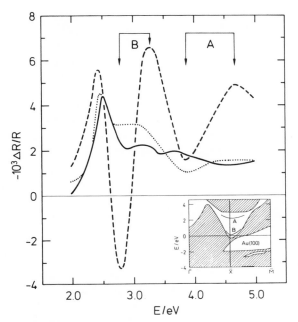

FIGURE 36. Normal-incidence ER spectra of Au(100)-(5 × 20) (——), Au(100)-(1 × 1) (- - -), and Au(111) (· · · ·) in 0.01 M HClO$_4$. Potential step from −0.2 to +0.3 V versus SCE. After Refs. 101 and 102. Inset: projected band structure and surface states A and B of the unreconstructed Au(100). After Ref. 86.

fields (about a factor of c/v_F), we can approximate the charge perturbation at the surface by a δ-function and use standard optics for the treatment of surface plasmons.[105]†

The transverse electric fields inside (−) and outside (+) the solid, which are coupled to the charge density perturbation, are described by (we define the outer normal as the $+z$-direction):

$$\mathbf{E}^{\pm}(x, z, t) = \mathbf{E}_0^{\pm} \exp\left[i(k_x \cdot x - \omega t)\right] \exp\left(k_z^{\pm} \cdot z\right) \qquad (48)$$

with

$$k_z^{+} = -[k_x^2 - (\omega/c)^2 \varepsilon_1]^{1/2} \qquad (48a)$$

$$k_z^{-} = [k_x^2 - (\omega/c)^2 \hat{\varepsilon}_m]^{1/2} \qquad (48b)$$

where ε_1 and $\hat{\varepsilon}_m$ are, respectively, the dielectric functions of the ambient medium (in our case the electrolyte) and the solid (in our case always a metal). The electric field of the surface plasmon represents a harmonic wave along

---

† In the large-k region of the surface plasmon dispersion curve, which is accessible only by excitation with electrons, this approximation is no longer valid and nonlocal optics must be applied.[49]

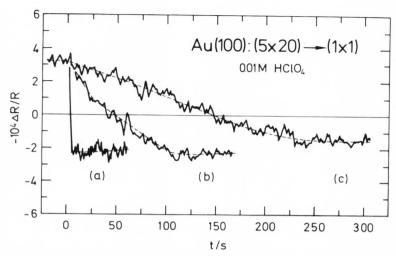

FIGURE 37. Normal-incidence ER effect monitoring the transition $(5 \times 20) \to (1 \times 1)$ of Au(100) in 0.01 M $HClO_4$. $\lambda = 400$ nm. At $t = 0$, the potential is stepped from 0 V, where the (5×20) structure is stable, to (a) +0.65 V, (b) +0.55 V, and (c) +0.50 V, where the transition into (1×1) is induced by anion adsorption. After Ref. 101.

the surface (x, y-plane) with frequency ω and wave vector k_x, parallel to the surface, while perpendicular to the surface, the field is exponentially decaying. $|k_z^+|$ is a measure for the penetration depth of the surface plasmon into the electrolyte (the reciprocal value, to be more precise).

The dispersion relation for surface plasmons is derived from the usual boundary conditions as:

$$\varepsilon_1 \cdot k_z^- - \hat{\varepsilon}_m \cdot k_z^+ = 0 \qquad (49)$$

which can be rearranged into the form (for $\omega < \omega_p$)[103]:

$$k_x = \left(\frac{\omega}{c}\right) \cdot \left(\frac{\varepsilon_1 \hat{\varepsilon}_m}{\varepsilon_1 + \hat{\varepsilon}_m}\right)^{1/2} \qquad (50)$$

Such a dispersion curve $\omega(\mathbf{k})$ is shown schematically in Figure 38. Excitation of surface plasmons is achieved with p-polarized light, where the normal component of the electric field vector induces a surface charge. However, special arrangements are required since direct excitation by light is still not possible because of lack of momentum (light line and SP dispersion curve do not intersect under normal reflection conditions; see Figure 38). Direct excitation of SPs by p-polarized light is only possible in an attenuated total reflection (ATR) configuration as sketched in Figure 38, where the light line can be inclined towards larger k values by a high refractive index of the ATR element so that it intersects with the dispersion curve.[106,107] SP excitation is then

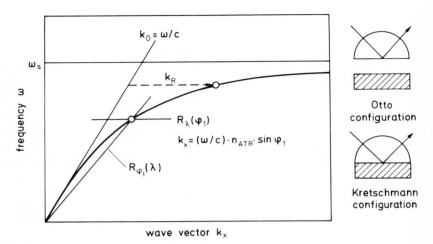

FIGURE 38. Surface plasmon dispersion curve $\omega(k_x)$. k_R: Wave vector due to roughness. The experimental arrangements due to Otto and Kretschmann for surface plasmon excitation in ATR are also shown.

recognized by a pronounced minimum in the reflectance. Alternatively, it has been shown that surface roughness acts as a momentum source and facilitates SP excitation.[108] Both ways will be discussed in the following section.

SP Excitation in the ATR Mode

SP excitation is often done under ATR conditions where, by choosing the correct angle of incidence, momentum conservation between light and SPs can be achieved. The two commonly employed experimental arrangements are those developed by Otto[106] and Kretschmann.[107] In the so-called Otto configuration, the solid under investigation by SPs is separated by a thin electrolyte gap (typically of the order of 500 to 1000 nm) from the ATR element (a prism or, preferably, a hemicylinder), while in the so-called Kretschmann configuration, a thin film of appropriate thickness ($\sim$50 nm) is evaporated directly onto the ATR element (Figure 38). In the first case, the electrolyte gap acts as a resonator for coupling light and surface plasmon; in the second case, it is the metal film proper. The resonance character of this phenomenon makes SP excitation very sensitive to perturbations in the surface region despite the fact that the penetration depth of the surface electromagnetic field is in many cases of the order of a few hundred nanometers and therefore comparable to the penetration depth of ordinary light reflected from a metal surface. Both methods, as well as the theoretical background for SP excitation, are well documented in the literature,[23,103-105] to which the interested reader is referred for more details.

In recent years, the Kretschmann configuration has been used nearly exclusively for SP studies because of easy preparation of thin metal films of a well-defined thickness on glass or quartz. Such films were of course polycrystalline. Despite the experimental difficulties in obtaining the exact electrolyte gap thickness, which ranges between 500 and 1000 nm depending on the wavelength, the Otto configuration has the distinct ability of studying single-crystal surfaces. In the author's laboratory, the "focused ATR" method[109] was employed, where instead of using a parallel light beam, the beam is *focused* onto the flat side of the ATR element, thus supplying a wide range of momenta (according to the range in angles of incidence because of $k_x \sim \sin \varphi_1$).[110] SP excitation is then seen in the reflected light cone by the appearance of a dark line. By either moving a photodetector across the light cone or imaging the cone onto an optical multichannel analyzer, the reflectance as a function of angle of incidence can be recorded. The gap thickness has been adjusted by a micrometer screw which moved the single-crystal electrode to and fro. The optimum gap thickness was found empirically by the appearance of a dark line in the reflected light cone.

SP excitation has been employed to determine the surface optical properties of an Ag single-crystal electrode and the dependence on the crystallographic orientation.[110,111] The question to be answered was whether differences in surface package and symmetry could be picked up in optical studies despite the relatively large penetration depth of the electromagnetic fields, and without applying modulation spectroscopy, which always yields relative changes rather than absolute numbers. The experimental results of such a study, the dispersion curves for various Ag single-crystal surfaces, are reproduced in Figure 39. Remarkably clear differences in the dispersion curves for the three low-index faces of Ag were found, which demonstrates beyond doubt that the differences in the surface optical constants for different crystallographic faces can be studied by SP excitation. Even the anisotropy in the surface optical constants of the (110) face is clearly detectable: for Ag(110), SP excitation in [00$\bar{1}$] and in [1$\bar{1}$0] direction occurs at different points in the ω–$\mathbf{k}$-plane, while no such directional dependences were found for Ag(111) and Ag(100).

From the dispersion curve, the real part of the complex dielectric function of the metal can be readily calculated. For Im $\hat{\varepsilon}_m \ll 1$ (i.e., small damping of the SP waves) and with $k_0 = (\omega/c)$, we obtain from Eq. (50):

$$\varepsilon_m' = \frac{\varepsilon_1 \cdot (k_x/k_0)^2}{\varepsilon_1 - (k_x/k_0)^2} \tag{50a}$$

An evaluation of the Ag dispersion curves (Figure 39) has revealed small, but distinct differences in ε_{Ag}' for the different crystallographic faces[105,110] [one should keep in mind that only two phases, metal and ambient, are considered in Eq. (50) and hence its use will yield an average over bulk and surface contributions to ε_m'], which correspond to differences in the reflectivity of the metal/electrolyte interface of only less than one percent. Such an effect would

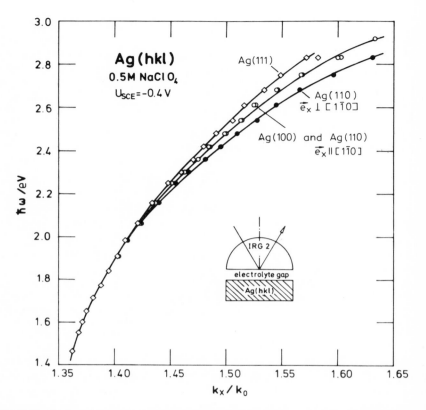

FIGURE 39. Surface plasmon dispersion curves for Ag single-crystal electrodes in 0.5 M NaClO$_4$. Note the anisotropy in the optical response for Ag(110). After Ref. 111.

be rather difficult to be picked up safely from normal-incidence reflectance measurements of the absolute reflectance values, while it is quite easily determined by SP excitation in the ATR mode. The latter technique requires the exact measurement of angles, which is easier to do than the determination of absolute reflectivities.

For completeness' sake, we briefly mention here also the equation for a three-phase system in the linear approximation ($d \ll \lambda$). With the usual notation for ambient (ε_1), thin film ($\hat{\varepsilon}_2$), and metal substrate ($\hat{\varepsilon}_3$), the change due to film formation in the SP wave vector Δk_x, parallel to the surface, is given by[107]:

$$\Delta k_x = \frac{(-\varepsilon_1 \hat{\varepsilon}_3)^{1/2}}{\varepsilon_1 - \hat{\varepsilon}_3} \cdot k_x^2 \cdot \left[1 - \left(\frac{k_x}{k_0}\right)^2 \left(\frac{1}{\hat{\varepsilon}_2} + \frac{\hat{\varepsilon}_2}{\varepsilon_1 \hat{\varepsilon}_3}\right) \right] \cdot d \qquad (51)$$

where k_x is the wave vector for the bare metal surface as given by Eq. (50).

A dependence of the SP excitation energy on the electrode potential has also been reported.[112-114] This is shown in Figure 40 for polycrystalline films

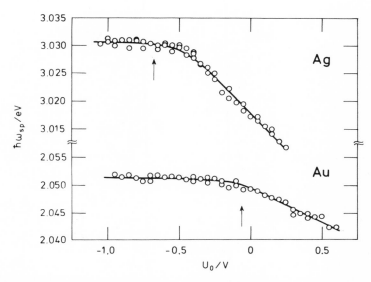

FIGURE 40. Shift in surface plasmon excitation energy with electrode potential (versus SCE) for polycrystalline Ag and Au in 0.5 M NaClO$_4$. $\varphi_1 = 60°$ (51°) for Ag (Au). The respective pzc's are indicated by arrows. After Ref. 112.

of Ag and Au measured in the Kretschmann configuration.[112] A rather asymmetric behavior of the SP excitation energy on electrode charging is found (note that the potentials of zero charge for the two metals are quite different as indicated by the arrows in Figure 40). When the electrode surface is charged positively, $\hbar\omega_{sp}$ decreases with electrode potential, the slopes being -25 and -13 meV/V for Ag and Au, respectively.[112] This follows the expected trend, since the electron density at the electrode surface is lowered at positive potentials ($\omega_{sp} \sim \omega_p \sim N^{1/2}$). On charging the electrode negatively, however, the shift of $\hbar\omega_{sp}$ ceases and $\hbar\omega_{sp}$ becomes independent of the potential. Model calculations have shown that a change in the molar refractivity of the Helmholtz layer with electrode potential cannot account for the observed shift at potentials anodic of the pzc. Measurements in electrolytes with ions of widely varying polarizability gave nearly identical results,[32] supporting the idea that the effect originates from changes in the metal surface with electrode potential rather than on the electrolyte side. It is believed that this observation offers some clues about the electron distribution in metal surfaces in contact with an electrolyte and about its potential dependence. For example, it has been speculated that the asymmetric response of $\hbar\omega_{sp}$ (Figure 40) reflects the different response of the evanescent tail of the electron density to positive or negative charging [e.g., different polarizability of the metal electrons at positive and negative potentials; for more details, see p. 314 of Ref. 105]. More experimental and theoretical work, however, seems necessary in this very promising field in order to unravel all the information which is still buried in SP excitation.

SP Excitation and Surface Roughness

As mentioned earlier, surface roughness can act as a momentum source and hence SP excitation by external reflection is allowed on rough surfaces.[108] Unlike a surface grating, which represents one well-defined wave vector, a randomly roughened surface supplies a whole spectrum of wave vectors (Fourier spectrum), most of them covering the large-k portion of the SP dispersion curve (Figure 38). Consequently, SP excitation by roughness usually occurs at relatively high energies, close to ω_s.[64,104,108] The latter is defined by the condition $\varepsilon'_m = -\varepsilon_1$, which for silver in contact with an aqueous electrolyte is fulfilled around 3.5 eV. Indeed, when an initially flat Ag surface [e.g., an Ag(111) electrode] is roughened by electrochemical oxidation–reduction cycles, a derivative-like structure appears in the ER spectrum at 3.5 eV, which we can assign to SP excitation (Figure 41a). The derivative-like structure in $\Delta R/R$ arises from the fact that $\hbar\omega_{sp}$ is potential dependent and hence the minimum in the static reflectance spectrum due to SP excitation is shifted along the $\hbar\omega$-axis with the potential variation.[64] Under external-reflection conditions, the shift of $\hbar\omega_{sp}$ with U has been reported to be about -0.2 eV/V for positive potentials,[64] which is roughly one order of magnitude larger than that found at small k values by ATR measurements (see above). The more pronounced potential influence on $\hbar\omega_{sp}$ at large k values can be understood by the smaller penetration depth ($\sim k_x^{-1}$) of the SPs, which makes them more surface sensitive than at small k values.

Since SP excitation is facilitated by a rough surface, this technique can be used in turn to study surface roughness and roughening. This is demonstrated in Figure 41a, b with an Ag(111) electrode, which had been subjected to various degrees of roughening by oxidation–reduction cycles, i.e., by electrochemical dissolution and redeposition of small amounts of Ag. The importance of these so-called activation cycles for the observation of surface-enhanced Raman scattering (SERS) has been discussed extensively in the literature[115] as well as the possible role in SERS of SP excitation via roughness.[116] When a perfectly flat Ag(111) electrode is subjected to oxidation–reduction cycles in perchlorate solution, the surface becomes immediately rough, as is seen by the appearance of the typical SP excitation features around 3.5 eV (in addition to an overall decrease in R due to light scattering): the derivative-like structure in the ER spectrum and the pronounced minimum in the corresponding static reflectance spectrum.[64] A totally different behavior is observed for Ag in chloride-containing electrolytes (Figure 41b). In such a solution, the oxidation of Ag leads to the formation of an AgCl film on the surface, which is reduced again on the cathodic scan of the potential cycle. Hence, Ag^+ is precipitated as AgCl at the surface rather than dissolved into

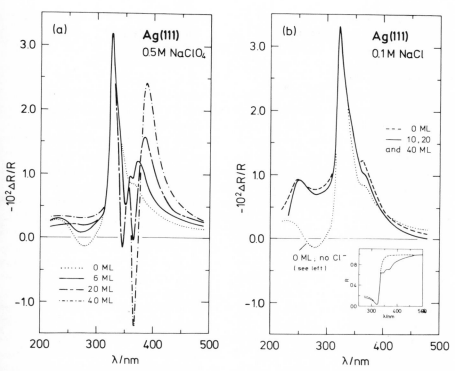

FIGURE 41. Normal-incidence ER-spectra of Ag(111) before and after electrochemical oxidation–reduction cycles (ORC) of various degrees in (a) 0.5 M NaClO$_4$ and (b) 0.1 M NaCl. Potential step from −0.55 to −0.05 V versus SCE. Inset: reflectivity spectrum of Ag(111) (- - -) before and after a 40-monolayer ORC in NaCl; (——) after an ORC of 40 monolayers in NaClO$_4$.

solution. Under such conditions, the Ag surface remains surprisingly smooth despite severe oxidation–reduction cycles. Ultimately, after moving back and forth the equivalent of about 100 monolayers, SP excitation becomes discernible, indicating that finally the surface has been roughened somewhat. This result should not be interpreted to indicate that the surface is still atomically flat; in fact, from Raman studies an atomic scale roughness was inferred (a roughness which is not detected by light but evidenced by an increased surface reactivity[115]). But large-scale roughness, which sustains SP excitation and causes light to scatter, can certainly be ruled out, a result to be considered in any SERS model which claims SP excitation as a major contributor to the enhancement factor[116] (note that most SERS experiments have been conducted in chloride-containing electrolytes). The conclusions on the surface morphology drawn from the SP excitation experiments were fully supported by *ex situ* post-electrochemical examinations of these surfaces by high-energy electron diffraction (RHEED).[117] The precipitation of the oxidized silver in a thin, uniform surface layer obviously prevents formation of larger clusters during the reduction process, in contrast to redeposition of Ag from the electrolyte phase. Monitoring the oxidation–reduction cycle for Ag in perchlorate solutions by reflectance measurements clearly reveals that surface roughening (as indicated by a decrease in R) occurs mainly during redeposition and not during dissolution.[117]

As mentioned already, the (random) roughness of a metal surface provides a large spectrum of k values for momentum conservation during SP excitation.[108] Consequently, excitation will occur over an extended energy range. If a more well-defined momentum source is desirable, surface gratings may be used for SP excitation.[118] In the following, we shall demonstrate that stepped single-crystal surfaces are very interesting in that respect because their regularly occurring steps supply a well-defined momentum, and this in one surface direction only. Figure 42 contains the ER spectra of an Ag(110) surface which was polished 3° off the [110] normal, with the steps along the [00$\bar{1}$]

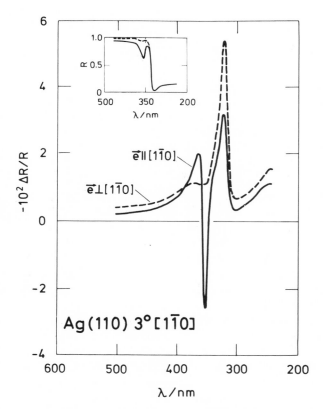

FIGURE 42. Normal-incidence ER spectra and static reflectance spectra (inset) of a stepped Ag(110) surface. 0.5 M NaClO$_4$. $\Delta U = 500$ mV. The steps lead in the [110] direction. After Ref. 70.

direction.[70] Assuming steps of atomic height, the terraces of such a surface are about 6 nm wide, which would correspond roughly to $|\mathbf{k}| = 1\,\text{nm}^{-1}$. The normal-incidence ER spectra of this surface, as shown in Figure 42, were recorded for $\mathbf{e}\|[00\bar{1}]$ and $\mathbf{e}\|[1\bar{1}0]$, which coincides with $\mathbf{e}$ being parallel and perpendicular to the steps (note that moving perpendicular to the steps would lead up and down the stairs). The difference in the two ER spectra (Figure 42) in the SP excitation region around 3.5 eV (350 nm) is indeed striking. For $\mathbf{e}\|[1\bar{1}0]$ the structure due to SPs is extremely sharp and pronounced, while in the other direction SP excitation is hardly seen at all in $\Delta R/R$. In the first case, the direction of the electric field vector is such that the steps can supply a well-defined $\mathbf{k}$-value for SP excitation, while in the second case a more or less perfectly flat surface is seen by the light. The same behavior is also noticed in the static reflectance spectra for the two polarization directions at normal incidence (see inset in Figure 42). In the first case ($\mathbf{e}\|[1\bar{1}0]$), a sharp dip in R is observed around 3.5 eV, indicative of strong and well-defined SP excitation, which is largely absent for the other direction of $\mathbf{e}$. We note in passing that outside the energy range for SP excitation, the ER spectra are nearly identical with those of perfectly flat Ag(110),[14,110] exhibiting the very same characteristic polarization anisotropy in $\Delta R/R$ for the two main crystallographic directions (see Section 4.1).

4.4. Double-Layer Contributions to Electroreflectance

The ER effect of the metal/electrolyte interface is usually dominated by contributions from the metal surface, especially in the interband transition region. The potential-induced changes in the optical properties of the Helmholtz layer proper, which constitutes a purely dielectric film in the photon energy range under study, are expected to give rise to $\Delta R/R$ values which are orders of magnitude smaller than those of the metal surface. This is readily demonstrated with the help of the linear approximation equations (29a) and (29b).

Attempts have been made in the past to unravel these two types of contributions for polycrystalline surfaces of Pb[119] and Au[120] and for Hg.[119] The separation procedure rested largely upon two simple correlations derived from the linear approximation equations for a three-phase model: (a) when the optical properties of the metal surface are only slightly different from those of the bulk (that is to say, the perturbation causing a reflectance change arises from the metal side), then at $\varphi_1 = 45°$, $(\Delta R/R)_p \approx 2(\Delta R/R)_s$ (as a direct consequence of Abeles' relation: $R_p = R_s^2$ at $\varphi_1 = 45°$); (b) if, on the other hand, a dielectric film on top of a metal substrate is the cause for a $\Delta R/R$ effect, we find $(\Delta R/R)_p \approx 0$ and $(\Delta R/R)_s \neq 0$ for $\varphi_1 = 45°$, while $(\Delta R/R)_p$ increases markedly for higher angles of incidence, say, 70°. It was therefore concluded that the ER effect at $\varphi_1 = 45°$ represents the metal contribution only (especially since one indeed often finds $(\Delta R/R)_p$ twice as large as $(\Delta R/R)_s$), while at $\varphi_1 = 70°$ both the metal and the double-layer contributions are included in $\Delta R/R$. On this basis, the double-layer effect was extracted from ER measurements at 45° and 70°. Because it was found for $\varphi_1 = 45°$ that $(\Delta R/R)_p$ varied linearly with surface charge, any deviation from linearity for such a plot at $\varphi_1 = 70°$ was blamed on the double-layer contribution. Clearly, such a model grossly oversimplifies the metal's role in ER, and therefore the results on the double-layer contribution to ER are likely to be wrong. In view of the fact that at least three different sources in the metal (namely, free and bound electrons and surface states) contribute markedly to the ER effect in a rather complicated and nonlinear fashion, the strict linearity of the ER effect as a function of surface charge at $\varphi_1 = 45°$ seems rather fortuitous. (Even if one is willing to believe that the nonlinear part of $\Delta R/R$ at $\varphi_1 = 70°$ originates from the double layer, it still needs to be proven whether the effect is due to double-layer water, as has been claimed in the literature,[119,120] or due to ions, especially specifically adsorbing anions. In the latter case, the cause of the corresponding reflectance change would be termed *film formation* (Section 5) rather than *electroreflectance*.)

An alternative route in the search for double-layer effects in ER is to exclude systematically the various contributions from the metal side by choosing the appropriate experimental conditions. For example, by extending the wavelength range for the optical studies of Ag into the near-IR region, i.e., far below the onset of the interband transitions, and by choosing the right crystallographic surface orientation where effects from surface states can be ruled out, we are left with the free-electron contribution of the metal. Under these conditions, this should give rise only to a smooth and unstructured background in $\Delta R/R$. This is fulfilled for Ag(111) around 1000 nm, which thus seems a good starting point for the search for double-layer contributions to the optical response of a metal/electrolyte interface. ER measurements under such conditions have revealed a derivative-like structure around the pzc, when $\Delta R/R$ was recorded as a function of electrode potential.[71] The result of such a study is shown in Figure 43, where a comparison is made between the reflectivity change with potential for Ag(111) in 0.5 M NaF at 1000 nm and the corresponding double-layer capacity C_{DL}. The former curve was obtained by integration of the ER signal, $-(\Delta R/R)/\Delta U$. The surprisingly close similarity between the curves $R(U)$ and $C_{DL}(U)$ strongly suggests that the derivative-like structure in the ER signal around the pzc indeed originates from the Helmholtz layer (note that the free-electron contribution to the ER would be directly proportional to C_{DL}: $\Delta R/R \sim C_{DL}$!). The sign of the optical response is such that the reflectivity of the interface is lowest at the pzc (more strictly speaking, at the maximum of C_{DL}; see Figure 43) and increases with potential on either side. Unfortunately, this behavior is still rather difficult to rationalize in a simple three-phase model with a purely dielectric film.[11] For example, at 1000 nm the effect in $\Delta R/R$ is too large to be explained solely by a change in the refractive index of the interfacial water, although the intimate

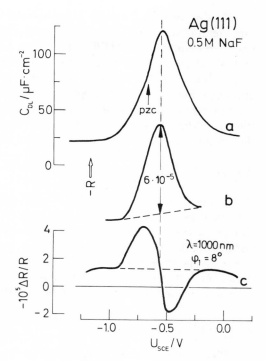

FIGURE 43. Double-layer capacity (a), integrated ER signal (b; see text) and ER signal (c) for Ag(111) in 0.5 M NaF as a function of potential. Normal incidence; $\lambda = 1000\,nm$; $\Delta U_{pp} = 112\,mV$; pzc: $-0.67\,V$. After Ref. 71.

relation between optical response of the interface and its double-layer capacity is eye-catching. Therefore, it may well be that the observed ER dependence (Figure 43) arises again from the metal side. It would have to be a contribution which is related to C_{DL}, such as the polarizability of the electron tail at the metal/electrolyte interface, which is assumed to decrease with surface charging in both directions. In conclusion, it looks as if there are a number of interesting observations[71,119,120] which suggest that a double-layer contribution to the optical response of the metal/electrolyte interface has been detected. Yet, we are still far from having definite answers and hence we have to await more experimental studies, preferably with well-defined single-crystal surfaces and in wavelength regions which extend further into the infrared or the ultraviolet (i.e., regions where the double-layer constituents become absorbing and therefore are more easily picked up in $\Delta R/R$), before we can draw any safe conclusions about the optical properties of the double layer proper.

5. Chemisorption and Film Formation

In the following, several selected examples are given for the study of submonolayer adsorbates and very thin films on metal electrodes. These include specifically adsorbed anions, dye molecules, oxides, metal adsorbates, and thin metal films. In by far the most cases, differential reflectance spectroscopy has been employed, where the normalized reflectance difference between bare and adsorbate-covered surface, $\Delta R/R = [(R(d) - R(0)]/R(0)$, is often determined, the adsorbate having a mean thickness of $d(\sim\theta)$. Less frequently, the adsorbate properties are studied by electroreflectance, i.e., by modulating the electrode potential *without* changing the coverage. As a typical example of the latter, the investigation of dye molecules, particularly of their Stark effect due to the electric field in the double layer, should be mentioned.

5.1. Oxides

Under appropriate conditions, many metals form thin oxide layers when subjected to anodic potentials. As is well known to the electrochemist, metals like platinum or gold form about a monolayer of chemisorbed oxygen (or a surface oxide) before oxygen evolution starts.[121,122] These monomolecular oxide layers were among the first systems to be studied optically in greater detail, mainly because they give rise to relatively large effects in $\Delta R/R$ of several percent and therefore are easily detected.[12,13]

Platinum Oxide

The oxide layers on polycrystalline Pt in acid media have been studied most thoroughly by Plieth *et al.*[35,37] The film optical constants have been evaluated for various thicknesses from $\Delta R/R$ spectra in the range of 1 to 6 eV by Kramers–Kronig analysis, first assuming isotropic optical constants and later anisotropic ones with uniaxial symmetry.[37] In both cases, the optical properties were indicative of a semiconducting film with a band gap of about 1.3 eV.

Later on, measurements were carried out with Pt single-crystal electrodes.[123] The $\Delta R/R$ spectra for oxide formation on the three low-index planes of Pt—(111), (110), and (100)—differ very little from each other, even for low oxygen coverages. Such a behavior, which is distinctly different from that seen in many other adsorption studies (see, e.g., Section 5.3 on metal adsorbates), obviously results from a strongly localized bond between oxygen and a Pt surface atom, with very little (if any) lateral interaction. It is therefore not surprising that oxygen adsorption (or oxide formation) on Pt(110) did not show the usual polarization anisotropy found with other adsorbates, i.e., $\Delta R/R$ at normal incidence did not differentiate between e∥[001] and e∥[110].[124] We mention in passing that the study of oxide formation on Pt single-crystal surfaces is a rather delicate problem, because this process tends to destroy the single crystallinity. As has been proven by *ex situ* LEED measurements, several anodic potential excursions into the oxide region with Pt single-crystal electrodes introduce a noticeable amount of steps and other surface defects,[56] which are readily noticed also in the corresponding hydrogen adsorption voltammograms.[125] Hence, this system is not amenable to repetitive potential cycling or stepping into the oxide region for determination of $\Delta R/R$ by the lock-in technique. In this respect, the use of a rapid-scan spectrometer in connection with signal averaging is very advantageous, as the $\Delta R/R$ spectra are obtained with a few potential steps only (see Section 3).

Gold Oxide

The $\Delta R/R$ spectra for oxygen adsorption (or oxide formation) on an Au(100) single-crystal surface are reproduced in Figure 44, the spectral shape being markedly influenced by the substrate reflectivity (the so-called $1/R$ effect). As in the case of Pt, it was found that the spectra do not show any pronounced dependence on the surface crystallographic orientation, the overall spectral shape in all cases being very similar to that for polycrystalline gold. Attempts have been made in the past to evaluate the optical constants of the thin oxygen layer on polycrystalline Au; these, however, led to contradictory results. In the early work of Kolb and McIntyre,[13] the film optical constants were evaluated by the simple three-phase model, and the result was indicative of strong absorptive properties with an absorption coefficient α of the order

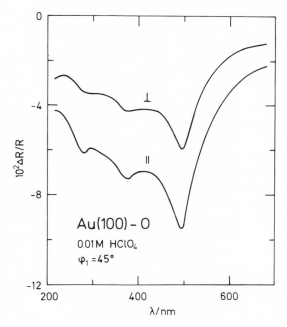

FIGURE 44. Normalized reflectance change, $\Delta R/R$, of Au(100) due to oxide formation. 0.01 M HClO$_4$. $\varphi_1 = 45°$. Potential step from +0.3 to +1.4 V vs. SCE. After Ref. 117.

of 10^5 cm^{-1}. Plieth *et al.*[126] used the experimental data of Kolb and McIntyre and applied their more sophisticated data evaluation which included uniaxial anisotropy of the film optical constants. Their result suggested that the film itself has negligible absorption, the $\Delta R/R$ being due mainly to the change in the bulk absorption caused by the formation of the thin oxide overlayer. The latter result is somewhat difficult to rationalize in view of the large reflectance change of several percent which is caused by a film of monolayer thickness only. Intuitively, one should expect that such a film would be strongly absorbing.† More experimental data, preferably with atomically flat gold surfaces, seem necessary to reach a conclusive answer and to avoid complications by surface roughness in the evaluation of the film anisotropy.

Iridium Oxide

While in the previous two examples we have discussed the initial stages of oxidation, we now shall focus our attention on the optical properties of thicker oxide layers. Anodically formed iridium oxide films have attracted particular attention because of their pronounced electrochromic effect.[128] When an Ir electrode is scanned anodically in 0.5 M H$_2$SO$_4$, oxidation starts at +0.6 V versus SCE. On the cathodic scan, however, the oxide layer is not reduced to the metallic state but to a low-conductivity hydroxide film facilitating further oxide formation with each anodic potential cycle. Continuous cycling of the iridium electrode between -0.25 and $+1.3$ V (SCE), at a frequency of 1 cps, therefore has been used as a standard treatment for the formation of thick anodic iridium oxide films.[129]

During such an experiment, it can be nicely observed by the naked eye that the oxide film undergoes a rapid and reversible change in color, from bleached (colorless) at cathodic potentials, say at 0 V, to bluish at anodic values, which is attributed to a change in the valence state of the Ir atom from +3 to +4: Ir(OH)$_3$ $\rightleftharpoons$ IrO(OH)$_2$ + H$^+$ + e^-.[129] Accordingly, the reflectivity of the electrode surface changes markedly with potential, as is demonstrated in Figure 45. The inset of this figure shows the spectral dependence of the reflectance change due to formation of the bluish film, which is characterized by two strong absorption bands around 2.2 eV (570 nm) and 4.1 eV (300 nm), the first one being responsible for the blue color. In addition to these two absorption bands, a marked background absorption is noted for the entire wavelength region under study. When the reflectivity of the oxide-covered iridium electrode is monitored as a function of potential (Figure 45), the transition from reduced (bleached) to oxidized (colored) state between 0.4 and 0.8 V is clearly seen. This transition is gradual and by no means complete at 0.8 V but continues on with potential. Measurements at different wavelengths

---

† Watanabe and Gerischer[127] have measured the anodic photocurrent for polycrystalline gold covered with approximately one monolayer of oxide. From the high quantum yield, an absorption coefficient α of about 10^5 cm^{-1} was indeed derived.

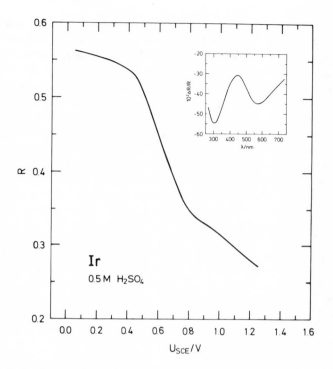

FIGURE 45. Reflectivity R of an oxide-covered iridium electrode in 0.5 M H_2SO_4 as a function of electrode potential. $\lambda = 570$ nm; $\varphi_1 = 45°$; p-polarization. Inset: corresponding $\Delta R/R$ spectrum due to potential stepping from +0.4 to +0.8 V versus SCE. After Ref. 117.

indicate that the background absorption has a similar potential dependence.

The electrochromic effect of the anodic iridium oxide film, which accompanies the change in oxidation state, has been explained by a potential-induced shift in the Fermi level causing a noticeable change in level occupancy.[129] In the bleached state, the Fermi level falls within the energy gap between the Ir $5d$-derived t_{2g} and e_g bands where the density of states is very low, which is the reason for the semiconducting properties of the bleached film with its low conductivity. With the transition from Ir^{3+} to Ir^{4+}, which is accompanied by the release of a proton, the Fermi level is shifted into the t_{2g} band, causing metallic-like properties of the bluish film. Because of the partial unfilling of the t_{2g} band, optical transitions in the near-IR and the visible region now become possible and are responsible for the blue color of the film.[129]

Ruthenium Oxidation

The following example deals with the spectroscopic identification of corrosion products which are generated during the oxygen evolution on Ru electrodes in acid media. Although this case does not fit exactly into the topic of this section, because—as we shall see in a minute—the corrosion product under study is soluble and is detected in the electrolyte rather than on the electrode surface, it demonstrates very nicely the molecular specificity of optical methods. It is well known that ruthenium and its four-valent oxide are excellent catalysts for chlorine and oxygen evolution; however, these reactions are accompanied by the corrosion of the metal.[130] From cyclic voltammetry under various stirring conditions, it has been inferred that the corrosion product is soluble, but a safe assignment of its valence state was difficult to reach solely on the grounds of electrochemical measurements. Since the absorption spectra of (soluble) Ru(VI), Ru(VII), and Ru(VIII) oxo species are so distinctly different from each other (note the inset of Figure 46[131]), the answer to this problem calls for the application of *in situ* reflectance spectroscopy. Indeed, in an optical study, RuO_4 was identified as the main, if not the only, corrosion product during the evolution of O_2 on a Ru electrode in sulfuric acid.[132] In Figure 46 are shown the differential reflectance spectra of a Ru electrode for which the potential was stepped from -0.1 V to $+1.17$ V versus SCE (solid line), where O_2 evolution and Ru corrosion take place, and to $+1.05$ V versus SCE (dotted line), which is not positive enough for these reactions to occur. Subtraction of these spectra (curve B in Figure 46) removes any

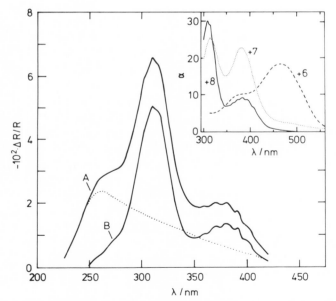

FIGURE 46. $\Delta R/R$ spectrum due to corrosion of a Ru electrode in 0.5 M H_2SO_4. ($\cdots$) -0.1 to $+1.05$ V versus SCE; no corrosion. (A) -0.1 to $+1.17$ V versus SCE, where corrosion takes place. Curve B was obtained by subtracting the dotted curve from curve A. After Ref. 132. The inset shows the absorption spectra of the various Ru-oxo species. After Ref. 131.

potential-induced change in the electrode's reflectivity and hence yields the unperturbed absorption spectrum of the corrosion product which is accumulated in front of the electrode and sampled by the light beam. The excellent agreement for the spectral shapes of curve B and the extinction coefficient for RuO_4 leaves no doubt about the nature of the corrosion product.[132]

5.2. Ions and Molecules

As we have pointed out already, an entity at the surface or in the electrochemical double layer in submonolayer amounts is readily picked up by reflectance spectroscopy only when it is either strongly absorbing (e.g., a dye molecule) or strongly interacting with the substrate (i.e., chemisorbed). In the latter case, the species is then optically detected either as an adsorbate-induced perturbation of the substrate's optical constants or because the interaction with the substrate changes the electronic (and hence optical) properties of the adsorbed species in such a way that it becomes absorptive, even if this species is nonabsorbing in solution (surface complex, surface compound).

Bromide Adsorption on Gold

Bromide ions are well known to adsorb specifically (that is to say, chemisorb) on gold electrodes.[133,134] While aqueous bromide solutions are transparent in the visible, Br^- ions obviously change their optical properties markedly upon adsorption on a metal surface. They give rise to considerable shifts of the surface plasmon dispersion curves[29,135] and to noticeably large $\Delta R/R$ values.[133] Differential reflectance spectra due to Br^- adsorption on polycrystalline gold were measured by Plieth[41] and the film optical constants evaluated by Kramers–Kronig analysis assuming uniaxial anisotropic film optical properties. Besides a marked anisotropy, which would be expected for a monolayer adsorbate, the tangential component of the dielectric surface excess ($\triangleq \hat{\varepsilon}_{2t}$, see Section 2.3) reveals two broad absorption bands at about 1.7 and 4.0 eV (Figure 47). This represents a dramatic change of the optical properties of the Br^- ion upon specific adsorption (e.g., the ion bcomes colored), which is indicative of a surface compound formation where the electronic states of adsorbate and substrate strongly mix to create new states.

Whenever the lateral interaction in an adlayer is weak, we observe a linear dependence of $\Delta R/R$ on coverage θ (or on the average film thickness d). Such a relation, which simply implies that the film dielectric constants, $\hat{\varepsilon}_2$, are independent of θ (Eq. 29), is conveniently used for a precise determination of coverages and hence of adsorption isotherms. Especially in cases like halide adsorption where the evaluation of surface excesses from charge measurements is hampered by an unknown electrosorption valency (unless a very elaborate concentration dependence study is made[134]), the spectroscopic method of determining coverages can indeed be advantageous. Adzić *et al.*[133] have

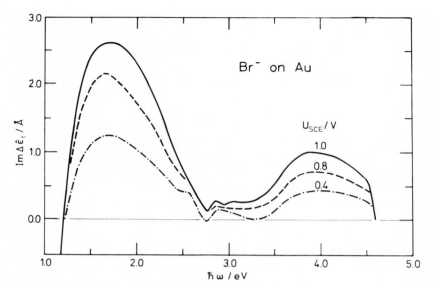

FIGURE 47. Imaginary part of the tangential component of the dielectric surface excess for a gold electrode covered with different amounts of Br^-. 0.5 M H_2SO_4 + 1 mM KBr. After Ref. 41.

studied Br^- ion adsorption on polycrystalline gold electrodes and determined by differential reflectance spectroscopy the adsorption isotherms $\Gamma_U = f(c)$, as well as the electrosorption valency $\gamma = -(1/F) \cdot (\partial q_m/\partial \Gamma)_U$. The strict linearity between $\Delta R/R$ and θ (or Γ) was tested by applying diffusion-controlled adsorption conditions and monitoring $\Delta R/R$ as a function of time; the measured values obeyed exactly the $t^{1/2}$ law (Sand equation). We should mention in this context that measurements of the surface resistance of thin film electrodes have also been used successfully to determine coverages with high precision. It was shown that the relative change in the electric resistance is indeed proportional to the amount of halide ions on the surface,[136] which allowed for an evaluation of electrosorption valencies for halides on gold.[136,137] The drawback of this elegant technique, which was mainly advocated by Hansen,[138] is the necessity of having film electrodes (usually 50-nm-thick metal films evaporated onto glass, quartz, or mica), which restrains its use in the study of single-crystal electrodes.[65]

Pyridine and Pyrazine on Ag(111)

Pyridine and pyrazine have received considerable attention in Raman spectroscopy because of their tremendously enhanced Raman scattering cross section when adsorbed on "activated" Ag surfaces.[95,115] The activation consisted of an oxidation–reduction cycle in the presence of the Raman-active molecule and, preferably, in chloride-containing electrolyes. By use of reflect-ance spectroscopy, it was possible to show that such a treatment leads to a surface complex, e.g., a pyridine molecule strongly bound to the Ag surface, which has optical properties distinctly different from those of the same molecule in solution.[139] While the aqueous solutions of both pyridine and pyrazine are transparent in the visible region with their first allowed optical transitions occurring in the near-UV region, the very same molecules adsorbed on Ag show pronounced absorption bands in the near-IR and visible wavelength range. Figure 48a contains the differential reflectance spectra, $\Delta R/R$, for pyridine and pyrazine adsorbed on Ag(111). To be precise, the reflectance change is due to an oxidation–reduction cycle in the presence of pyridine or pyrazine, during which monolayer amounts of Ag are oxidized and precipitated at the surface as AgCl and reduced to metallic Ag again. Because the same treatment in pyridine/pyrazine-free solutions yields no significant structures in $\Delta R/R$, the pronounced absorption bands (Figure 48a) can be safely assigned to a surface complex formed between molecules and Ag surface as a con-sequence of their strong interactions. Similar conclusions were reached by surface plasmon investigations.[141] Figure 48b shows the ER spectrum for pyrazine on Ag(111). The derivative-like structure in $\Delta R/R$ indicates a shift of the absorption band at 600 nm with electrode potential, which is about −25 meV/V.

These examples certainly demonstrate how much the properties of an entity can be changed upon chemisorption, or how different a molecule can

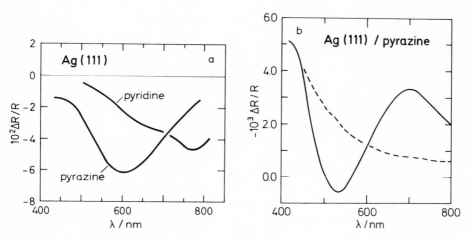

FIGURE 48. (a) $\Delta R/R$ spectra for Ag(111) in 0.1 M KCl due to an oxidation–reduction cycle of about 5 monolayers in the presence of pyridine and pyrazine, respectively. $\varphi_1 = 66°$; p-polarization. After Refs. 139 and 140. (b) ER spectrum for Ag(111) in 0.1 M KCl + 0.05 M pyrazine before (– – –) and after (——) the oxidation–reduction cycle. Potential step from −0.5 to 0.0 V versus SCE. $\varphi_1 = 66°$; p-polarization.

be in the free and the chemisorbed state. A further example of this kind, just to mention it briefly, has been given by Lamy *et al.*,[142] who studied CO chemisorbed on a Pt electrode. Adsorption of CO in submonolayer amounts from CO-saturated acid solutions causes a remarkably large reflectance change at 360 nm, a wavelength at which CO in solution does not absorb at all. The large $\Delta R/R$ values of 1 to 2% due to CO adsorption on Pt are again indicative of a drastic change in the molecule's electronic structure upon interacting with the Pt surface.

Dye Molecules Adsorbed on Metal Electrodes

The study of dye adsorption on electrode surfaces by optical methods seems particularly interesting and rewarding for more than one reason. (i) Dye molecules often have a transition moment of a well-defined direction, permitting a determination, by polarization and angle-dependence studies, of the orientation of the adsorbed molecule with respect to the electrode surface. (ii) Because the dye's absorption bands are usually sharp and well pronounced, even relatively weak interactions with the substrate are readily detectable in the corresponding reflectance spectra by a band broadening and by peak shifts.[143] (iii) The influence of strong electric fields on the optical transitions of dye molecules has been frequently studied in the past, and energy shifts of absorption bands with electric field have been firmly established (electrochromism).[144] It is therefore hoped to use adsorbed dye molecules as a local probe for the precise determination of the electric field in the double layer.

Electrochromism studies of dye molecules adsorbed on electrode surfaces have been reported by Plieth *et al.*[145-147] Two of their results are reproduced in Figure 49, showing the potential-modulated reflectance spectra of *p*-aminonitrobenzene (*p*-ANB)[147] and of *p*-dimethylaminonitrostilbene[146] adsorbed on polycrystalline Pt electrodes. In the case of *p*-ANB, the bands at 436 and 620 nm have been attributed to monomer and dimer absorption, respectively. The appearance of dimers (or higher aggregates) in the absorption spectra of adsorbed dyes seems a widespread phenomenon and obviously results from the high surface concentration of dye molecules.[148] The monomer band was found to shift to shorter wavelengths with increasingly positive potentials by 0.13 eV/V, while the dimer band showed an opposite shift of about the same magnitude.[147] From the polarization dependence—an effect in $\Delta R/R$ has been observed only for *p*-polarized light (Figure 49)—it was concluded that the molecule is oriented perpendicular to the surface. From the shift of the 436-nm absorption peak with electrode potential, an electric field strength of 5.6×10^7 V cm^{-1} at the position of the chromophore was derived.[147] This value is lower than expected for the field strength in the (compact) double layer, and it indicates that the dye molecule "sees" the residual potential drop in the diffuse part of the double layer. Obviously,

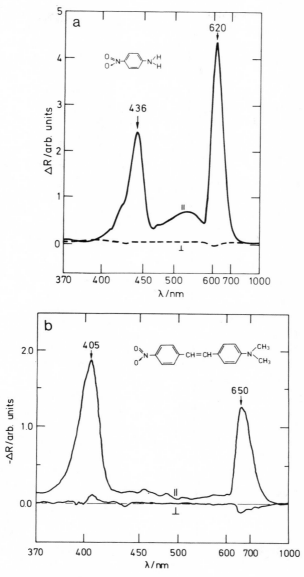

FIGURE 49. Modulated reflectance (ER) spectra for (a) p-aminonitrobenzene and (b) p-dimethylaminonitrostilbene on polycrystalline Pt. 0.5 M Na$_2$SO$_4$ + 5 × 10^{-5} M dye. φ_1 = 65°; p- and s-polarized light; ΔU_{pp} = 100 mV. After Refs. 146 and 147.

the dye molecules in general are too large to act as a probe for the compact double layer. Small molecules, which would fit into the double layer, naturally absorb in the UV range and therefore are not yet readily accessible by conventional light sources. In this respect, the use of synchrotron radiation in combination with suitable thin-layer cells may be a promising way to go.

Finally, we come back to the striking polarization dependence, which is shown in Figure 49, where a potential-induced reflectance change is only observed for p-polarized light. It suggests that the dye molecule is oriented with its dipole moment perpendicular to the electrode surface. We have to bear in mind, however, that *two* requirements must be met in order to obtain a potential-induced reflectance change. Firstly, the scalar product of the light electric vector and the transition moment has to be nonzero; and secondly, the transition moment has to have a component in the direction of the modulating electric field, which is the cause for a ΔR and which is always normal to the surface. Since for s-polarization the light electric vector and the modulating electric field are always normal to each other, a molecule with its transition moment either parallel *or* perpendicular to the surface would not be detected in $\Delta R / R$ with s-polarized light.

5.3. Metal Adsorbates

The formation of metal adsorbates is most conveniently studied at the metal/electrolyte interface when the so-called underpotential deposition (upd) occurs.[149] It has been well known for many years that a large number of metals are deposited in submonolayer amounts, usually up to one monolayer, onto foreign metal electrodes at underpotentials, that is, at potentials which are positive of the respective Nernst potential for bulk deposition.[149,150] This is simply a consequence of the fact that in these cases, the strength of the substrate–adsorbate bond exceeds that of the adsorbate–adsorbate bond. Upd is readily demonstrated by cyclic voltammetry, which contains direct information on the adatom binding energy as a function of coverage. In Figure 50 are shown the anodic scans of cyclic current–potential curves for various Pt single-crystal electrodes covered with a Cu monolayer.[151] Such curves are often called "electrochemical desorption spectra," as they are in some respects comparable to thermal desorption spectra in UHV,[152] but have the distinct advantage over thermal desorption spectroscopy of being performed under equilibrium conditions. Integration of the $I-U$ curves (Figure 50) yields directly the corresponding adsorption isotherms. Distinct differences in the bond energies are seen for Cu adsorption on the three low-index faces of Pt. Particularly for Pt(110), we note that Cu monolayer formation occurs in two, energetically quite different, steps. From a careful analysis of such desorption spectra and by comparison of the adsorption isotherms with the results of structural investigations performed in vacuum, the conclusion was reached

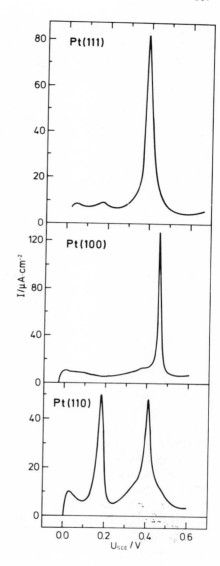

FIGURE 50. Electrochemical desorption spectra for a Cu monolayer on Pt(hkl). Scan rate: $10 \, \text{mV s}^{-1}$. $0.5 \, M \, H_2SO_4 + 1 \, mM \, CuSO_4$.

that superstructures (i.e., ordered adsorption) were formed during upd on single-crystal surfaces.[153-155]

Because of the strong interaction of the upd layer with the substrate, the adsorbate's optical properties are expected to differ markedly from those of the corresponding bulk material. The $\Delta R / R$ spectra (i.e., the reflectance change due to metal deposition) for a Cu monolayer on Pt single-crystal electrodes are shown in Figure 51.[151] Not only do these spectra reveal that the optical

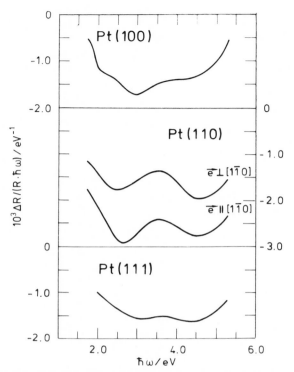

FIGURE 51. Normalized reflectance change $\Delta R/R$ for a Cu monolayer on Pt(hkl). Normal incidence. Note the polarization anisotropy for Pt(110). After Ref. 151.

properties of the Cu monolayer indeed differ markedly from those of bulk Cu, they also show a pronounced dependence on the crystallographic orientation of the substrate.[151] Furthermore, we note from Figure 51 that for Cu on Pt(110), the normal-incidence $\Delta R/R$ spectra show a pronounced anisotropy parallel to the surface, which reflects the twofold symmetry of the substrate.

The optical properties (the imaginary part of $\hat{\varepsilon}$, to be precise) of a Cu monolayer on Pt(110) are shown in Figure 52, as evaluated from the respective $\Delta R/R$ spectra by Kramers–Kronig analysis for the two main crystallographic directions.[151] The corresponding curve for bulk Cu also is shown for comparison. For both surface directions, a strong absorption band is seen around 2.5 eV, but the sharp rise of ε'' at 2.0 eV due to the onset of the interband transition, which is typical for bulk Cu, is not yet present.

In studying the coverage dependence of $\Delta R/R$, we can obtain information about lateral interactions in the adsorbate layer and hence we may obtain some clues to the geometric structure of the adlayer. The following case study is based on the assumption that with coverage θ the adsorbate–adsorbate interaction is changed rather than the adsorbate–substrate interaction. On these

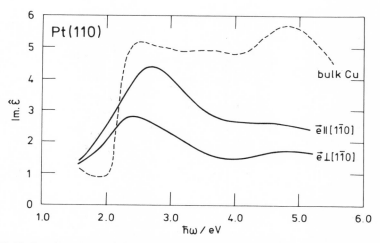

FIGURE 52. Imaginary part of the dielectric function of a Cu monolayer on Pt(110) for the electric field vector in the two main crystallographic directions; (– – – –) corresponding values for bulk Cu. After Ref. 151.

grounds, $\hat{\varepsilon}_2(\theta)$ should reflect mainly the lateral interaction in the adlayer. From the linear approximation equation (29a) and by substituting d by $\theta \cdot d_0$, we conclude that $\Delta R/R(\theta)$ should be a straight line through the origin for θ-independent optical constants of the adsorbate (no or constant lateral interaction: strongly localized bond or island growth). When we envisage a reconstruction of the adlayer at a certain coverage θ, i.e., the adlayer rearranges itself so that all adatoms suddenly occupy new sites, then a change from one straight line through the origin to another one with different slope is expected at that coverage. Finally, we could think of an occupation of new sites at a certain coverage, e.g., after a certain ordered structure had been completed. This filling in of new sites with different optical properties should cause a slope change in $\Delta R/R(\theta)$. These three cases, which seem to be the most likely ones for monolayer formation, are depicted schematically in Figure 53. A systematic investigation was performed for Cu on Pt(hkl), the results of which are reproduced in Figure 54.[151] The strikingly close correlation of experimental observations and model predictions permits some straightforward conclusions to be drawn about the structure of Cu adsorbates on Pt electrodes. For example, for Cu on Pt(111), a restructuring of the adlayer at medium coverages seems obvious. An intriguing piece of information is given by $\Delta R/R(\theta)$ for Cu on Pt(110). Besides the lateral anisotropy in the optical response for $e\|[001]$ and $e\|[1\bar{1}0]$, which we already mentioned, we note a slope change in $\Delta R/R(\theta)$ for $e\|[001]$ at $\theta = 0.5$, indicating a change in the adlayer's properties in this crystallographic direction. We have learnt already from cyclic voltammetry (Figure 50) that there are two energetically different

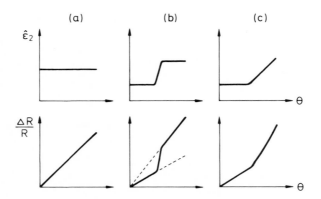

FIGURE 53. Schematic diagram of the coverage dependence of $\Delta R/R$ for various types of adsorbates. (a) Coverage-independent optical properties; (b) rearrangement in the adsorbate; (c) gradual change in lateral interaction as the monolayer is completed. After Ref. 151.

adsorption sites for $\theta < 0.5$ and $\theta > 0.5$, respectively. This difference in adsorption sites, however, is noted only for light with $\mathbf{e}\|[001]$, while no such change is observed with $\mathbf{e}\|[1\bar{1}0]$. From these observations, it was concluded that Cu is preferentially deposited in the substrate's grooves along the [110]-direction as densely packed chains, but for $\theta < 0.5$ only every second groove is occupied. Such a model would account for the observed change in the lateral interaction at $\theta = 0.5$ along [001], as the adsorbate–adsorbate distance in the [001]-direction would be reduced by a factor of 2 when each row becomes filled for $\theta \geq 0.5$. It may be noteworthy that in gas-phase studies with (110) substrates, the filling of every second row at low coverages as a result of long-range order has indeed been observed.[156] Despite the very interesting information obtained by differential reflectance spectroscopy, definite conclusions on the structure of the Cu adlayer cannot be gained. The above-described model, although plausible, needs confirmation by more structure-sensitive techniques, such as *ex situ* LEED or RHEED.[157]

For adatoms smaller than the substrate atoms, the complete monolayer is usually in registry with the substrate surface, while adatoms larger than the substrate atoms often form close-packed monolayers, regardless of the substrate surface symmetry.[153] The first case is true for Cu on Pt, and indeed the lateral anisotropy in $\Delta R/R$ for Cu on Pt(110), which reflects the substrate's twofold symmetry, is seen up to the complete monolayer. For Pb on Ag(110), a system which falls into the second category, a distinctly different behavior was reported: only at medium coverages was a clear polarization anisotropy observed, while the anisotropy disappeared for the completed monolayer.[124] This suggests that at medium coverages, Pb forms an adlayer which is in registry with the Ag substrate and hence acquires the twofold symmetry of the substrate, while at high coverages, the Pb atoms arrange themselves into a

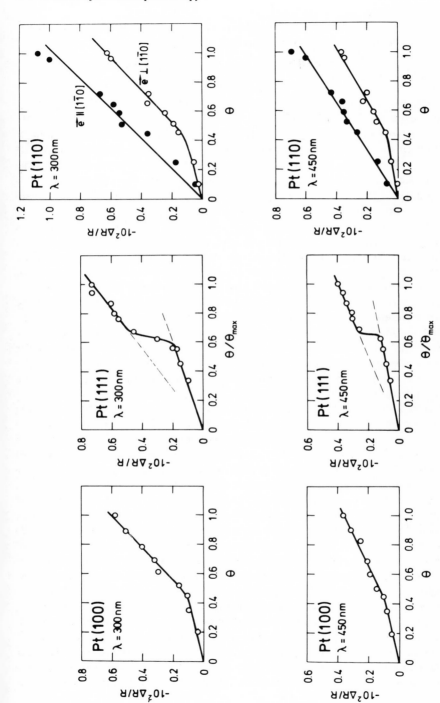

FIGURE 54. Normalized reflectance change $\Delta R/R$ due to Cu deposition onto Pt(hkl) as a function of Cu coverage. Normal incidence, two different wavelengths. Note the anisotropy in the optical response for Cu on Pt(110). After Ref. 151.

hexagonal close-packed overlayer which does not show polarization anisotropy anymore because of the sixfold symmetry.

5.4. Metal Film Formation

The optical properties of metal monolayers differ markedly from those of the respective bulk material, as we have just seen in the previous section. Hence, it will be interesting to study the gradual transition from monolayer to bulk properties and to determine the minimum film thickness which is necessary for the deposit to acquire bulk optical properties. Besides this, reflectance measurements may be employed to determine the structure and the quality of thin metal overlayers, e.g., whether a thin film is continuously deposited on a substrate or whether it consists of clusters.

By monitoring the reflectance of a Pt electrode, onto which Ag was deposited from solution at constant (diffusion-limited) rate, McIntyre and Kolb[12] could demonstrate that a thickness of about 3 monolayers was required before the optical response was that expected for bulk Ag. An evaluation of the dielectric constants for metal monolayers deposited at underpotentials often reveals [but not always; see Refs. (80, 158)] a positive ε_2' throughout the entire wavelength region under study,[57,117] whereas the real part of a metal's dielectric function is usually negative in this region (Figures 3–5). A Kramers–Kronig analysis of $\Delta R/R$ spectra for Cu deposited onto Pt revealed that for a 3 monolayer-thick Cu film, ε_2' is already negative (at least for $\hbar\omega \leq 3.5$ eV) and a steep rise with $\hbar\omega$ in ε_2'' around 2 eV indicates the onset of bulklike interband transitions.[32]

In a systematic study, the deposition of thin Ag overlayers on a Cu(111) electrode was investigated by electroreflectance and differential reflectance spectroscopy, employing an optical thin-layer cell (Figure 20d).[47] Figure 55 contains the ER spectra of a Cu(111) electrode covered with Ag overlayers of various thicknesses. We remember that the bulk optical properties of Ag are markedly influenced by the plasmon excitation at 3.85 eV. This plasmon excitation, which causes the reflectivity of bulk Ag to drop sharply from about unity by nearly two orders of magnitude (note the inset of Figure 4), is clearly seen in the ER spectra of massive Ag(111) as a pronounced dip in $-(\Delta R/R)$ for p-polarized light (Figure 23b). It can be discerned from the sequence of curves in Figure 55 that the bulk plasmon begins to show up around coverages of 2.5 monolayers as evidenced by the dip in $-\Delta R/R$ developing at 3.8 eV, which is clearly seen as such for $\theta = 3.2$ monolayers (ML). Since the low-energy bulk plasmon at 3.8 eV is very characteristic of the electronic properties of massive Ag, the appearance of the plasmon excitation feature in the ER spectrum indicates definitively at which film thickness bulk behavior is reached. We again find that a film thickness of 3 monolayers suffices to yield bulk optical properties. It should be noted, however, that in all the above-cited cases, metal deposition started by the growth of a uniform monolayer (by

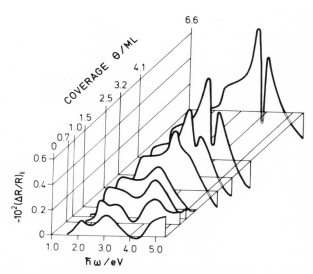

FIGURE 55. ER spectra for Cu(111) covered with various amounts of Ag. The Ag coverage is given in units of monolayers. 0.5 M H$_2$SO$_4$. $\Delta U_{pp} = 300$ mV; p-polarization at 45° angle of incidence. After Ref. 47.

underpotential deposition), which obviously is a prerequisite for the deposition of thin metal overlayers of uniform thickness. We mention in passing that no upd is observed for Cu on Ag.[149] In that case, the deposit grows from the very beginning in three-dimensional islands, and a continuous Cu overlayer on Ag is not reached up to a deposition equivalent to many tens of Cu monolayers.[47]

We now return to the question about the role of nonlocal optics in the study of metal electrodes. In Figure 56a, the differential reflection spectra are shown for a Cu(111) electrode covered with a thin Ag overlayer; i.e., $\Delta R/R$ represents the reflectance change of the Cu(111) electrode due to deposition of a 5-monolayer-thick Ag film.[47] These experimental spectra can then be compared with model calculations for a 1.2-nm-thick Ag film on Cu, assuming bulk optical constants for the Ag overlayer. The result, Figure 56b, convincingly shows that the experimental data for p-polarized light can only be reproduced satisfactorily when spatial dispersion is included in the calculation. The use of classical optics, on the other hand, leads to marked differences in the spectral shapes, which could be misinterpreted to mean that the film has optical properties still substantially different from those of bulk Ag. This, however, is not true as is demonstrated by the almost perfect agreement between experiment and theory when spatial dispersion is included in the calculation and bulk optical constants are used for the 1.2-nm-thick Ag overlayer.[47]

The last and perhaps most dramatic example to illustrate the importance of spatial dispersion is presented in Figure 57, where for various wavelengths,

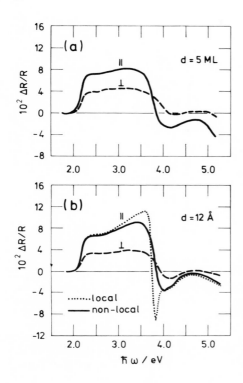

FIGURE 56. Normalized reflectance change $\Delta R/R$ due to deposition of an Ag overlayer onto Cu(111). (a) Experimental result for a 5-monolayer-thick film; $\varphi_1 = 45°$. (b) Calculated spectra with (——) and without ($\cdots$) inclusion of longitudinal waves. After Ref. 47.

the ER effect of an Ag-covered Au(111) electrode is shown as a function of silver film thickness D. For this purpose, Ag was deposited under diffusion-controlled conditions, i.e., at constant rate, onto the (very flat) Au(111) electrode and the ER effect recorded continuously as a function of time. Except for the thickness range below 1 nm, where $\Delta R/R$ varies rapidly with D for s- and p-polarized light because of a marked dependence of the film optical properties on D for the first three monolayers (see above), the ER effect for s-polarization rises smoothly with film thickness and eventually reaches a limiting value. For p-polarization, however, pronounced oscillations in $\Delta R/R$ are seen for photon energies close to $\hbar\omega_p$. It has been shown that these oscillations are caused by standing plasma waves in the Ag overlayer, which acts as a resonator at certain thicknesses.[159] From the distance of two succeeding maxima in $-(\Delta R/R) = f(D)$, the plasma wavelength is readily obtained as about 4 nm (compared to about 320 nm for the transverse wave in this energy region). Hence, such measurements allow a direct and reliable determination of the bulk plasmon dispersion curve $\omega(\mathbf{k})$ for Ag.[159] The oscillations in $\Delta R/R$ for p-polarization (Figure 57) have been reproduced in a model calculation with nonlocal optics and assuming (thickness-independent) bulk optical properties of the Ag film.[159] It is obvious that a calculation of the

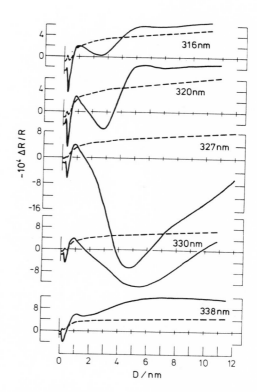

$-10^4 \Delta R/R$

316nm

320nm

327nm

330nm

338nm

D / nm

FIGURE 57. ER effect for Au(111) during deposition of an Ag overlayer of thickness D. $\varphi_1 = 45°$; (——) p- and (- - -) s-polarization; $\Delta U_{pp} = 100\,mV$. After Refs. 71 and 159.

film optical constants from the data in Figure 57 using standard optics would yield a completely wrong result, namely, oscillating optical properties of the Ag film up to large D values, which is absolutely not the case.

The above-described observation of standing plasma waves in thin films certainly requires very homogeneous overlayers of uniform thickness [actually, the degradation of the oscillations at larger thickness values (Figure 57) is ascribed to an increased thickness variation within the film, or to the appearance of surface roughness as the deposition continues]. In this respect, the occurrence of upd as the initial step in metal deposition obviously seems an important supposition for obtaining thin metal films of rather uniform thickness. Upd of metals, so far, has only been reported for metal electrodes as substrate but not for semiconductor electrodes. For the latter, metal deposition has always been found to start by three-dimensional nucleation and subsequent cluster growth,[160] although under UHV conditions monolayer formation has indeed been observed for evaporation of Pd and Ni onto ZnO.[161,162]

The deposition of metals onto semiconductor electrodes is sensitively monitored by differential reflectance spectroscopy because of the vast differences in the optical constants of metals and semiconductors.[60,163,164] Figure 58 illustrates the reflectance changes, $\Delta R/R$, for a ZnO electrode (which is an

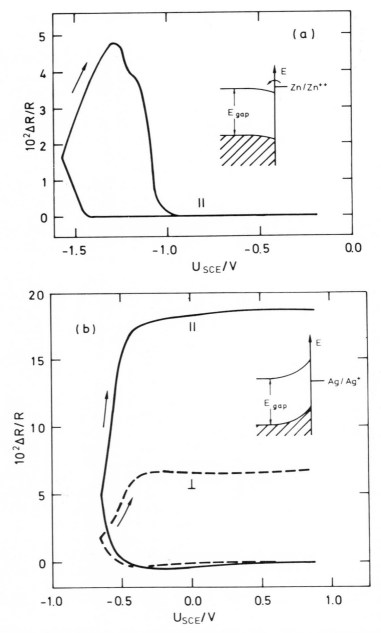

FIGURE 58. Monitoring the deposition of (a) Zn and (b) Ag on a ZnO electrode via the resulting reflectance change $\Delta R/R$. $\lambda = 500$ nm. $\varphi_1 = 45°$. Note that in contrast to the Zn deposit, Ag is not oxidatively stripped again at positive potentials, because the substrate cannot take up the electrons (see inset). After Ref. 163.

n-type material) onto which Zn and Ag are deposited from solution. On the cathodic scan, the metal deposition is clearly seen in both cases by an increase in reflectance. Except for a small overvoltage, Zn is deposited at the expected potential value for the Zn deposition reaction. This is negative of the substrate's flat-band potential, and therefore electrons are available at the surface for the discharge of the Zn^{2+} ions. The redox potential for Ag deposition, however, is positive of the flat-band potential, and no electrons are at the surface due to the upward bending of the energy bands.[163] Silver deposition therefore starts only at much more negative potentials, when electrons are accumulated at the ZnO surface. For similar reasons, the Zn deposit is readily desorbed on the anodic potential scan, whereas the Ag deposit clearly remains on the surface, even at potentials positive of the respective Nernst potential. The reason for this behavior lies in the very positive redox potential of Ag (see inset of Figure 58), which causes the energy level for the Ag/Ag^+ reaction to fall within the bandgap of the ZnO electrode, so that the electron transfer from the deposit to the substrate is inhibited. Only illumination with light of $\hbar\omega \geq \hbar\omega_{gap}$ facilitates photooxidation of the Ag deposit.

As we mentioned above, metal deposition on semiconductor electrodes usually starts with three-dimensional nucleation, which leads to discontinuous (clustered) overlayers. Such films have apparent optical properties which can be distinctly different from those of the metal clusters proper and which are determined also by the refractive index of the embedding medium (in our case, the electrolyte), the shape of the particles, and their volume fraction (filling factor f_m). According to the Maxwell–Garnett theory, for spherical particles the optical response is[165]:

$$\frac{\hat{\varepsilon}_a - 1}{\hat{\varepsilon}_a + 2} = f_m \cdot \frac{\hat{\varepsilon}_m - 1}{\hat{\varepsilon}_m + 2} + (1 - f_m)\frac{\varepsilon_1 - 1}{\varepsilon_1 + 2} \tag{52}$$

where $\hat{\varepsilon}_a$ is the apparent dielectric function of the aggregated film, $\hat{\varepsilon}_m$ is the dielectric function for the metal particles proper (usually the bulk values of the metal), and f_m represents the fractional volume of the metal particles of which the aggregated film consists. Optical studies on thin evaporated metal films on glass showed that aggregated (island) films exhibit "nonmetallic" apparent optical properties (i.e., $n_a > k_a$) at very low film thicknesses, which could be rationalized by the Maxwell–Garnett theory.[166]

When Zn is deposited from solution onto a ZnO electrode and the reflectivity is monitored as a function of film thickness (i.e., as a function of charge), the optical response depends noticeably on the deposition potential. For example, it was reported that for a deposition potential just negative of the respective Nernst potential, the reflectivity decreased continuously with increasing amount of Zn deposited, while the opposite optical behavior was observed for deposition at high overvoltages. Obviously, a film of better quality (higher reflectivity) was formed in the latter case, which is easily understood from the correlation between nucleation behavior and overpotential. Since the

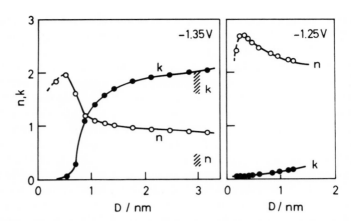

FIGURE 59. Optical constants, *n* and *k*, of a Zn deposit on ZnO as a function of mean film thickness *D*, formed at two different deposition potentials. The optical constants of bulk Zn are indicated by the shaded areas. After Ref. 163.

number of nuclei increases exponentially with overpotential η, deposition at large η causes the nuclei to coalesce much faster than at low η and hence leads to smoother films. Figure 59 shows an evaluation of the apparent film optical constants as a function of film thickness *D* for deposition of Zn onto ZnO at high and low overpotentials.[163] While the low-reflectivity film, deposited at −1.25 V versus SCE, retains its anomalous, nonmetallic behavior up to several nanometers of average thickness, the high-reflectivity film (−1.35-V deposition potential) exhibits a transition to metallic behavior around *D* = 1 nm. An average film thickness of 3 nm suffices to reach optical properties very close to those reported for bulk Zn.[163] A similarly smooth and continuous Zn film on ZnO was obtained by cathodic decomposition of the ZnO electrode, where the thin Zn overlayer was formed by reducing the Zn^+ ions of the ZnO lattice.[60] Optical experiments were also performed for Tl on ZnO[164] and Cd on CdS, the latter by deposition of Cd from solution as well as by photodecomposition of the CdS electrode proper.[60,164]

6. Summary and Outlook

It is the author's firm belief that *in situ* reflectance spectroscopy in the visible and near-UV region can still contribute in an important way towards a better understanding of the structure and the properties of the electrochemical interface and of electrode processes. Its virtues are the molecular specificity and its nondestructive and easy use. The wavelength limitations set forth by the electrolyte, which are presently still encountered and which restrict the routinely performed studies to the range of about 1 to 6 eV, should be removed in the near future by the construction of appropriate thin-layer cells, similar to those developed for *in situ* infrared spectroscopy.[167] Extension of the wavelength region into the near-vacuum-UV range with the use of synchrotron radiation would be very valuable for a complete characterization of the optical properties of bare and adsorbate-covered surfaces.

Unlike Raman or infrared spectroscopy, with their well-defined capability of determining vibrational properties, the application of optical spectroscopy in the UV–visible range is less focused. This is because UV–visible reflectance spectroscopy intrinsically is to yield electronic excitation spectra, which unfortunately for solids are often broad and unstructured and hence not very informative as such. Therefore, this technique has increasingly been used for other tasks also, such as the determination of coverages, the assessment of structural information, and the study of surface roughness. The aim of this chapter was to demonstrate this versatility.

Besides the variety of techniques which are now available for *in situ* reflectance studies, such as electroreflectance, differential reflectance, surface plasmon excitation, and external or internal reflection, there are other techniques described in the literature which operate on the same physical basis (e.g., absorption of light) and hence yield similar information. They should have been discussed jointly with reflectance spectroscopy but were omitted because of time and space limitations. These include ellipsometry,[168] photocurrent spectroscopy,[169] photoemission into electrolytes,[170] and, somewhat remotely, surface conductivity.[171] The intimate correlation between reflectance spectroscopy and ellipsometry is obvious, both techniques being capable of the determination of optical constants of solids. The principles of ellipsometry are well covered in the literature,[168] and a number of groups have used this method for the study of the very same problems addressed in this chapter (e.g., for the study of metal adsorbates formed at underpotentials).[80,158,172] Ellipsometry and reflectance spectroscopy have even been combined to determine all three unknown parameters of a three-phase system, namely, real and imaginary parts of the complex refractive index and thickness of the film phase.[173] Similarly, photocurrent spectroscopy is a powerful technique for the study of semiconducting overlayers and has been used extensively to investigate oxide layers on metal electrodes.[169] Again it is obvious that the photocurrent response is closely connected with the absorption coefficient of the layer under study and hence a cross-correlation with the results of reflectance studies is possible. Less frequently used in electrochemical studies is electrode resistance monitoring, an elegant technique to investigate adsorbates.[136,171] This method, although restricted to the study of thin film electrodes, has been successfully used to determine adsorption isotherms, especially in those cases where purely electrochemical studies were hampered by large faradaic currents.[137] Combining surface resistance measurements with reflectance spectroscopy seems particularly intriguing because of the different aspects of both techniques.[136] With all these techniques at hand, which are applicable *in situ* and which deal with the interaction of UV–visible light with electrode surfaces, a wide variety of problems can be solved as they provide the means of studying the reactions at the electrochemical interface at a much more detailed and specific level than classical electrochemical techniques ever can.

Appendix I. How to Calculate the Nonlocal Fresnel Reflection Coefficient for a Three-Phase-System†

Consider a three-phase system with ambient (ε_1), thin film $(\hat{\varepsilon}_2)$ of thickness d, and substrate $(\hat{\varepsilon}_3)$, where phases 2 and 3 are metals, i.e., spatially dispersive media. The dielectric function $\hat{\varepsilon}_j$ of phase j is given by (Eq. 20)

$$\hat{\varepsilon}_j = \hat{\varepsilon}_{jb} - \frac{\omega_{nj}^2}{\omega(\omega + i\gamma_j)} \tag{20a}$$

with the usual separation into bound- and free-electron contributions. Note that a different sign convention has been used in Ref. 50, and is used in this appendix $(\hat{n} = n + ik)$, which leads to a different sign for the dampling factor in Eq. (20a). Also note the unusual definition of the wave vectors, k_z and κ (see below), which includes the imaginary unit i (e.g., $\exp(-k_z d)$ is already a plane wave).

We first write down the expressions for the normal component, k_z, of the wave vector of the transverse waves in media 1, 2, and 3 (we consider p-polarization only, with x and z defining the plane of incidence):

$$k_z^{(1)} = (k_x^2 - \varepsilon_1 \omega^2/c^2)^{1/2}$$

$$k_z^{(2)} = (k_x^2 - \hat{\varepsilon}_2 \omega^2/c^2)^{1/2}$$

$$k_z^{(3)} = (k_x^2 - \hat{\varepsilon}_3 \omega^2/c^2)^{1/2}$$

and of the longitudinal waves in media 2 and 3:

$$\kappa^{(2)} = \left[k_x^2 - \hat{\varepsilon}_2 \omega^2 \left(1 + \frac{i\gamma_2}{\omega} \right) \middle/ (\hat{\varepsilon}_{2b}\beta_2) \right]^{1/2}$$

$$\kappa^{(3)} = \left[k_x^2 - \hat{\varepsilon}_3 \omega^2 \left(1 + \frac{i\gamma_3}{\omega} \right) \middle/ (\hat{\varepsilon}_{3b}\beta_3) \right]^{1/2}$$

with $\beta = \frac{3}{5} \cdot v_F^2$ [$v_F = (\hbar/m) \cdot (3\pi^2 N)^{1/3}$ is the Fermi velocity] and $k_x = (2\pi/\lambda)\sqrt{\varepsilon_1} \sin \varphi_1$ (take $\text{Re}(k_z^{(j)}, \kappa^{(j)}) > 0$ and $\text{Im}(k_z^{(j)}, \kappa^{(j)}) < 0$). We then

† From Ref. 50.

introduce the following abbreviations:

$$a = \exp(-k_z^{(2)} \cdot d); \qquad b = \exp(-\kappa^{(2)} \cdot d); \qquad \delta = \hat{\varepsilon}_{2b}/\hat{\varepsilon}_{3b}$$

$$H = 8k_x^2 k_z^{(2)}(\hat{\varepsilon}_3\delta - \hat{\varepsilon}_2)/\kappa^{(3)}$$

$$G = [\hat{\varepsilon}_{3b}(\hat{\varepsilon}_3\delta - \hat{\varepsilon}_2)]/[\hat{\varepsilon}_3(\hat{\varepsilon}_2 - \hat{\varepsilon}_{2b})]$$

$$C = k_z^{(1)}\hat{\varepsilon}_2/\varepsilon_1 \pm k_z^{(2)} \pm k_x^2(\hat{\varepsilon}_2 - \hat{\varepsilon}_{2b})/(\kappa^{(2)}\hat{\varepsilon}_{2b})$$

$$D = \left(1 + \frac{\kappa^{(2)}}{\kappa^{(3)}}\delta\right)(k_z^{(2)}\hat{\varepsilon}_3 \pm k_z^{(3)}\hat{\varepsilon}_2) + G[k_z^{(2)}\hat{\varepsilon}_3 \pm k_z^{(3)}\hat{\varepsilon}_2 \pm H/(8k_z^{(2)})]$$

$$F = \left(1 - \frac{\kappa^{(2)}}{\kappa^{(3)}}\delta\right)(k_z^{(2)}\hat{\varepsilon}_3 \pm k_z^{(3)}\hat{\varepsilon}_2) + G[k_z^{(2)}\hat{\varepsilon}_3 \pm k_z^{(3)}\hat{\varepsilon}_2 \pm H/(8k_z^{(2)})]$$

The Fresnel reflection coefficient $r_{\parallel123}^{\text{n.l.}}$ for a three-phase system with nonlocal optics included is then given by:

$$r_{\parallel123}^{\text{n.l.}} = [D(\oplus) \cdot C(\ominus\ominus) + D(\ominus) \cdot C(\oplus\ominus) \cdot a^2 + F(\oplus) \cdot C(\ominus\oplus) \cdot b^2$$

$$+ F(\ominus) \cdot C(\oplus\oplus) \cdot a^2b^2 - Hab]$$

$$\times [D(\oplus) \cdot C(\oplus\oplus) + D(\ominus) \cdot C(\ominus\oplus) \cdot a^2 + F(\oplus) \cdot C(\oplus\ominus) \cdot b^2$$

$$+ F(\ominus) \cdot C(\ominus\ominus) \cdot a^2b^2 + Hab]^{-1}$$

The two sign symbols in the parentheses of C indicate the signs to be taken in the expression of C [e.g., $C(\oplus\ominus) = k_z^{(1)}\hat{\varepsilon}_2/\varepsilon_1 + k_z^{(2)} - k_x^2 \cdots$]. The sign symbol given with D or F indicates that this sign should be chosen throughout the whole expression for D or F.

Neglect of spatial dispersion in the above-described formalism must yield the well-known expressions of standard optics. This is done by setting $\beta = 0$ and $\kappa^{(2)}$, $\kappa^{(3)} \to \infty$.

Appendix II. Dielectric Constants of Some Selected Materials

ℏω (eV)	Au ε'	Au ε''	Ag[a] ε'	Ag[a] ε''	Cu ε'	Cu ε''	Pt ε'	Pt ε''	Pd ε'	Pd ε''	Fused quartz ε'	H
1.6	−22.1	1.1	−29.0	0.4	−23.3	1.1	−23.1	25.1	−19.0	21.5	2.11	1.
1.7	−18.6	1.0	−25.8	0.4	−19.6	1.0	−21.3	22·6	−17.4	19.8	2.12	1.
1.8	−15.5	0.9	−22.1	0.4	−16.4	0.9	−19.4	20.2	−15.8	17.9	2.12	1.
1.9	−12.9	1.0	−19.5	0.5	−13.6	0.9	−17.9	18.5	−14.5	16.6	2.12	1.
2.0	−10.6	1.1	−17.4	0.5	−10.9	0.9	−16.5	16.7	−13.3	15.2	2.12	1.
2.1	−8.7	1.2	−15.5	0.5	−8.7	1.4	−15.2	15.4	−12.3	14.2	2.13	1.
2.2	−6.8	1.5	−13.8	0.4	−6.1	3.3	−13.9	14.1	−11.2	13.1	2.13	1.
2.3	−5.4	1.9	−12.3	0.4	−5.3	4.7	−12.9	13.1	−10.4	12.3	2.13	1.
2.4	−3.7	2.3	−10.9	0.4	−5.0	5.1	−11.8	12.1	−9.7	11.4	2.14	1.
2.5	−2.1	3.7	−9.7	0.3	−4.8	5.2	−11.0	11.4	−9.0	10.8	2.14	1.
2.6	−1.3	4.2	−8.6	0.3	−4.5	5.1	−10.2	10.5	−8.3	10.1	2.14	1.
2.7	−1.1	4.9	−7.6	0.3	−4.1	5.1	−9.5	9.9	−7.8	9.6	2.15	1.
2.8	−1.1	5.4	−6.7	0.2	−3.6	5.0	−8.9	9.3	−7.3	9.0	2.15	1.
2.9	−1.1	5.6	−5.9	0.2	−3.2	5.0	−8.3	8.8	−6.8	8.6	2.15	1.
3.0	−1.0	5.8	−5.2	0.2	−2.8	4.9	−7.7	8.3	−6.3	8.1	2.16	1.
3.1	−1.0	5.8	−4.5	0.2	−2.4	4.9	−7.3	7.9	−6.0	7.8	2.16	1.
3.2	−0.9	5.8	−3.8	0.2	−2.1	4.8	−6.8	7.4	−5.6	7.3	2.17	1.
3.3	−0.8	5.8	−3.2	0.2	−1.8	4.9	−6.4	7.1	−5.2	7.1	2.17	1.
3.4	−0.6	5.7	−2.6	0.2	−1.5	4.9	−6.0	6.7	−4.8	6.7	2.18	1.
3.5	−0.4	5.7	−2.0	0.3	−1.3	4.9	−5.6	6.4	−4.5	6.5	2.18	1.
3.6	−0.3	5.9	−1.4	0.3	−1.1	4.9	−5.3	6.1	−4.3	6.2	2.18	1.
3.7	−0.4	6.0	−0.9	0.3	−1.0	4.9	−5.0	5.9	−4.0	6.0	2.19	1.
3.8	−0.5	6.0	−0.1	0.4	−0.9	4.8	−4.6	5.6	−3.8	5.8	2.19	1.
3.9	−0.6	5.9	0.8	0.8	−0.7	4.8	−4.4	5.4	−3.5	5.6	2.20	1.
4.0	−0.8	5.9	0.9	1.5	−0.5	4.8	−4.1	5.1	−3.3	5.4	2.21	1.
4.1	−0.9	5.7	0.9	2.4	−0.3	4.8	−3.9	4.9	−3.1	5.3	2.21	1.
4.2	−0.9	5.6	0.7	3.0	−0.1	4.9	−3.6	4.7	−2.9	5.1	2.22	1.
4.3	−1.0	5.4	0.5	3.4	0.0	5.0	−3.4	4.5	−2.7	5.0	2.22	1.
4.4	−1.1	5.2	0.3	3.6	0.0	5.2	−3.2	4.3	−2.6	4.9	2.23	1.
4.5	−1.1	5.0	0.2	3.8	0.0	5.4	−3.0	4.2	−2.4	4.8	2.24	1.
4.6	−1.1	4.8	0.0	3.8	−0.1	5.5	−2.8	4.0	−2.3	4.7	2.25	1.
4.7	−1.0	4.5	−0.1	3.8	−0.3	5.6	−2.6	3.9	−2.3	4.6	2.25	1.
4.8	−1.0	4.3	−0.2	3.7	−0.5	5.7	−2.4	3.7	−2.2	4.5	2.26	1.
4.9	−0.9	4.2	−0.2	3.7	−0.8	5.7	−2.2	3.6	−2.1	4.3	2.27	1.
5.0	−0.8	4.0	−0.2	3.7	−1.0	5.6	−2.0	3.5	−2.1	4.2	2.28	1.
5.1	−0.6	3.9	−0.2	3.6	−1.2	5.4	−1.9	3.4	−2.0	4.1	2.29	1.
5.2	−0.5	3.7	−0.2	3.6	−1.4	5.3	−1.7	3.3	−2.0	3.9	2.29	1.
5.3	−0.4	3.6	−0.2	3.5	−1.5	5.0	−1.5	3.2	−1.8	3.8	2.30	1.
5.4	−0.3	3.5	−0.2	3.4	−1.6	4.8	−1.4	3.1	−1.8	3.6	2.31	1.
5.5	−0.3	3.5	−0.2	3.4	−1.7	4.6	−1.2	3.0	−1.7	3.6	2.32	1.

[a] Ref. 20.

List of Symbols

D	dielectric displacement vector
d	film thickness
E	energy
E_F	Fermi energy
E	electric field vector
e	unit vector of electric field
ER	electroreflectance
I	light intensity
Im (x)	imaginary part of function x
k	extinction coefficient
k	wave vector
l_{TF}	Thomas–Fermi screening length
N	electron density/number of oscillators
$\hat{n}$	complex refractive index
n	refractive index
pzc	potential of zero charge
R	reflectivity
$R(0) \equiv R_{13}$	reflectivity of bare surface
$R(d) \equiv R_{123}$	reflectivity of surface covered with a film of thickness d
r	Fresnel reflection coefficient
SP	surface plasmon
U	electrode potential
v_F	Fermi velocity
α	absorption coefficient
$\tilde{\alpha}$	polarizability
$\beta = (3/5) \cdot v_F^2$	coefficient of spatial dispersion
γ	damping constant
δ^r	phase change due to reflection
$\hat{\varepsilon}$	complex dielectric function
ε'	real part of complex dielectric function
ε''	imaginary part of complex dielectric function
λ	vacuum wavelength of light
$\xi_j = \hat{n}_j \cdot \cos \varphi_j$	~normal component of wave vector in phase j
θ	coverage
φ_j	angle of incidence for phase j
φ_1	(real) angle of incidence in ambient phase (1)
σ	optical conductivity
$\tau = \gamma^{-1}$	lifetime
χ	electric susceptibility ($\chi = \varepsilon - 1$)
ω	angular frequency of light
ω_n	plasma frequency of free-electron metal $[\omega_n = (4\pi Ne^2/m^*)^{1/2}]$

ω_p plasma frequency of nearly-free-electron metal $(\omega_p = \omega_n/\sqrt{\varepsilon_b'})$

ω_{sp} surface plasma frequency

ω_s maximum value of ω_{sp} (frequency limit of dispersion curve)

Indices:

$s, \perp$	perpendicular to plane of incidence
$p, \parallel$	parallel to plane of incidence
n, t	normal, tangential to surface
t, l	transverse, longitudinal (nonlocal optics)
$1, 2, 3$	ambient, film, substrate
m	metal
f, b	free-, bound-electron contribution

References

1. T. E. Furtak, K. L. Kliewer, and D. W. Lynch (eds.), Non-Traditional Approaches to the Study of the Solid-Electrolyte Interface, *Surface Sci.*, Vol. 101 (1980).
2. W. N. Hansen, D. M. Kolb, and D. W. Lynch (eds.), Electronic and Molecular Structure of Electrode–Electrolyte Interfaces, *J. Electroanal. Chem.*, Vol. 150 (1983).
3. H. Gerischer and D. M. Kolb (eds.), Structure and Dynamics of Solid/Electrolyte Interfaces, *Ber. Bunsenges. Phys. Chem.* **91**, 262–496 (1987).
4. E. Yeager, A. Homa, B. D. Cahan, and D. Scherson, *J. Vac. Sci. Technol.* **20**, 628 (1982).
5. D. M. Kolb, *J. Vac. Sci. Technol.* **A4**, 1294 (1986).
6. W. Schmickler, *J. Electroanal. Chem.* **150**, 19 (1983).
7. D. F. A. Koch and D. E. Scaife, *J. Electrochem. Soc.* **113**, 302 (1966).
8. J. Feinleib, *Phys. Rev. Lett.* **16**, 1200 (1966).
9. T. Takamura, K. Takamura, W. Nippe, and E. Yeager, *J. Electrochem. Soc.* **117**, 626 (1970).
10. T. Takamura, Y. Sato, and K. Takamura, *J. Electroanal. Chem.* **41**, 31 (1973).
11. J. D. E. McIntyre and D. E. Aspnes, *Surface Sci.* **24**, 417 (1971).
12. J. D. E. McIntyre and D. M. Kolb, *Symp. Faraday Soc.* **4**, 99 (1970).
13. D. M. Kolb and J. D. E. McIntyre, *Surface Sci.* **28**, 321 (1971).
14. T. E. Furtak and D. W. Lynch, *Phys. Rev. Lett.* **35**, 960 (1975).
15. J. D. E. McIntyre, in : Advances in Electrochemistry and Electrochemical Engineering (R. H. Muller, ed.), Vol. 9, p. 61, Wiley, New York (1973).
16. F. Wooten, *Optical Properties of Solids*, Academic Press, New York (1972).
17. R. E. Hummel, *Optische Eigenschaften von Metallen und Legierungen*, Springer-Verlag, Berlin (1971).
18. J. A. Stratton, *Electromagnetic Theory*, McGraw-Hill, New York (1941).
19. See, e.g., M. Cardona, in: *Optical Properties of Solids* (S. Nudelman and S. S. Mitra, eds.), p. 137, Plenum Press, New York (1969).
20. P. B. Johnson and R. W. Christy, *Phys. Rev. B* **6**, 4370 (1972).
21. M. Cardona, *Modulation Spectroscopy*, Academic Press, New York (1969).
22. See, e.g., Ref. 16, p. 61.
23. H. Raether, *Excitation of Plasmons and Interband Transitions by Electrons*, Springer Tracts in Modern Physics, Vol. 88, Springer, Berlin (1980).

24. H. Ehrenreich and H. R. Philipp, *Phys. Rev.* **128**, 1622 (1962).
25. W. R. Hunter, *J. Opt. Soc. Am.* **55**, 1197 (1965).
26. W. N. Hansen, *J. Opt. Soc. Am.* **58**, 380 (1968).
27. J. K. Sass and H. Laucht, *J. Electroanal. Chem.* **67**, 260 (1976).
28. See, e.g., P. J. Hyde, C. J. Maggiore, A. Redondo, S. Srinivasan, and S. Gottesfeld, *J. Electroanal. Chem.* **186**, 267 (1985).
29. D. M. Kolb, *J. Phys. (Paris)* **38**, C5-167 (1977).
30. D. M. Kolb, D. Leutloff, and M. Przasnyski, *Surface Sci.* **47**, 622 (1975).
31. T. Takamura, K. Takamura, and F. Watanabe, *Surface Sci.* **44**, 93 (1974).
32. R. Kötz, Thesis, Technical University, Berlin (1978).
33. E. D. Palik (ed.), *Handbook of Optical Constants of Solids*, Academic Press, Orlando (1985).
34. W. J. Plieth and K. Naegele, *Surface Sci.* **50**, 53 (1975).
35. K. Naegele and W. J. Plieth, *Surface Sci.* **50**, 64 (1975).
36. See, e.g., Ref. 33, pp. 56-58 and references cited therein.
37. K. Naegele and W. J. Plieth, *Surface Sci.* **61**, 504 (1976).
38. W. J. Plieth and K. Naegele, *Surface Sci.* **64**, 484 (1977).
39. W. J. Plieth, *Dechema Monographie* **90**, 69 (1981).
40. M. J. Dignam, M. Moskovits, and R. W. Stobie, *Trans. Faraday Soc.* **67**, 3306 (1971).
41. W. J. Plieth, *Isr. J. Chem.* **18**, 105 (1979).
42. A. J. Bennett and D. Penn, *Phys. Rev. B* **11**, 3644 (1975).
43. F. Forstmann and R. R. Gerhardts, in: *Festkörperprobleme (Adv. in Solid State Phys.)*, Vol. XXII, p. 291, Vieweg, Braunschweig (1982).
44. K. L. Kliewer, *Surface Sci.* **101**, 57 (1980).
45. P. J. Feibelman, *Progr. Surface Sci.* **12**, 287 (1982).
46. R. Kötz, D. M. Kolb, and F. Forstmann, *Surface Sci.* **91**, 489 (1980).
47. R. Kötz and D. M. Kolb, *Surface Sci.* **97**, 575 (1980).
48. F. Forstmann, K. Kempa, and D. M. Kolb, *J. Electroanal. Chem.* **150**, 241 (1983).
49. F. Forstmann and H. Stenschke, *Phys. Rev. Lett.* **38**, 1365 (1977); *Phys. Rev. B* **17**, 1489 (1978).
50. F. Forstmann and R. R. Gerhardts, *Metal Optics near the Plasma Frequency*, Springer Tracts in Modern Physics, Vol. 109, Springer, Berlin (1986).
51. K. L. Kliewer, *Phys. Rev. B* **14**, 1412 (1976).
52. H. Petersen and S. B. M. Hagström, *Phys. Rev. Lett.* **41**, 1314 (1978).
53. K. Kempa and F. Forstmann, *Surface Sci.* **129**, 516 (1983).
54. D. E. Aspnes and N. Bottka, in: *Semiconductors and Semimetals* (R. K. Willardson and A. C. Beer, eds.), Vol. 9, p. 457, Academic Press, New York (1972).
55. A. Bewick and A. M. Tuxford, *Symp. Faraday Soc.* **4**, 114 (1970).
56. F. T. Wagner and P. N. Ross, *J. Electroanal. Chem.* **150**, 141 (1983).
57. D. M. Kolb and R. Kötz, *Surface Sci.* **64**, 698 (1977).
58. J. D. E. McIntyre, in: *Optical Properties of Solids—New Developments* (B. O. Seraphin, ed.), p. 555, North-Holland, Amsterdam (1976).
59. See, e.g., R. Dornhaus, M. B. Long, R. E. Benner, and R. K. Chang, *Surface Sci.* **93**, 240 (1980).
60. D. M. Kolb and H. Gerischer, *Electrochim. Acta* **18**, 987 (1973).
61. T. Takamura, K. Takamura, and E. Yeager, *J. Electroanal. Chem.* **29**, 279 (1971).
62. D. Dickertmann, F. D. Koppitz, and J. W. Schultze, *Electrochim. Acta* **21**, 967 (1976).
63. P. O. Nilsson and D. E. Eastman, *Phys. Scri.* **8**, 113 (1973).
64. D. M. Kolb and R. Kötz, *Surface Sci.* **64**, 96 (1977).
65. M. S. Zei, Y. Nakai, G. Lehmpfuhl, and D. M. Kolb, *J. Electroanal. Chem.* **150**, 201 (1983).
66. J. O'M. Bockris, M. A. Devanathan, and K. Müller, *Proc. R. Soc. Ser. A* **274**, 55 (1963).
67. N. D. Lang and W. Kohn, *Phys. Rev. B* **7**, 3541 (1973).
68. W. N. Hansen and A. Prostak, *Phys. Rev.* **174**, 500 (1968).
69. J. D. E. McIntyre, *Surface Sci.* **37**, 658 (1973).
70. R. Kötz and D. M. Kolb, *Z. Phys. Chem. N.F.* **112**, 69 (1978).
71. D. M. Kolb, *J. Phys. (Paris)* **44**, C10-137 (1983).

72. D. W. Lynch, *Surface Sci.* **103**, 289 (1981).
73. T. E. Furtak and D. W. Lynch, *J. Electroanal. Chem.* **79**, 1 (1977).
74. C. N. van Huong, C. Hinnen, J. LeCoeur, and R. Parsons, *J. Electroanal. Chem.* **92**, 239 (1978). Note that the crystallographic directions for Au(110) have been interchanged.
75. R. Kofman, R. Garrigos, and P. Cheyssac, *Surface Sci.* **101**, 231 (1980).
76. R. Kötz and H. J. Lewerenz, *Surface Sci.* **78**, L233 (1978).
77. R. Smoluchowski, *Phys. Rev.* **60**, 661 (1941).
78. J. D. E. McIntyre and W. F. Peck, *Faraday Discuss. Chem. Soc.* **56**, 122 (1973).
79. K. Kempa, *Surface Sci.* **157**, L323 (1985).
80. F. Chao and M. Costa, *Surface Sci.* **135**, 497 (1983).
81. W. Shockley, *Phys. Rev.* **56**, 317 (1939).
82. See, e.g., G. Chiarotti, S. Nannarone, R. Pastore, and P. Chiaradia, *Phys. Rev. B* **4**, 3398 (1971).
83. K. M. Ho, B. N. Harmon, and S. H. Liu, *Phys. Rev. Lett.* **44**, 1531 (1980).
84. D. M. Kolb, W. Boeck, K. M. Ho, and S. H. Liu, *Phys. Rev. Lett.* **47**, 1921 (1981).
85. K. M. Ho, C. L. Fu, S. H. Liu, D. M. Kolb, and G. Piazza, *J. Electroanal. Chem.* **150**, 235 (1983).
86. S. H. Liu, C. Hinnen, C. N. van Huong, N. R. DeTacconi, and K. M. Ho, *J. Electroanal. Chem.* **176**, 325 (1984).
87. A. Goldmann, V. Dose, and G. Borstel, *Phys. Rev. B* **32**, 1971 (1985).
88. W. Jacob, V. Dose, U. Kolac, T. Fauster, and A. Goldmann, *Z. Physik* **B63**, 459 (1986).
89. W. Boeck and D. M. Kolb, *Surface Sci.* **118**, 613 (1982).
90. R. Kofman, R. Garrigos, and P. Cheyssac, *Thin Solid Films* **82**, 73 (1981).
91. P. Heimann, H. Neddermeyer, and H. F. Roloff, *J. Phys. C* **10**, L17 (1977).
92. B. Reihl, R. R. Schlittler, and H. Neff, *Phys. Rev. Lett.* **52**, 1826 (1984).
93. B. Reihl, K. H. Frank, and R. R. Schlittler, *Phys. Rev. B* **30**, 7328 (1984).
94. G. Binnig, K. H. Frank, H. Fuchs, N. Garcia, B. Reihl, H. Rohrer, F. Salvan, and A. R. Williams, *Phys. Rev. Lett.* **55**, 991 (1985).
95. See, e.g., J. Billmann and A. Otto, *Solid State Commun.* **44**, 105 (1982).
96. G. Piazza and D. M. Kolb, unpublished.
97. S. Trasatti, *J. Electroanal. Chem.* **33**, 351 (1971).
98. K. M. Ho, private communication.
99. See, e.g., G. A. Somorjai, *Chemistry in Two Dimensions*, Cornell University, Ithaca, New York (1981), pp. 143–146.
100. M. A. Van Hove, R. J. Koestner, P. C. Stair, J. P. Biberian, L. L. Kesmodel, I. Bartos, and G. A. Somorjai, *Surface Sci.* **103**, 189, 218 (1981).
101. D. M. Kolb and J. Schneider, *Surface Sci.* **162**, 764 (1985).
102. D. M. Kolb and J. Schneider, *Electrochim. Acta* **31**, 929 (1986).
103. A. Otto, in: *Optical Properties of Solids—New Developments* (B. O. Seraphin, ed.), p. 677, North-Holland, Amsterdam (1976).
104. H. Raether, in: *Physics of Thin Films* (G. Hass, ed.), Vol. 9, p. 145, Academic Press, New York (1977).
105. D. M. Kolb, in: *Surface Polaritons* (V. M. Agranovich and D. L. Mills, eds.), p. 299, North-Holland, Amsterdam (1982).
106. A. Otto, *Z. Physik* **216**, 398 (1968).
107. E. Kretschmann, *Z. Physik* **241**, 313 (1971).
108. H. Raether, in: *Surface Polaritons* (V. M. Agranovich and D. L. Mills, eds.), p. 331, North-Holland, Amsterdam (1982).
109. E. Kretschmann, *Opt. Commun.* **26**, 41 (1978).
110. A. Tadjeddine, D. M. Kolb, and R. Kötz, *Surface Sci.* **101**, 277 (1980).
111. A. Tadjeddine and D. M. Kolb, unpublished.
112. R. Kötz, D. M. Kolb, and J. K. Sass, *Surface Sci.* **69**, 359 (1977).
113. F. Abeles, T. Lopez-Rios, and A. Tadjeddine, *Solid State Commun.* **16**, 843 (1975).
114. A. Tadjeddine and D. M. Kolb, *Proc. 4th Int. Conf. on Solid Surfaces*, Cannes, 1980, Vol. I, p. 615, Supplement "Le Vide, les Couches Minces" No. 201.

115. A. Otto, *Appl. Surface Sci.* **6**, 309 (1980).
116. M. Moskovits, *Solid State Commun.* **32**, 59 (1979).
117. D. M. Kolb, unpublished.
118. A. Girlando, M. R. Philpott, D. Heitmann, J. D. Swalen, and R. Santo, *J. Chem. Phys.* **72**, 5187 (1980).
119. A. Bewick and J. Robinson, *J. Electroanal. Chem.* **60**, 163 (1975), **71**, 131 (1976).
120. C. Hinnen, N. van Huong, A. Rousseau, and J. P. Dalbera, *J. Electroanal. Chem.* **95**, 131 (1979).
121. D. Dickertmann, J. W. Schultze, and K. J. Vetter, *J. Electroanal. Chem.* **55**, 429 (1974).
122. H. Angerstein-Kozlowska, B. E. Conway, and W. B. A. Sharp, *J. Electroanal. Chem.* **43**, 9 (1973).
123. W. Boeck, Thesis, Technical University, Berlin (1981).
124. D. M. Kolb, R. Kötz, and D. L. Rath, *Surface Sci.* **101**, 490 (1980).
125. K. Yamamoto, D. M. Kolb, R. Kötz, and G. Lehmpfuhl, *J. Electroanal. Chem.* **96**, 233 (1979).
126. W. J. Plieth, H. Bruckner, and H. J. Hensel, *Surface Sci.* **101**, 261 (1980).
127. T. Watanabe and H. Gerischer, *J. Electroanal. Chem.* **117**, 185 (1981).
128. D. N. Buckley and L. D. Burke, *J. Chem. Soc. Faraday Trans. 1* **71**, 1447 (1975).
129. R. Kötz and H. Neff, *Surface Sci.* **160**, 517 (1985).
130. S. Trasatti and W. E. O'Grady, in: *Advances in Electrochemistry and Electrochemical Engineering* (H. Gerischer and C. W. Tobias, eds.), Vol. 12, p. 177, Wiley, New York (1981).
131. R. E. Connick and C. R. Hurley, *J. Am. Chem. Soc.* **74**, 5012 (1952).
132. R. Kötz, S. Stucki, D. Scherson, and D. M. Kolb, *J. Electroanal. Chem.* **172**, 211 (1984).
133. R. Adzić, E. Yeager, and B. D. Cahan, *J. Electroanal. Chem.* **85**, 267 (1977).
134. C. N. van Huong, C. Hinnen, and A. Rousseau, *J. Electroanal. Chem.* **151**, 149 (1983).
135. See, e.g., J. G. Gordon and S. Ernst, *Surface Sci.* **101**, 499 (1980).
136. W. N. Hansen, *Surface Sci.* **101**, 109 (1980).
137. D. L. Rath and W. N. Hansen, *Surface Sci.* **136**, 195 (1984).
138. W. J. Anderson and W. N. Hansen, *J. Electroanal. Chem.* **43**, 329 (1973).
139. B. Pettinger, U. Wenning, and D. M. Kolb, *Ber. Bunsenges. Phys. Chem.* **82**, 1326 (1978).
140. J. Schneider, D. M. Kolb, and A. Otto, unpublished.
141. A. Tadjeddine and D. M. Kolb, *J. Electroanal. Chem.* **111**, 119 (1980).
142. N. Collas, B. Beden, J. M. Leger, and C. Lamy, *J. Electroanal. Chem.* **186**, 287 (1985).
143. H. Gerischer, *Faraday Discuss. Chem. Soc.* **58**, 219 (1974).
144. H. Labhart, *Adv. Chem. Phys.* **13**, 179 (1967); W. Liptay, *Angew. Chem.* **81**, 195 (1969).
145. W. J. Plieth, P. Gruschinske, and H. J. Hensel, *Ber. Bunsenges. Phys. Chem.* **82**, 615 (1978).
146. P. Schmidt and W. J. Plieth, *J. Phys. (Paris)* **44**, C10-171 (1983).
147. P. H. Schmidt and W. J. Plieth, *J. Electroanal. Chem.* **201**, 163 (1986).
148. R. Memming, *Faraday Discuss. Chem. Soc.* **58**, 261 (1974).
149. See, e.g., D. M. Kolb, in: *Advances in Electrochemistry and Electrochemical Engineering* (H. Gerischer and C. W. Tobias, eds.), Vol. 11, p. 125, Wiley, New York (1978).
150. W. J. Lorenz, H. D. Hermann, N. Wüthrich, and F. Hilbert, *J. Electrochem. Soc.* **121**, 1167 (1974).
151. D. M. Kolb, R. Kötz, and K. Yamamoto, *Surface Sci.* **87**, 20 (1979).
152. See, e.g., E. Bauer, H. Poppa, G. Todd, and F. Bonczek, *J. Appl. Phys.* **45**, 5164 (1974).
153. J. W. Schultze and D. Dickertmann, *Surface Sci.* **54**, 489 (1976).
154. K. Jüttner and W. J. Lorenz, *Z. Phys. Chem. N.F.* **122**, 163 (1980).
155. T. Takayanagi, D. M. Kolb, K. Kambe, and G. Lehmpfuhl, *Surface Sci.* **100**, 407 (1980).
156. W. Heiland, F. Iberl, E. Taglauer, and D. Menzel, *Surface Sci.* **53**, 383 (1975).
157. D. Aberdam, R. Durand, R. Faure, and F. El Omar, *Surface Sci.* **162**, 782 (1985).
158. F. Chao and M. Costa, *Thin Solid Films* **82**, 3 (1981).
159. G. Piazza, D. M. Kolb, K. Kempa, and F. Forstmann, *Solid State Commun.* **51**, 905 (1984).
160. P. Bindra, H. Gerischer, and D. M. Kolb, *J. Electrochem. Soc.* **124**, 1012 (1977).
161. W. Gaebler, K. Jacobi, and W. Ranke, *Surface Sci.* **75**, 355 (1978).
162. D. Schmeisser and K. Jacobi, *Surface Sci.* **88**, 138 (1979).

163. D. M. Kolb, *Ber. Bunsenges. Phys. Chem.* **77**, 891 (1973).
164. D. M. Kolb, *Faraday Disc. Chem. Soc.* **56**, 138 (1973).
165. O. S. Heavens, *Optical Properties of Thin Solid Films*, Butterworths, London (1955), p. 177.
166. P. L. Clegg, *Proc. Phys. Soc. (London)* **65B**, 774 (1952).
167. D. K. Roe, J. K. Sass, D. S. Bethune, and A. C. Luntz, *J. Electroanal. Chem.* **216**, 293 (1987).
168. R. H. Muller, in: *Advances in Electrochemistry and Electrochemical Engineering* (R. H. Muller, ed.), Vol. 9, p. 167, Wiley, New York (1973).
169. L. M. Peter, *Ber. Bunsenges. Phys. Chem.* **91**, 419 (1987).
170. Yu. Ya. Gurevich, Yu. V. Pleskov, and Z. A. Rotenberg, *Photoelectrochemistry*, Consultants Bureau, New York (1980).
171. D. L. Rath, *J. Electroanal. Chem.* **150**, 521 (1983).
172. See, e.g., R. H. Muller and J. C. Farmer, *Surface Sci.* **135**, 521 (1983).
173. J. Horkans, B. D. Cahan, and E. Yeager, *Surface Sci.* **46**, 1 (1974).

5

Infrared Reflectance Spectroscopy

Bernard Beden and Claude Lamy

1. Introduction and Historical Survey

The structure of the electrode/electrolyte interface plays an important, growing role in electrochemistry and electrocatalysis. On one hand, the distribution of charged particles and dipolar molecules in the double layer (i.e., the transition region between the surface of the metal electrode and the bulk of the electrolyte solution) under the combined influence of diffusion and potential gradients determines the potential barrier, which strongly influences the rate of electrochemical reactions, i.e., reactions involving charge transfer through the interface.[1] On the other hand, many electrode reactions proceed through adsorbed intermediates, which are produced during chemisorption of the reacting molecules (the so-called electroactive species) on the electrode surface. The mechanisms of these electrocatalytic reactions will therefore greatly depend on the nature of the electrode material, which determines the structure of the adsorbed intermediates.[2]

Before the eighties, most of the *in situ* investigations concerning the structure of the electrode/electrolyte interface or the nature of adsorbed intermediates were performed using electrochemical measurements, which give electrical currents as a response of the interface perturbed by a carefully chosen potential program.[3] A typical example, linear sweep voltammetry (or cyclic voltammetry), consists in applying an electrode potential E as a single (or repetitive) triangular waveform of sufficient amplitude, i.e., of the order of a few volts, and in recording the electrical current, I, flowing through the interface versus E (voltammogram). The reader may refer to Figure 25, which displays the voltammogram of a platinum electrode in contact with 0.5 M H_2SO_4. By numerical integration of the current–potential (time) relationship, one obtains the quantity of electricity associated with the electrochemical process under

Bernard Beden and Claude Lamy • Laboratory of Chemistry I—Electrochemistry and Interactions, University of Poitiers, U.A. CNRS No. 350, 86022 Poitiers, France.

study, since, during a single sweep, E is proportional to time t ($v = dE/dt$ is called the sweep rate). Then, knowing either the number of molecules involved (e.g., by radiotracer methods) or the number of active sites on which the molecule is adsorbed, one may calculate the number of electrons per molecule (epm) involved or the number of electrons per site (eps), respectively. Finally, with the help of these data, one may attempt to guess the nature and the structure of the intermediates involved. But the accuracy of such determinations is rather poor and 50 to 100% errors are usual. Therefore, it is extremely difficult, if not impossible, to determine the reaction mechanisms only by electrical measurements on a macroscopic scale, which explains numerous controversies in the literature.[4]

The need for experimental methods allowing investigations of the electrode/electrolyte interface at the molecular level has become acute during the last two decades, in which period many sophisticated techniques were developed for the solid/vacuum and solid/gas interfaces. More precisely, it has become apparent that physicochemical methods have to be used in conjunction with electrochemical methods, or at least with an electrochemical control of the electrode surface by applying a suitable potential program. This last point is of major importance in obtaining experimental data which are significant for the electrochemical experiments. Furthermore, these physicochemical methods, which must operate *in situ*, should be able to provide detailed information in the following areas:

• the properties of the electrode surface
• the structure and constitution of the double layer
• the nature and the structure of adsorbed species
• their quantity at the electrode surface, i.e., the coverage of the electrode surface
• their interaction between themselves, and with the substrate

Most of the powerful techniques developed at the solid/low-pressure gas interface, such as low-energy electron diffraction (LEED), grazing incidence high-energy electron diffraction (RHEED), Auger electron spectroscopy (AES), X-ray photoelectron spectroscopy (XPS), and electron energy loss spectroscopy (EELS), are not directly applicable for *in situ* investigations of the electrode/electrolyte interface because they need high- or ultrahigh-vacuum conditions. However, some of these *in vacuo* techniques may be used in electrochemical studies by means of transfer experiments. However, these experiments become very delicate and highly sophisticated, and the question arises of whether the structure of the electrode/electrolyte interface is perturbed or not, either by the ultrahigh-vacuum environment or by contamination of the electrode surface during transfer.[5]

Of the different remaining experimental methods that may be used for *in situ* investigations of the electrode/electrolyte interface, such as electron spin resonance (ESR),[6] Mössbauer spectroscopy,[7] or optical spectroscopies,[8] the latter are actually the most convenient. These techniques, which are extremely sensitive, are able to follow changes in surface coverage down to a few percent of a monolayer. They can be very selective, because light is absorbed only by molecules or species whose difference in energy levels matches its frequency. Finally, in certain circumstances, the response time is very small (below 1 ms), thus allowing kinetic studies.

Due to its very nature, the electrode/electrolyte interface may conveniently be studied by reflection–absorption spectroscopy. The first attempts in the infrared wavelength range were made with internal reflection spectroscopy. This allows multiple reflections at the electrode surface to increase the signal, which was otherwise too weak for direct measurement.[9,10] However, due to inherent difficulties of this method (e.g., the need for a transparent substrate, the necessity for a thin metal layer as electrode), specular external reflection spectroscopy now is preferred for the *in situ* investigation of electrode processes.

Specular reflection spectroscopy in the UV–visible wavelength range has been developed since the mid-sixties for *in situ* studies of the electrode/electrolyte interface. The technique, which uses potential modulation,[11] has reached a degree of development now that makes it possible to detect very small changes in reflectivity, of the order of 10^{-6}, due to the presence of a surface film or an adsorbed species.[12] Moreover, the use of a rapid-scan spectrometer (e.g., the Harrick RSS) permits rapid changes in concentration to be followed, and thus the rates of chemical and/or electrochemical reactions occurring in the layer adjacent to the electrode surface can be monitored.[13] However, UV–visible spectroscopy suffers major drawbacks due to its inability to identify adsorbed intermediates which do not absorb light in this spectral region and the difficulty in determining the electronic structure of the electrode surface without a model. Conversely, specular reflection spectroscopy in the infrared range appears to be an ideal tool for the identification of molecules adsorbed at the electrode/electrolyte interface. This is because the molecules in the adsorbed state retain their infrared fingerprints, which arise from the characteristic frequencies associated with the vibrational modes of the functional groups that they contain.

As early as 1975, infrared specular reflection spectroscopy was recognized as a powerful technique for investigating adsorbed molecules at the solid/gas interface.[14] Using a polarization modulation technique, high sensitivity is achieved, typically, 10^{-3} to 10^{-5} absorbance units, sufficient to detect as low as 10^{-3} of a monolayer for adsorbed species with a strong absorption band (such as CO).[15] This ability to study adsorbates on single-crystal surfaces under controlled conditions thus provides a link between the highly sophisticated *in vacuo* techniques and vibrational spectroscopy investigations that affords fundamental information at the molecular level about the structure of solid/gas interfaces.[16]

The development of infrared reflection spectroscopy for studying the electrode/electrolyte interface is more recent, since it was believed that the presence of a solvent—particularly water, which displays strong IR absorption bands—precluded the use of such techniques.[17] However, the ability of modulation techniques to recover very small signals buried in a large-amplitude background makes it possible to detect signals of the magnitude expected for infrared spectra of adsorbed molecules at the electrode/electrolyte interfaces. Besides the usual modulation techniques available for the solid/gas interface, such as polarization modulation or wavelength modulation, modulation of the electrode potential confers on infrared reflection spectroscopy a unique ability to study *in situ* electrode/electrolyte interfaces. Potential modulation is particularly easy to achieve when the interface is controlled electrochemically by a technique such as voltammetry.

The first demonstration of the correctness of this approach was given quite recently by Bewick *et al.*, who reported infrared spectra of adsorbed hydrogen and of water molecules in the double-layer region of a platinum electrode in contact with 1 M H_2SO_4.[18] Since this pioneering work, realized with potential modulation and a dispersive spectrometer, the development of this technique has been tremendous. In particular, two directions have evolved:

- The improvement of the methodology, namely, the use of other modulation techniques such as polarization modulation and Fourier transform spectroscopy.
- The investigation of various aqueous and nonaqueous systems, including adsorption of molecules and ions at the electrode/electrolyte interface and oxide films on a metal substrate.

The first part of this chapter aims to provide a theoretical background to specular reflection spectroscopy at metal surfaces and to its application to *in situ* studies of the structure of the electrode/electrolyte interface. In the second part, the different techniques and experimental equipment employed will be described, with particular attention given to the design of the spectrochemical cell. Rather than reviewing all of the results recently obtained in this growing field, the third part will present some typical examples in detail. These examples will deal with adsorption of several important species (such as H, H_2O, CO, (CH_3OH, HCOOH) at the electrode/aqueous electrolyte interface, together with the adsorption of some organic molecules (TBAF, TCNE, etc.) from nonaqueous solvents.

Semiconductor electrodes will not be considered here, because the properties of the semiconductor/electrolyte interface are influenced by the existence of a space charge layer inside the semiconductor. Reflectance spectroscopy in the infrared range was applied early to the study of the semiconductor/electrolyte interface to determine the characteristics of this space charge layer (free carriers, surface states, etc.).[19] The reader interested in the status of this field is referred to the work of Seraphin.[20]

Besides infrared reflection spectroscopy, Raman spectroscopy, particularly surface-enhanced Raman spectroscopy (SERS), is able to provide information about the structure of the electrode/electrolyte interface through the identification of adsorbed species from their vibrational spectra.[21,22] However, Raman spectroscopy is not of as general use as infrared spectroscopy, and the basis of the enhancement mechanism, which is observed only for a limited number of systems (e.g., adsorption of pyridine on a silver electrode), is not yet very clear.[23] Surface-enhanced Raman scattering is discussed by Birke and Lombardi in Chapter 6 of this book.

2. Theory of Reflection–Absorption Spectroscopy

Infrared reflectance spectroscopy, like UV–visible reflectance spectroscopy, is based on the specular reflection of the incident light on the substrate surface. It is therefore relevant to give briefly, in this section, the basic equations describing the propagation of an electromagnetic plane wave in an absorbing medium and its reflection at the boundary which separates two contiguous phases of different optical properties.

2.1. Propagation of an Electromagnetic Plane Wave[24]

The propagation of an electromagnetic plane wave *in vacuo* is defined by its wave vector $\mathbf{k}_0 = (2\pi/\lambda_0)\,\mathbf{u} = 2\pi\,\bar{\nu}\mathbf{u}$, where $\mathbf{u}$ is the unit vector in the propagation direction, λ_0 the wavelength *in vacuo*, and $\bar{\nu} = 1/\lambda_0$ the wavenumber, and by its angular frequency $\omega = (2\pi/T) = 2\pi\nu$, where T is the period and ν the frequency. The wavelength and the frequency, which describe the periodicity of the electric field vector (and the magnetic field vector) in space and time, respectively, are related by the dispersion relations $\lambda_0 = cT$, or $\omega = ck_0$, where c is the velocity of light *in vacuo* ($c = 2.997925 \times 10^8 \text{ m s}^{-1}$). Typical values of λ_0, $\bar{\nu}$, ν, and the energy involved, E ($E = h\nu$, where $h = 6.62618 \times 10^{-34}$ J s is Planck's constant) are given in Table I for the ultraviolet, visible, and infrared regions.

TABLE I. Values for λ_0, $\bar{\nu}$, ν, and E

Region	λ (nm)	$\bar{\nu}$ (cm^{-1})	ν (Hz)	E (eV)^a
Ultraviolet				
Far	10–200	1×10^6–5×10^4	3×10^{16}–1.5×10^{15}	124–6.20
Near	200–380	5×10^4–2.63×10^4	1.5×10^{15}–7.9×10^{14}	6.20–3.26
Visible	380–780	2.63×10^4–1.28×10^4	7.9×10^{14}–3.8×10^{14}	3.26–1.59
Infrared				
Near	780–2500	1.28×10^4–4000	3.8×10^{14}–1.2×10^{14}	1.59–0.496
Middle	2500–5×10^4	4000–200	1.2×10^{14}–6×10^{12}	0.496–0.0248
Far	5×10^4–1×10^6	200–10	6×10^{12}–3×10^{11}	0.0248–1.24×10^{-3}

^{a} 1 eV ~ 1.6022×10^{-19} J ~ 96.485 kJ mol^{-1} ~ 8065.5 cm^{-1} ~ 2.4180×10^{14} Hz.

The instantaneous value of the oscillating electric field vector $\mathbf{E}$, obtained as a solution of Maxwell's equations, may be represented at a position vector $\mathbf{r}$ of space and at time t by the following expression:

$$\mathbf{E} = \mathbf{E}^0 \exp\left[i(\omega t - \mathbf{k}_0 \cdot \mathbf{r} + \delta)\right] \tag{1}$$

where $\mathbf{E}^0$ (components E_x^0, E_y^0, E_z^0) is the wave amplitude, δ is the phase angle usually taken equal to zero, and $i = \sqrt{-1}$. For convenience, $\mathbf{E}$ is treated as a complex function of the space coordinates x, y, z and time t, but one has to take the real part to deal with numerical solutions. The magnetic field vector $\mathbf{B}$ has a similar expression, but is not considered here, since in optical spectroscopy the interaction between the electromagnetic field and the molecules proceeds through the electric dipole moment $\mathbf{p}$ with the interaction energy $W = -\mathbf{p} \cdot \mathbf{E}$.

Electromagnetic waves travel through homogeneous, isotropic, nonabsorbing media with the phase velocity v, which is related to the index of refraction n by $n = c/v = (\varepsilon\mu)^{1/2}$, where ε is the optical dielectric constant and μ the magnetic permeability of the medium. For nonmagnetic materials, as is the usual case, μ can be taken equal to unity, and thus $n = \sqrt{\varepsilon}$. Since the phase velocity is always smaller than the light velocity in vacuum, the refractive index is greater than unity. The wavelength is modified according to $\lambda = vT = (c/n)T = \lambda_0/n$ (i.e., the dispersion relation becomes $\omega = vk$) and the wave vector $\mathbf{k}_0$ in Eq. (1) must be replaced by

$$\mathbf{k} = \frac{2\pi}{\lambda}\mathbf{u} = n\frac{2\pi}{\lambda_0}\mathbf{u} = n\mathbf{k}_0$$

If the medium is absorbing, the radiation propagates as an exponentially damped wave represented by the expression:

$$\mathbf{E} = \mathbf{E}^0 \exp\left[i(\omega t - n\mathbf{k}_0 \cdot \mathbf{r})\right] \exp\left[-k(\mathbf{k}_0 \cdot \mathbf{r})\right] \tag{2}$$

where k is the extinction coefficient $(k > 0)$. Introducing a complex refractive index $\hat{n} = n - ik$, Eq. (2) may be written by analogy with Eq. (1) as follows:

$$\mathbf{E} = \mathbf{E}^0 \exp\left[i(\omega t - \hat{n}\mathbf{k}_0 \cdot \mathbf{r})\right] \tag{3}$$

Since the refractive index and the dielectric constant are related by $n = \sqrt{\varepsilon}$, or $\varepsilon = n^2$, the dielectric constant $\hat{\varepsilon}$ of an absorbing medium is also complex, i.e.:

$$\hat{\varepsilon} = \hat{n}^2 = \varepsilon' - i\varepsilon''$$

with

$$\varepsilon' = n^2 - k^2 \quad \text{and} \quad \varepsilon'' = 2nk \tag{4}$$

The variables ε' and ε'' (like n and k) are not independent, but are linked together by the Kramers–Kronig relations. Taking the positive z-direction $0z$ as the propagation direction, the electric field vector $\mathbf{E}$ perpendicular to the propagation direction is characterized at the origin of the coordinates ($\mathbf{r} = \mathbf{0}$) by its two components E_x and E_y, since $E_z = 0$:

$$E_x = E_x^0 \cos (\omega t + \delta_x^0)$$
$$E_y = E_y^0 \cos (\omega t + \delta_y^0)$$

(5)

where δ_x^0 and δ_y^0 are the respective phase angles at $t = 0$ (and $\mathbf{r} = \mathbf{0}$). If either one of the wave amplitudes, E_x^0 or E_y^0, is equal to zero or if the phase angles δ_x^0 and δ_y^0 are equal (modulo π), linearly polarized light will result. If a constant phase angle difference, $\delta_x^0 - \delta_y^0 = \delta_0$, exists between the two components E_x and E_y, the light is elliptically polarized, since the extremity of the electric field vector traces an ellipse in the $x0y$ plane. For $E_x^0 = E_y^0$ and $\delta_x^0 - \delta_y^0 = \pi/2$ (modulo π), circularly polarized light will result. If the two components of $\mathbf{E}$ have a random phase relationship, random or unpolarized light results, as is the case for natural light.

Any electromagnetic wave with random polarization can be represented by the superposition of two plane waves, such as E_x and E_y, linearly polarized in two perpendicular directions.

2.2. Fundamentals of Absorption Spectroscopy. Selection Rules[25,26]

Classically, the light intensity is given by the flux of the Poynting vector, i.e., the light intensity I of a linearly polarized plane wave is proportional to the mean square of the electric field strength, $\langle \mathbf{E}^2 \rangle$. Using either Eq. (2) or (3), this gives the light intensity in the propagation direction $0z$, as follows:

$$I = I^0 \exp (-2kk_0 z) = I^0 \exp \left(-\frac{4\pi k}{\lambda_0} z \right)$$

where I^0 is the light intensity at the origin ($z = 0$). This is exactly the Beer-Lambert law:

$$I = I^0 \exp (-\alpha z)$$

(6)

Thus, the absorption coefficient α is related to the extinction coefficient k according to:

$$\alpha = 2kk_0 = \frac{4\pi k}{\lambda_0}$$

(7)

From the quantum mechanical viewpoint, light absorption occurs when a molecule initially in a state of energy E_1 is excited to an upper state of energy E_2 by a photon of energy $h\nu_0$, thereby satisfying the Planck relation:

$$h\nu_0 = \Delta E = E_2 - E_1$$

This gives a spectral line, at a frequency ν_0, the intensity of which is equal to the rate of energy absorption, i.e.:

$$I = \frac{dE}{dt} = h\nu(N_2 - N_1)\rho(\nu)B_{12}$$

(8)

where N_1 and N_2, the numbers of molecules in states with energy E_1 and E_2, respectively, are linked by the Boltzmann distribution, i.e.:

$$N_2 = N_1 \exp\left[-(E_2 - E_1)/k_B T\right]$$

where $k_B = 1.38066 \times 10^{-23} \text{ J K}^{-1}$ is the Boltzmann constant, T is the absolute temperature, $\rho(\nu)$ is the energy density of radiation, given by the Planck distribution:

$$\rho(\nu) = \frac{8\pi h\nu^3}{c^3}\left[\exp\left(\frac{h\nu}{k_B T}\right) - 1\right]^{-1}$$

and B_{12} is the Einstein coefficient for absorption, which is proportional to the transition probability per unit time. B_{12} can be calculated by resolving the time-dependent Schrödinger equation by the time-dependent perturbation method, i.e.:

$$B_{12} = \frac{4\pi^2}{h^2}\left|W_{12}\right|^2 \quad \text{with } W_{12} = \int \psi_1 \hat{W}\psi_2 \, d\tau$$

where W_{12} is the matrix element, taken on the basis of unperturbed quantum states ψ_1, ψ_2 of energy E_1, E_2, of the perturbation $\hat{W}$ due to the interaction of the electromagnetic field with the electric dipole moment ($\mathbf{p} = -e_0\mathbf{r}$) of the molecule. The perturbation energy is given by:

$$W = -\mathbf{p} \cdot \mathbf{E} = e_0\mathbf{r} \cdot \mathbf{E}$$

so that the transition probability involves the transition dipole moment $\mathbf{D}_{12} = \int \psi_1 \, e_0\mathbf{r}\psi_2 \, d\tau$, since $|W_{12}|^2 = |D_{12}|^2|\mathbf{E}|^2$, and the light intensity is proportional to $\langle \mathbf{E}^2 \rangle$.

The energy level E of a quantum state ψ is obtained by solving the time-independent Schrödinger equation in the Born–Oppenheimer approximation, where the energy contributions of electronic, vibrational, and rotational motions can be separated:

$$E = E_{\text{elec}} + E_{\text{vib}} + E_{\text{rot}} \tag{9}$$

The energy levels involved in the infrared range are associated with the vibrational energy E_{vib}, which for a harmonic oscillator is given by:

$$E_{\text{vib}} = (v + \tfrac{1}{2})h\nu_e \quad \text{with } v = 0, 1, 2, 3 \ldots \text{ an integer quantum number,}$$

and for an anharmonic oscillator by:

$$E_{\text{vib}} = (v + \tfrac{1}{2})h\nu_e - (v + \tfrac{1}{2})^2 x_e h\nu_e + \cdots \tag{10}$$

where $\nu_e = (1/2\pi)(F/\mu)^{1/2}$ is the fundamental frequency of a normal mode of vibration with the force constant F and reduced mass μ. The anharmonicity constant, $x_e = h\nu_e/4D_e$, is related to the spectroscopic heat of dissociation D_e (measured from the bottom of the potential energy curve). Some typical vibrational frequencies are given in Tables IIa and IIb.

TABLE IIa. Vibrational Wavenumbers
of Some Diatomic Molecules

Molecule	$\bar{\nu}$ (cm^{-1})
HF	3958
HCl	2885
HBr	2559
HI	2230
HD	3817
CO	2143
NO	1876

TABLE IIb. Vibrational Wavenumbers of Some Group
Vibrational Modes

Mode of vibration	Range of wavenumbers, $\bar{\nu}$ (cm^{-1})
C—H stretching	2850–2960
C—H bending	1340–1465
C—C stretching	700–1250
C=C stretching	1620–1680
C≡C stretching	2100–2260
O—H stretching	3590–3650
C=O stretching	1640–1780
C≡N stretching	2215–2275
N—H stretching	3200–3500
H-bonds	3200–3570

The number of normal modes of vibration is equal to the number of degrees of freedom in which the molecule can vibrate, i.e., $3N-6$ for a polyatomic molecule with N atoms, and $3N-5$ for a linear molecule.

Fine structure of spectra due to rotational motion is not considered here, since in both the condensed state and the adsorbed state, rotation of the molecule is restricted.

The selection rules are easily derived from Eq. (8), which states that the light intensity will be nonzero only if B_{12} is nonzero. This means that the electrical dipole moment must be nonzero and that the matrix element must be different from zero (selection rules). For a pure vibrational motion, this implies a relation between the quantum number of vibration, v, of the two energy states involved, i.e., $\Delta v = v_2 - v_1 = \pm 1$ for a harmonic oscillator (fundamental frequency), with overtone transitions $\Delta v = \pm 2, \pm 3, \ldots$, of much lower intensities for an anharmonic oscillator.

One last point worth mentioning is the variation of the absorbed intensity with the light frequency. This can be seen from Eq. (8), which shows that I is proportional to the number of molecules N and varies as ν^2, i.e., light intensity in the infrared range is about 10^2 lower (for $\lambda = 5$ μm or $\bar{\nu} =$ 2000 cm^{-1}) than in the visible range ($\lambda = 500$ nm, $\bar{\nu} = 2 \times 10^4$ cm^{-1}). This emphasizes one of the difficulties in obtaining infrared spectra of adsorbed molecules because these are present at the surface in number $N \approx 10^{15}$ molecules or less per cm^2, which is much lower than for bulk molecules (about $N \approx 10^{21}$ molecules per cm^3 of a 1 M solution).

2.3. Specular Reflection. Application to Reflection–Absorption Spectroscopy. Surface Selection Rules[27-29]

When light is reflected at the surface separating two homogeneous, isotropic, semi-infinite phases, each one being characterized by its optical constants, (n_1, k_1) and (n_2, k_2), respectively, both the intensity and the state of polarization are affected.

The direction of propagation, parallel to the wave vector **k**, and the unit vector **n** perpendicular to the plane of reflection P define the plane of incidence P' (Figure 1). Both the electric field vectors, $\mathbf{E}_i$ for the incident plane wave and $\mathbf{E}_r$ for the reflected plane wave, can be decomposed into two independent perpendicular components, linearly polarized, respectively, in the plane of incidence, $\mathbf{E}_p$ (p-polarization), and perpendicular to the plane of incidence, $\mathbf{E}_s$ (s-polarization). For these two components, there is no change in the polarization state upon reflection, as seen by symmetry considerations, but a change in amplitude and phase results.

The reflectivity coefficient R of the interface is defined as the ratio between the intensity I_r of the reflected light and the intensity I_i of the incident light:

$$R = \frac{I_r}{I_i} = \frac{\langle \mathbf{E}_r^2 \rangle}{\langle \mathbf{E}_i^2 \rangle}$$

The expression of R for p-polarization and s-polarization can be obtained from the Fresnel coefficients as follows:

$$R_p = \frac{tg^2(\hat{\varphi}_1 - \hat{\varphi}_2)}{tg^2(\hat{\varphi}_1 + \hat{\varphi}_2)} \tag{11a}$$

$$R_s = \frac{\sin^2(\hat{\varphi}_1 - \hat{\varphi}_2)}{\sin^2(\hat{\varphi}_1 + \hat{\varphi}_2)} \tag{11b}$$

where $\hat{\varphi}_1$, the angle of incidence, and $\hat{\varphi}_2$, the angle of refraction, are related by Descarte's law (or Snell's law):

$$\hat{n}_1 \sin \hat{\varphi}_1 = \hat{n}_2 \sin \hat{\varphi}_2 \tag{12}$$

Both relations (11) and (12) hold for absorbing media, characterized by a complex refractive index $\hat{n} = n - ik$, so that the angles of incidence and of refraction, as well as the reflectivity coefficients, are complex.

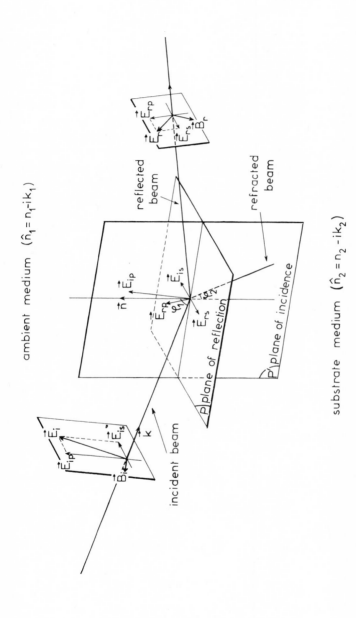

FIGURE 1. Reflection of an electromagnetic wave (**E**, **B**) of wave vector **k** arriving at an angle of incidence φ_1 at the boundary between two homogeneous media of different refractive index. (This figure corresponds to the case $n_1 < n_2$ and $\varphi_1 + \varphi_2 > \pi/2$.)

For normal incidence, assuming medium (1) to be nonabsorbing ($k_1 = 0$), relations (11) reduce to:

$$R = \frac{(n_2 - n_1)^2 + k_2^2}{(n_2 + n_1)^2 + k_2^2} \tag{11c}$$

for both s- and p-polarization.

The variation of the reflectivity coefficients with the angle of incidence is given in Figure 2 for various values of n and k.[30]

R_p can reach zero when tg $(\varphi_1 + \varphi_2)$ becomes infinite only for nonabsorbing media ($k_2 \approx 0$). This condition corresponds to the Brewster angle of incidence, defined by tg $\varphi_1^B = n_2/n_1$ (for the air/glass interface, where $n_1 = 1$ and $n_2 = 1.52$, this angle is $\varphi_1^B \approx 57°$). For absorbing media, R_p exhibits a minimum at the pseudo-Brewster angle. Figure 2 shows that the reflectivity coefficients of the two linearly polarized components are different, except for $\varphi_1 = 0°$ and $\varphi_1 = 90°$, so that the state of polarization of the incident light is

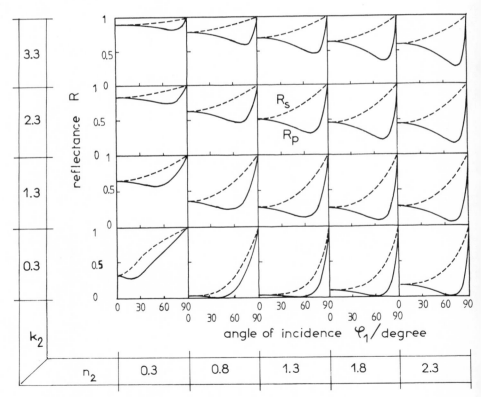

FIGURE 2. Variation of the reflectivity coefficients for p-polarization, R_P (——), and s-polarization, R_S (- - -), with the angle of incidence φ_1 for different sets of optical constants (n_2, k_2) of a substrate in contact with air ($n_1 = 1$, $k_1 = 0$). After Hunter.[30]

TABLE III. Changes in Phase Angles upon Reflection

	Change in phase angle for:	
	$\varphi_1 + \varphi_2 < \pi/2$	$\varphi_1 + \varphi_2 > \pi/2$
p-Polarization	π	0
s-Polarization	π	π

changed upon reflection. The degree of polarization of an unpolarized incident light after reflection at the interface may be defined by the ratio:

$$\Delta = \frac{I_{rs} - I_{rp}}{I_{rs} + I_{rp}} = \frac{R_s - R_p}{R_s + R_p} \tag{13}$$

which is maximum for the pseudo-Brewster angle (since R_p is minimum). These features are used in spectrometers with polarization modulation techniques, for example, in infrared reflection–absorption spectroscopy (Section 3.1.3).

Phase angles also are changed upon reflection. The change in phase depends on the angle of incidence and on the optical constant. Some values are given in Table III for the usual case considered here ($n_1 < n_2$).

If we consider now the presence of an adsorbed layer, or a film of thickness d, at the surface separating the two media considered above, one may use the three-phase model developed by Hansen[31] (Figure 3). In this model, the incident light, arriving from the first phase (solution with $\hat{n}_1 = n_1$), is absorbed in the second phase (the film of thickness d and with $\hat{n}_2 = n_2 - ik_2$) and then reflected at the interface between the second phase and the third phase (substrate with $\hat{n}_3 = n_3 - ik_3$). The reflected light has an intensity $I(d)$ lower

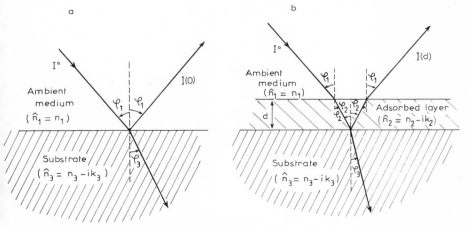

FIGURE 3. Two-phase (a) and three-phase (b) systems, after Hansen.[31]

than the intensity $I(0)$ in the absence of any absorbing film. Therefore, the normalized reflectance change will be defined by:

$$\frac{\Delta R}{R} = \frac{R(d) - R(0)}{R(0)} = \frac{I(d) - I(0)}{I(0)} \qquad (14)$$

which usually is negative for an absorbing film.

In the second phase, absorption of light results from interaction between the electric dipole moment of the molecule and the electric field vector, $\mathbf{E} = \mathbf{E}_i + \mathbf{E}_r$, resulting from the simultaneous presence of both the incident electric field vector $\mathbf{E}_i$ and the reflected electric field vector $\mathbf{E}_r$. This is illustrated for p-polarization in Figure 4, where $\mathbf{E}_p$ is decomposed into two components, $E_{p\perp}$, perpendicular to the plane of reflection, and $E_{p\parallel}$, parallel to the plane of reflection.

For angles of incidence smaller than the pseudo-Brewster angle φ_1^B ($\varphi_1 + \varphi_2 < \pi/2$), both p- and s-polarization display a change of phase angle of π (Table 3), so that the incident and the reflected lights tend to cancel out. However, for angles of incidence greater than φ_1^B, the angle of phase for p-polarization does not change, and both $\mathbf{E}_{ip}$ and $\mathbf{E}_{rp}$ add in phase. The resulting p-polarization, $\mathbf{E}_p = \mathbf{E}_{ip} + \mathbf{E}_{rp}$, will be thus maximum near the grazing incidence ($\varphi_1 \approx 90°$), whereas the resulting s-polarization, $\mathbf{E}_s = \mathbf{E}_{is} + \mathbf{E}_{rs}$, will be close to zero for any angle of incidence, as shown in Figure 5. This figure shows that $E_{p\parallel}$ is also very small, whereas $E_{p\perp}$ approaches a maximum (about twice the amplitude $E_i°$ of the incident light), for high angles of incidence. This is particularly true in the infrared range, where most of the metals behave as highly reflective surfaces due to a very high electrical conductivity ($\sigma > 10^7\ \Omega^{-1}\ m^{-1}$). This leads to high values of the extinction coefficient ($k > 10$). The angle of incidence corresponding to the maximum of $E_{p\perp}$, i.e., the

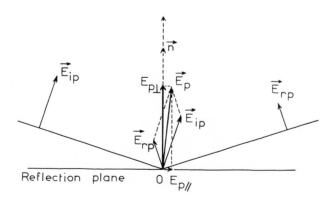

FIGURE 4. Vector $\mathbf{E}_p$ resulting from the sum of the p-components of the incident and reflected electric field vectors, $\mathbf{E}_{ip} + \mathbf{E}_{rp}$, at the reflection plane of a metal surface.

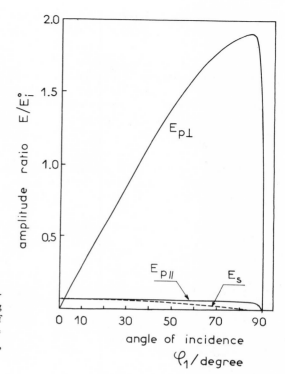

FIGURE 5. Variation of the amplitude of the components of the resulting electric field vector with the angle of incidence, at a bare metal surface ($n_3 = 3$, $k_3 = 30$) in contact with air ($n_1 = 1$, $k_1 = 0$). After Pritchard.[29]

maximum of light absorption, is therefore close to the grazing incidence, since it reaches about 88° at $\lambda = 5 \, \mu\text{m}$ (Table IV).

Therefore, the interaction with the adsorbed molecules will only be important for molecules having their electric dipole moment perpendicular to the separation plane. This restriction constitutes the so-called "surface selection rule," which is not in fact a true selection rule because no quantum numbers are involved.

Another consequence of these features is that the maximum absorption of the infrared beam will occur for p-polarized light. This would justify the use of a polarizer in reflectance spectroscopy.

Using the three-phase model and assuming that the thickness, d, of the film (or of the adsorbed layer) is much smaller than the wavelength, λ, McIntyre and Aspnes were able to calculate the normalized reflectivity changes for p- and s-polarizations to first order in d/λ.[32] For p-polarization, which is the only component to be considered here, this gives:

$$\left(\frac{\Delta R}{R}\right)_p = \frac{8\pi d \sqrt{\varepsilon_1} \cos \varphi_1}{\lambda} \text{Im}\left\{\left(\frac{\hat{\varepsilon}_2 - \hat{\varepsilon}_3}{\varepsilon_1 - \hat{\varepsilon}_3}\right)\left[\frac{1 - (\varepsilon_1/\hat{\varepsilon}_2\hat{\varepsilon}_3)(\hat{\varepsilon}_2 + \hat{\varepsilon}_3) \sin^2 \varphi_1}{1 - (1/\hat{\varepsilon}_3)(\varepsilon_1 + \hat{\varepsilon}_3) \sin^2 \varphi_1}\right]\right\} \quad (15)$$

TABLE IV. Optical Constants and Optimal Angle of Incidence, φ_1^{max}, Corresponding to the Maximum of Absorption for a Single Reflection on a Metal Surface in the Infrared Range

Metal	$\bar{\nu}$ (cm^{-1})	Optical constants		φ_1^{max} (deg)
		n	k	
\multicolumn{5}{c}{Data at different wavenumbers $\bar{\nu}^a$}
Ag	4444	0.77	15.4	86
	3333	1.65	20.1	87
	2222	4.49	33.3	88
Au	5000	0.47	12.5	85
	2500	0.80	24.5	87
	2000	1.81	33	88
Cu	4444	1.03	11.7	85
	2500	1.87	21.3	87
	1818	3.16	28.4	88
Pt	5000	5.70	1.70	85
	3333	7.7	1.59	86
	2000	11.5	1.37	87
\multicolumn{5}{c}{Data at $\bar{\nu} = 2100$ cm$^{-1\,b}$}
Al		6.8	32.0	88
Cr		3.8	14.7	86
Co		3.8	15.7	86
Cu		3.5	30.0	88
Au		4.2	27.6	88
Fe		5.3	15.5	85
Mn		6.2	6.7	77
Mo		4.2	18.2	87
Ni		5.4	18.6	86
Nb		5.1	18.0	86
Pd		6.3	15.3	86
Pt		5.0	20.0	87
Ag		2.9	33.7	88
Ta		6.0	12.7	85
Ti		5.5	10.5	83
W		4.7	17.6	86
V		5.1	18.2	86
Zr		4.3	10.2	84

[a] After J. Corset, LASIR, CNRS Thiais, France, personal communication.
[b] After R. G. Greenler, cited in Ref. 29.

where Im means that one has to take the imaginary part of the expression in braces, φ_1 is the angle of incidence, and ε_1 is the dielectric constant of medium 1, i.e., the electrolyte, taken as real. Equation (15) can be simplified for highly conductive metals ($k_3 > 10$, i.e., $\hat{\varepsilon}_3 \gg \hat{\varepsilon}_2 > \varepsilon_1$ and taking $\varepsilon_1 = 1$), as follows:

$$\left(\frac{\Delta R}{R}\right)_p \approx \frac{8\pi d \sin \varphi_1 \, \text{tg} \, \varphi_1}{\lambda} \, \text{Im} \, \{-1/\hat{\varepsilon}_2\} \tag{15a}$$

which can be further simplified for relatively small extinction coefficients k_2 for the film, i.e.:

$$\left(-\frac{\Delta R}{R}\right)_p \approx \frac{16\pi k_2 d \sin \varphi_1 \, \text{tg} \, \varphi_1}{\lambda n_2^3} = \frac{4 \sin \varphi_1 \, \text{tg} \, \varphi_1}{n_2^3} \, \alpha_2 d \tag{15b}$$

where α_2 is the absorption coefficient of the film—see the Beer–Lambert law and Eq. (7). For high angles of incidence, the factor $4 \sin \varphi_1 \, \text{tg} \, \varphi_1/n_2^3$ can reach values greater than unity and, thus, behaves as an enhancement factor for reflectance spectroscopy as compared to transmission spectroscopy.

This quantitative theory fails for a monolayer or submonolayer of adsorbed molecules, because it is impossible to define a complex dielectric constant $\hat{\varepsilon}_2$ for a film of molecular thickness since the layer is not a continuum. However, it is possible, in a first approximation, to treat the monolayer like the film and to define an effective index of refraction (or dielectric constant) for a sub-monolayer and an effective thickness $\bar{d} = \theta d_0$, where d_0 is the thickness of a full monolayer and θ is the degree of coverage of the surface by adsorbed species. It turns out that if $\bar{d}$ in Eq. (15) is replaced by θd_0, the normalized reflectivity change $\Delta R/R$ will be, in a first approximation, proportional to the degree of coverage θ.

Using Eq. (15b) and taking $\alpha_2 = 1200 \, \text{mol}^{-1}$ liter cm^{-1} [33] and $n_2 = 1.1$ for CO, the relative reflectivity change $\Delta R/R$ due to a full monolayer (of thickness $d_0 = 3 \, \text{Å}$) of CO adsorbed on a platinum surface is estimated to be about 0.3% at the maximum absorption ($\bar{\nu} \approx 2100 \, \text{cm}^{-1}$ and $\varphi_1 \approx 88°$). Such low reflectivity changes necessitate the recovery of the useful signal from noise by signal treatment techniques (modulation techniques followed by lock-in detection and/or signal averaging techniques).

3. Experimental Techniques

It is only recently that infrared vibrational spectroscopy has been success-fully used as an *in situ* technique to investigate the electrode/electrolyte interface. However, now that the first applications have been published, there is no doubt that, with a little improvement, the technique will become a routine tool for obtaining vibrational data of adsorbed species or of species in the double layer.

Two main methods were adopted in the past nine years. In the so-called "external infrared reflectance spectroscopy" method, the beam passes through a window, crosses a very thin layer of the electrolytic solution, and makes a single reflection at the electrode surface (Figure 6a). In the "internal infrared reflectance spectroscopy" method, the beam reaches the surface from the inside, i.e., through an optically transparent substrate on which the thin-layer electrode is deposited, and makes one or several reflections in order to improve the signal-to-noise ratio (Figure 6b).

The slow development of external infrared reflectance spectroscopy at metal/solution interface can certainly be attributed to a general idea, commonly shared by most spectroscopists until the late 1970s, that the presence of a layer of aqueous solvent in contact with the electrode would be an insuperable obstacle preventing the infrared beam from reaching the electrode surface. However, by using enhancement techniques like phase-sensitive detection coupled with averaging methods, as well as the signal processing ability of microcomputers, it becomes possible to improve greatly the sensitivity and to overcome the difficulties due to the weakness of the signal.

The problem is not as acute for internal infrared reflectance spectroscopy, since the beam does not pass through the solution, and therefore this technique would seem to be more promising. However, there is a serious restriction due to the use of thin-layer deposited electrodes. The technique has proved to work well, insofar as the catalytic properties of the electrode surface are not considered. However, for applications in electrocatalysis, it must be taken into account that a thin-layer deposited electrode may behave very differently from a smooth bulky electrode.

At the moment, since both techniques have been demonstrated to be feasible, there is considerable activity in the area of infrared spectroscopy at

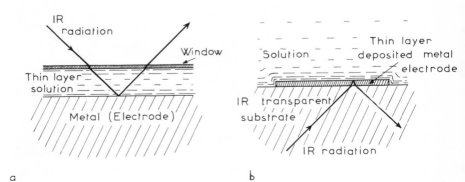

FIGURE 6. Schematic diagram showing the difference between (a) external reflection and (b) internal reflection.

the electrode/electrolyte interface. After the first publications, by Bewick *et al.*[18,34] for external reflection and/or Neugebauer *et al.*[35] for internal reflection, the techniques have developed in related but distinct directions, according to the type of spectrometer utilized (either dispersive or Fourier transform) and the design of the electrochemical cell (for external or internal reflection).

The various experimental methods are described in the next section, starting with dispersive spectrometers, then followed by Fourier transform spectrometers. This classification is not based on performance, which, at the present state of development, is roughly the same for both types of spectrometer.

3.1. Dispersive Spectrometers

3.1.1. Optical Components Used in Infrared Spectrometers Specially Designed for External Reflectance Spectroscopy

Various dispersive systems can be used as monochromators in infrared spectrometers. Gratings were used for the original developments of EMIRS (electrochemically modulated infrared reflectance spectroscopy)[36] and LPSIRS (linear potential sweep infrared reflectance spectroscopy)[37] techniques, but a simple wedge was employed in the case of IRRAS (infrared reflection–absorption spectroscopy).[38] In the initial development of the latter technique, poor resolution was thought to be less important than the resulting high-energy throughput.

Several spectrometer designs have been described in the literature, and a constant preoccupation is the maximization of energy throughput.

The first EMIRS experiment was realized in Southampton by Bewick *et al.*,[18,34] using a standard infrared spectrometer with a $\frac{1}{4}$-meter, $f/8$ grating monochromator. Due to the large number of mirrors, the modified optical path, and the characteristics of the grating, the luminosity was low and the equipment not optimal for EMIRS experiments. However, the results were encouraging. A second spectrometer was then designed by Bewick *et al.* and constructed by Anaspec Ltd.[36] The optical layout is given in Figure 7. The source is a Nernst filament. The $\frac{1}{2}$-meter $f/4$ monochromator is fitted with a three-position grating mount, and the large-surface 84 mm × 84 mm gratings are chosen to encompass the usual range of wavelengths. A combination of several filters, F, are available for order sorting. Two detectors can be used. One is a Golay-type Pye-Unicam IR 50 pneumatic detector; the other is a liquid-nitrogen-cooled HgCdTe photoconductor detector. A gold grid transmission polarizer, P, is positioned behind the exit slit of the monochromator to remove the undesirable *s*-polarization perpendicular to the incident plane.

A third-generation spectrometer, designed in Southampton but produced by HI-TEK, accomplishes a still improved energy throughput. This remarkable

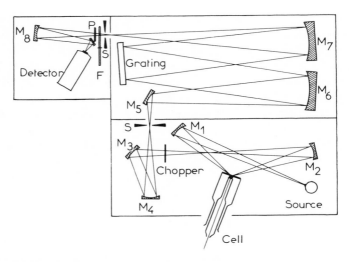

FIGURE 7. Schematic diagram of the EMIRS spectrometer produced by Anaspec Ltd. After Bewick *et al.*[36]

result was obtained by avoiding the use of three mirrors (M_3, M_4, and M_5 of Figure 7) and by focusing the emerging beam of M_2 directly onto the entrance slit of the monochromator.

A novel grating spectrometer for EMIRS was also tested by Kunimatsu in Sapporo and manufactured by Ritsu Ohyo Kogaku K.K. in Japan. Its originality is due to a design in which the electrochemical cell is located after the monochromator (Figure 8). An incorporated laser helps in alignment procedures. A special mount permits rotation of the cell without losing the

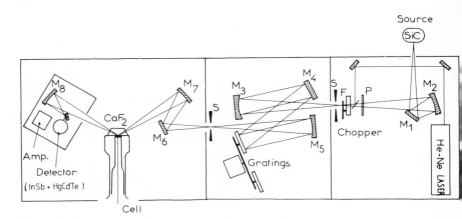

FIGURE 8. Schematic diagram of the IR spectrometer produced by Ritsu Applied Optics. After Kunimatsu.[39]

focus on the electrode surface. A two-color cooled HgCdTe–InSb detector is employed in order to achieve the maximum detectability[39] over a large range of wavelengths.

3.1.2. Signal Detection and Processing

Radiation Detectors. The question of how to compare different radiation detectors, especially those which work by different principles, remains difficult. Usually, several parameters are given by manufacturers. These include:

- The responsivity $R = V_S/(HS_D)$, which is defined as the ratio of the output voltage V_S to the product of the irradiance H and the sensitive area S_D. The units of R are V W^{-1} if V_S is in volts, H in watts per cm^2, and S_D in cm^2.
- The time constant τ, which characterizes the speed of response of the detector. It is generally defined as the time needed for the detector to reach $1 - e^{-1} \approx 0.63$ of the final value after its illumination.
- The detectivity D, in W^{-1}, which is defined as $D = R/V_n$, where V_n is the noise of the detector. Each detector has a certain noise level, which depends on the temperature and on the frequency bandwidth, Δf. Theoretically, the white noise can be written as proportional to the square root of the bandwidth, i.e., $V_n = k\sqrt{\Delta f}$ (Figure 9). It is necessary to cool some types of detectors to improve their performance.
- The specific detectivity D^*, in cm Hz$^{1/2}$ W^{-1}, which is calculated as the product of D and the square root of the factor $(S_D \times \Delta f)$. As it is less dependent on the bandwidth, D^* seems to be the most appropriate factor to use in comparing different detectors. However, for a full comparison, it is necessary also to take into account both the spectral region (which depends on the chosen window material) and the wavelength, λ_{peak}, at which there is the maximum response.

Figure 10 illustrates the variation of D^* with λ for the three common types of detectors used in EMIRS and IRRAS experiments. A special, two-color

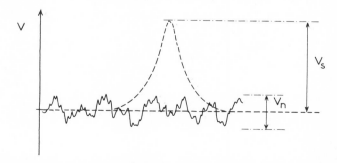

FIGURE 9. Comparison between the output voltage V_s of an IR detector and its noise voltage V_n.

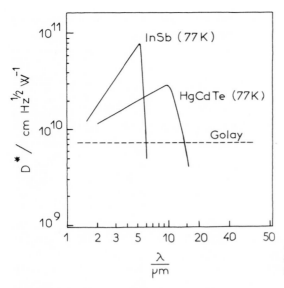

FIGURE 10. Variation of the specific detectivity D^* with the wavelength λ for three types of IR detectors.

InSb–HgCdTe detector is used by Kunimatsu,[39] with the advantage of a wider spectral range. Performances of the three types of detectors also are given in Table V.

Enhancement of the Signal. A large contribution to the success of infrared reflectance spectroscopy at the electrode/electrolyte interface results from the coupling of spectroscopic and electrochemical techniques. Furthermore, due to the strong absorption of the solvent and the subsequent weakness of the emerging beam, it is indispensable to use extremely powerful additional techniques to extract the signals buried in noise.

Therefore, in addition to the spectrometer itself and controlling microcomputer, equipment is required firstly to monitor the electrochemical processes occurring at the metal/solution interface (e.g., a potentiostat and waveform generator) and secondly to enhance the signal prior to its storage and processing. Usually the signals in phase with a reference modulation are detected

TABLE V. Characteristics of IR Detectors

	Detector		
	HgCdTe (77 K)	InSb (77 K)	Pneumatic Golay
Spectral range (μm)	2 to 15–20	1.5 to 5.5	0.4 to 1000
λ_{peak} (μm)	From 12 to 17	5	
Time constant, τ	0.1–0.2 μs	0.1–1.0 μs	30 ms
$10^{-10}\ D^*(cm\ Hz^{1/2}\ W^{-1})$	0.5–2.0	5–8	0.7

with a phase-sensitive detector and/or an averaging system, followed by suitable processing with microcomputer programs.

Phase-Sensitive Detection. The purpose of a phase-sensitive detector (PSD) is to extract a weak signal buried in noise. The signal which is to be analyzed (generally a mixture of various frequencies, ω_i) is first modulated at a suitable reference frequency ω_0, then input to the phase-sensitive detector and compared with the reference signal (Figure 11). Voltages which differ in frequency are mostly rejected, whereas those at frequency ω_0 pass through a maximum when they are in phase with the reference signal. Thus, the phase-sensitive detector allows the signals in phase and frequency to pass and rejects the noise. This is a very powerful method which is widely employed in spectroscopic techniques. In EMIRS, the practice is to modulate the electrode potential, whereas in IRRAS it is usual to modulate the polarization state of the reflected beam.

Signal Averaging. Another way to improve the sensitivity has been well known by spectroscopists since the development of Fourier transform spectroscopy. It consists in using ensemble averaging and signal processing by means of computers or sophisticated digital oscilloscopes. The signal-to-noise ratio S/N improves with the number of coadded and averaged spectra. More exactly, S/N varies as the square root of the number of samples, n. This means that 4 samples are necessary to improve S/N by a factor of 2, 16 samples to

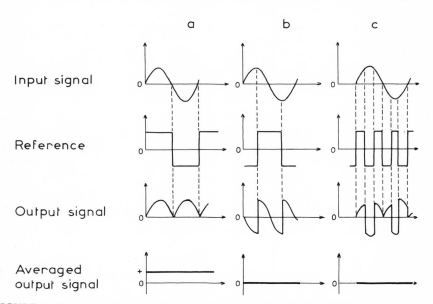

FIGURE 11. Schematic diagrams of the output voltage of a phase-sensitive detector: (a) The input and reference signals have the same frequency and are in phase; (b) the input and reference signals have the same frequency but are out of phase; (c) the input and reference signals differ in frequency and in phase.

improve it by an additional factor of 2, and 64 samples to improve it by a third factor of 2. Theoretically, there is no limit to the averaging and improvement process within instrumental resolution. However, some restrictions arise from both the necessity to keep constant all the experimental parameters and the difficulty in avoiding drift of the electronic devices.

Shape of the Signals. The shape of the absorption band depends on the type of detector and on the way the signal is processed. For well-controlled experimental conditions, the shape usually is not particularly important, especially if only qualitative information is required. However, it is necessary to be very careful for quantitative analysis, and a complete interpretation of the infrared spectra presumes a good knowledge of the electrochemical behavior of the system.

Electrochemically Modulated Infrared Reflectance Spectroscopy (*EMIRS*). Due to the synchronous detection, the EMIRS signal is at each time proportional to the difference in intensity of the reflected radiation at the two potential limits of the pulse applied to the electrode. The pulses may range from a few tens to hundreds of millivolts, with a frequency of a few hertz to several tens of hertz, depending on the response of the electrochemical system. Various theoretical forms are to be expected for the EMIRS signal depending on whether the species which absorb the radiation do so only at one or at two potential limits (Figure 12). According to the various limit cases, the EMIRS signal may appear as a single band (with either a positive or a negative sign), a bipolar band (with positive and negative lobes), or even not at all if at constant coverage of the adsorbed species, the change of potential does not sufficiently affect the force constant of the bond. The way the species is adsorbed on the surface (i.e., flat or perpendicular) is also a dominant factor governing whether absorption of the radiation occurs.[36]

The first EMIRS experiment[40,41] immediately showed the importance of the bands with a bipolar shape. In fact, a bipolar band has a pseudo-derivative shape, the origin of which had been clearly established.[36] It remains to correlate quantitatively the peak-to-peak intensity to the superficial concentration of adsorbed species: in other words, the question is how to obtain quantitative information from the peak-to-peak intensity of a bipolar EMIRS band. An attempt has been made assuming a Lorentzian profile for the peak.[42] The mathematical expression of such a peak is given by:

$$Y(\bar{\nu}) = \frac{Y_{max}}{1 + \left(\dfrac{\bar{\nu} - \bar{\nu}_{max}}{0.5\, \delta\bar{\nu}_{1/2}}\right)^2} \tag{16}$$

where Y is the ordinate corresponding to the wavenumber $\bar{\nu}$, Y_{max} is the maximum value for $\bar{\nu}_{max}$, and $\delta\bar{\nu}_{1/2}$ is the linewidth at half-height. It is shown in Figure 13a that it is easy to construct a bipolar band, C, by subtraction of

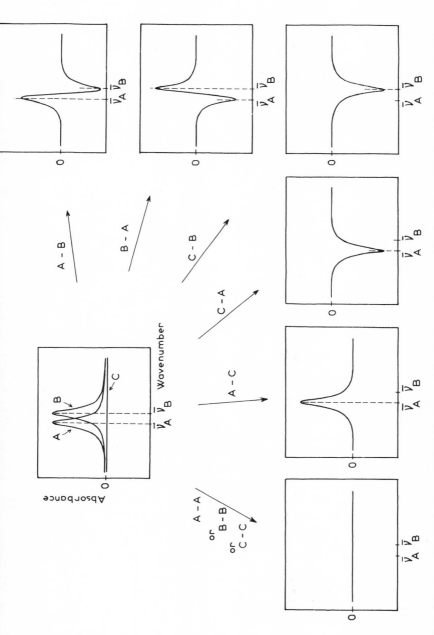

FIGURE 12. Examples of possible shapes of EMIRS bands according to the way the normal IR absorption bands A and B, at the two limiting potentials of the modulation, are combined together, or combined with the signal background C.

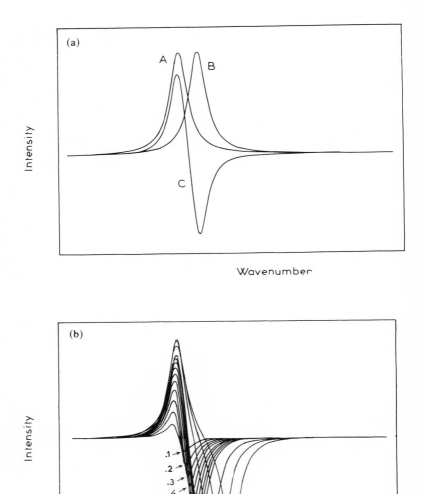

FIGURE 13. Example of the construction of a bipolar EMIRS band: (a) By subtracting two absorption bands, A and B, of the same intensity and close together; (b) by subtracting two absorption bands at various relative peak separations $s/\delta\bar{\nu}_{1/2}$.

the two peaks A and B. Thus, band C would correspond to a model in which the absorption is of the same amplitude at both potential limits.

In Figure 13b, it is shown how the shift, s, in the peak separation of the bands A and B can alter the shape and the peak-to-peak amplitude of the signal C. For $s < \delta\bar{\nu}_{1/2}$, the shape of the EMIRS signal resembles that of a pure derivative; for $s > \delta\bar{\nu}_{1/2}$, a shoulder appears due to the greater separation of A and B. Such a shoulder is generally not observed experimentally when the modulation amplitude is small, which leads to the conclusion that for most EMIRS bands, $s < \delta\bar{\nu}_{1/2}$. However, at highest amplitudes, s increases, and it is interesting to detect the limiting cases for which $s = \delta\bar{\nu}_{1/2}$ (which is reached just before the shoulder becomes noticeable). Thus, if $\bar{\nu} - \bar{\nu}_{max} = s = \delta\bar{\nu}_{1/2}$, expression (16) becomes

$$Y_S = \frac{Y_{max}}{1 + \left(\dfrac{\delta\bar{\nu}_{1/2}}{0.5\,\delta\bar{\nu}_{1/2}}\right)^2} = \frac{Y_{max}}{5}$$

It is thus possible to calculate the peak amplitude of one lobe of the bipolar band, Y_{0p}, as:

$$Y_{0p} = Y_{max} - \frac{Y_{max}}{5} = \frac{4}{5}Y_{max}$$

and if $(Y_{pp})_s$ is the peak-to-peak amplitude of the bipolar band for $s = \delta\bar{\nu}_{1/2}$, it follows that:

$$(Y_{pp})_s = 2Y_{0p} = \tfrac{8}{5}Y_{max}$$

This relation can be used to estimate the intensity of the real infrared band by measuring Y_{pp} of the EMIRS band in the limiting case where the shift of the band center approaches the value of the linewidth at half-height.

Infrared Reflection–Absorption Spectroscopy (*IRRAS*). In this method, the signal is obtained as a true infrared band. A useful comparison between EMIRS and IRRAS signals was made by Russell *et al.*[43] for the case of the adsorption of CO on a smooth platinum electrode in acid solution. The result (Figure 14) is a good demonstration of the origin of the particular shape of the EMIRS bands.

Linear Potential Sweep Infrared Reflectance Spectroscopy (LPSIRS). Each signal in this technique is obtained as a reflectogram, i.e., a curve which gives the dependence between potential and the relative reflectivity at a fixed wavelength λ, $(\Delta R/R)_\lambda = f(E)$. To obtain the spectrum itself in the usual form $[(\Delta R/R), \lambda]_E$, many reflectograms recorded at close wavelengths (the separation of which determines the spectral resolution at least as much as the slit width) are required. Figure 15a shows how, by an appropriate section at a fixed potential E of the three-dimensional diagram $(\Delta R/R, \lambda, E)$, it is possible to reconstruct the spectrum (Figure 15b). This method is rather long; however, it appears to be a very useful technique for investigating in detail a

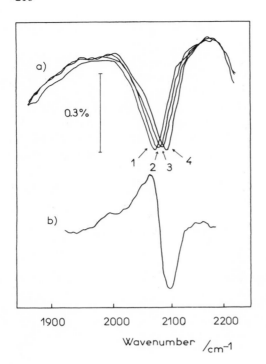

FIGURE 14. Relation between (a) the IRRAS spectra of CO adsorbed on Pt in 1 M HClO$_4$ at different constant potentials (versus SHE): (1) 50 mV; (2) 250 mV; (3) 450 mV; (4) 650 mV; and (b) the EMIRS spectrum of adsorbed CO modulated between 50 and 450 mV. After Russell *et al.*[43]

particular absorption peak under full electrochemical control of the electrode surface. In a variant of the method, it is also possible to superimpose a small modulation signal to the potential ramp delivered by the waveform generator and applied to the electrode by means of the potentiostat. Using this modulation as reference makes a synchronous detection possible. The sensitivity is increased very much but the signal appears as a derivative.

3.1.3. Techniques for External Reflectance Spectroscopy

EMIRS. This technique is a fruitful extension of the modulated electroreflectance spectroscopy (MERS) previously developed in the UV–visible range.[8,12,13,44] Basically, the method consists in studying the interaction of incident radiation with an adsorbed layer on a metallic surface of a smooth plane electrode, the potential of which is modulated at a given frequency ω_0 (generally close to 10 Hz). The limits of the pulses ΔE are chosen in order to induce a perturbation in either the degree of coverage of a given adsorbed species (in which case the amount of adsorbed species varies with the potential) or in the force constant of the intramolecular bonds of the species or of the bond which links the species to the surface (in which case the coverage remains

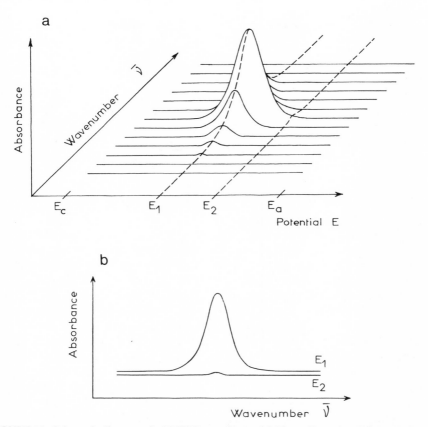

FIGURE 15. Schematic diagram of a LPSIRS experiment: (a) Three-dimensional diagram giving the absorbance versus the potential E during a sweep between E_c and E_a, for different wavenumbers $\bar{\nu}$ held constant; (b) reconstructed absorption spectra for two different potentials, E_1 and E_2.

constant). After a single reflection from the electrode surface, the beam is collected and focused on the detector. The further, synchronous analysis at ω_0 of the output signal permits rejection of the unmodulated information, in particular, absorption due to species in the bulk electrolytic solution, and thus distinguishes the absorption bands due to vibrations of the adsorbed species. The spectra are obtained as $[\Delta R/R, \lambda]_{\Delta E}$, and due to the modulation potential ΔE, the bands may appear with a complex form (refer to Section 3.1.2). The changes in relative reflectivity $\Delta R/R$ expected to be detected range in magnitude from 10^{-3} to 10^{-6}; these are obviously too weak for conventional equipment.

Experimental details for EMIRS have been described by Bewick *et al.*[36] Figure 16 gives the layout of the equipment for modulating the potential and processing the signal.

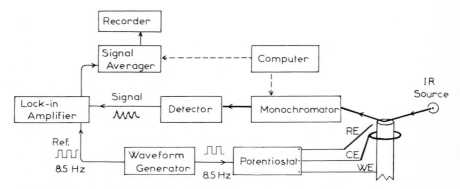

FIGURE 16. Block diagram of the complete EMIRS spectrometer. After Bewick et al.[36]

LPSIRS. In this method, originally developed by Kunimatsu,[37] a monochromatic incident light is reflected from the surface of a smooth electrode, the potential of which is varied rapidly between two suitable limits by conventional linear sweep cyclic voltammetry (LSCV). The minor changes which ensue in the absorption of the infrared radiation are detected by using spectral accumulation techniques and lead to the so-called "reflectograms" $[\Delta R/R, E]_\lambda$. By recording enough experiments at various close wavelengths λ and by subsequent transformations, it is possible to reconstruct the spectra in the usual form $[\Delta R/R, \lambda]_E$ (see Section 3.1.2 for treatment).

The advantage of LPSIRS is that repetitive potential sweeps continuously control the electrode surface by means of LSCV. This technique is particularly useful when unstable systems are considered. However, the successive spectral accumulations may require a rather long time (several hours), and the method is therefore restricted to investigations in narrow wavelength ranges. The equipment is the same as for EMIRS and need not be described further.

IRRAS. This technique is similar to one which was first applied to studies at the solid/gas interface.[38] It is based on Greenler's theory and on the surface selection rules which have been discussed in Section 2.3, and accordingly, adsorbed molecules (more precisely those with a nonzero dipole moment perpendicular to the electrode surface) are able only to absorb p-polarized radiation and not the s-polarized component. In contrast, due to their random orientation, molecules in the bulk absorb both the p and s components. Thus, in principle, it is possible to take advantage of this different behavior to discriminate between molecules at the surface and those in the bulk solution.

In the technique of Golden *et al.*,[38] the infrared radiation first is chopped at a frequency ω_c and focused on the electrode surface (Figure 17). The reflected beam passes then through a photoelastic modulator and through a polarizer P. The radiation is thus alternately modulated between s and p states. A double synchronous detection, which is necessary to demodulate the reflected

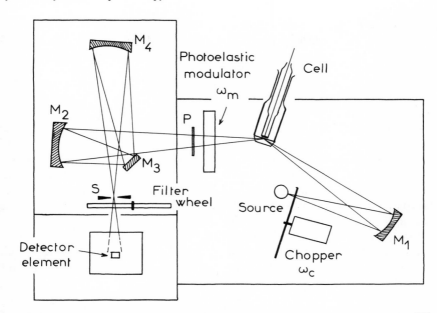

FIGURE 17. Schematic diagram of the IRRAS spectrometer designed by Golden *et al.*[(38)]

beam, allows the alternate measurement of the intensities I_s and I_p. In fact, because of the extremely weak variations of intensity and because of the large changes in reflectivity with respect to the wavelength, it is necessary to normalize the signals according to the relative expression

$$(\Delta)_\lambda = \frac{I_s - I_p}{I_s + I_p} \tag{17}$$

Under the experimental conditions used by Russell *et al.*[(43,45,46)] for the *in situ* investigation of the electrode/electrolyte interface, two series of measurements are performed: one, A, with the electrolytic solution alone; the other, B, with the electroactive species in the solution. Calculations of the expression $[\Delta_B - \Delta_A]_\lambda$ at each wavelength λ and for a given potential allow the rejection of the information due to the bulk solution and the establishment of the difference spectra of the adsorbed species at a fixed potential $[\Delta_{B-A}, \lambda]_E$.

For several experimental reasons, the IRRAS technique is somewhat hard to use. One of the main problems remains a possible shift of the baseline. However, even though IRRAS is presently less sensitive than EMIRS, the technique has an undeniable advantage: it gives true absorption bands and thus allows direct quantitative estimations.

3.1.4. Internal Reflection Spectroscopy

A few authors have utilized a grating spectrometer in conjunction with an electrochemical cell specially designed for attenuated total reflection (ATR).

By modulating the potential of the electrode, a synchronous detection is possible, and the sensitivity which is achieved is similar to that of EMIRS experiments. Only the cell design will be discussed therefore (Section 3.3.2).

3.2. Fourier Transform Infrared Spectroscopy (FTIRS)

3.2.1. Principle of FTIR Spectrometers

In a grating spectrometer, the light is dispersed by the monochromator, which means that the detector receives each wavelength separately and that, consequently, the intensity is low. In Fourier transform spectroscopy, on the other hand, all wavelengths are collected and received simultaneously by the detector. This high energy throughput results in a high signal-to-noise ratio.

It is beyond the scope of this chapter to describe Fourier transform infrared spectroscopy in detail. Recent papers by Pons *et al.*,[47] Pons,[48] and Habib and Bockris[49] have reported the use of FTIRS for *in situ* spectroelectrochemical studies.

The fundamental part of the instrument is the so-called Michelson interferometer. It consists of a beam splitter and two mirrors, one fixed, the other movable (Figure 18). The infrared radiation, emitted by a broadband source,

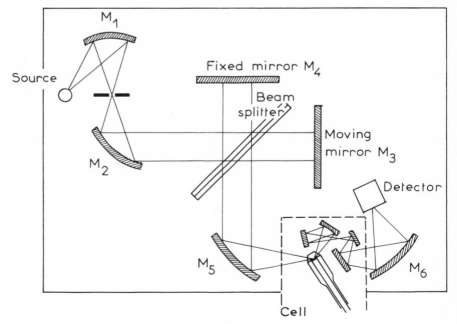

FIGURE 18. Schematic diagram of the FTIR spectrometer used by Pons.[48]

enters the interferometer and is split into two beams of equal intensity. Both beams are reflected back by the mirrors to the beam splitter. Depending on the movable mirror's relative displacement x, the beams are recombined either constructively or destructively. The resulting intensity is thus of $\sin x/x$ type. The interferogram, which is produced by the sum of the interferences, has its maximum at $x = 0$, because all wavelengths interact constructively, giving rise to the "center burst." On both sides of the center burst (i.e., when the movable mirror is displaced), the amount of constructive interference decreases rapidly (Figure 19).

The intensity, which depends on x, is given by

$$I(x) = \tfrac{1}{2}I(0) + \int_{-0}^{+\infty} I(\bar{\nu}) \cos (2\pi x\bar{\nu}) \, d\bar{\nu} \tag{18}$$

where $I(\bar{\nu})$ is the intensity at each wavenumber $\bar{\nu}$ and $I(0)$ its value at $x = 0$.

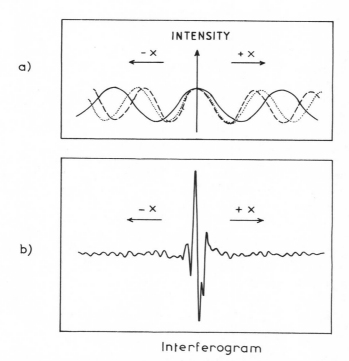

Interferogram

FIGURE 19. Principle of obtaining an interferogram with a Michelson interferometer: (a) Light intensity of the different monochromatic waves issued from the IR source as a function of the relative displacement x of the moving mirror; (b) constructive and destructive recombination of the reflected beams, leading to an interferogram with a maximum intensity when the waves are in phase ($x = 0$).

The second term of this expression is known as the *interferogram function*:

$$F(x) = \int_{-0}^{+\infty} I(\bar{\nu}) \cos{(2\pi x \bar{\nu})} \, d\bar{\nu} \qquad (19)$$

By performing a Fourier transformation on $F(x)$, it is possible to obtain the spectrum:

$$I(\bar{\nu}) = \int_{-\infty}^{+\infty} F(x) \cos{(2\pi \bar{\nu} x)} \, dx \qquad (20)$$

However, two important problems arise, which can both be solved by computer calculations. The first arises from the fact that the integral (20) of the interferogram function (19) is taken from $-\infty$ to $+\infty$, whereas the experimental measurement is limited by the maximum displacement of the movable mirror. (This correction is known as the apodization correction.) The second problem is due to the phase error, which makes the interferogram not totally symmetrical about $x = 0$.

A Fourier transform infrared spectrometer possesses several advantages over conventional grating spectrometers:

- simplicity and reliability
- speed (Fellgett's advantage)
- high energy throughput (Jacquinot's advantage), because no slit is utilized
- accuracy in frequency determination (Conne's advantage) due to the use of a subsidiary internal laser

3.2.2. Use for External Reflection Measurements

Commercial FTIR spectrometers are designed for transmission experiments. It is thus necessary to use a reflectance attachment to deviate the beam and to focus it on the surface of the working electrode, in the electrochemical cell.

In fact, as clearly noted by Pons et al.,[47] sophisticated FTIR spectrometers are required for work at the electrode/solution interface. In particular, they must allow alternating access to several memories for coadding the interferograms at two different potential limits that are repetitively changed. They must allow also the further spectral subtractions and divisions by the background which are necessary to generate the final spectrum of the detected species.

Subtractively Normalized Interfacial Fourier Transform Infrared Spectroscopy (SNIFTIRS). Pons et al.[47,48] have developed the SNIFTIRS technique, which consists in collecting successive series of four to eight interferograms at each of two potential limits, E_1 and E_2. These potentials are chosen according to the electrochemical behavior of the system studied. E_1 is the reference potential and E_2 a potential of interest at which an electrochemical process occurs. The step between E_1 and E_2 is repeated until the desired signal-to-noise ratio is obtained (an average of 100 scans is generally sufficient). If R_1 and R_2 are the reflectivities measured at E_1 and E_2, it is easy to calculate the relative change of reflectivity $\Delta R/R$ by

$$\frac{\Delta R}{R} = \frac{R_2 - R_1}{R_1} = \frac{R_2}{R_1} - 1 \qquad (21)$$

The SNIFTIRS technique has proved to be quite suitable for the *in situ* detection of either electrogenerated intermediates in the double layer or species adsorbed at the electrode surface.

Polarization Modulation in FTIR Spectroscopy (PM-FTIRS). Recently, Kunimatsu, Golden, and coworkers[50,51] have coupled the polarization modulation technique with Fourier transform infrared spectroscopy and applied the new method to electrochemical studies (Figure 20). In their work, a double modulation technique is employed, using a photomodulator working at high frequency (78 kHz) and a polarizer, both inserted in the optical path.

3.2.3. Use for Internal Reflection

As already pointed out, the FTIR technique is suitable for internal reflection experiments, on condition that reflectance accessories are used. This method was first developed in Austria by Neugebauer, Brinda-Konopik, and

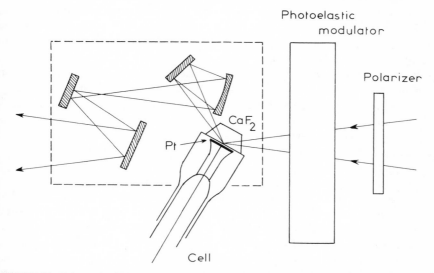

FIGURE 20. Schematic diagram of a polarization-modulated Fourier transform infrared spectrometer. After Kunimatsu *et al.*[84]

coworkers.[35,52,53] The description of the electrochemical cell is given below in Section 3.3.2.

3.3. Design of the Spectroelectrochemical Cell

The electrochemical cell design is of primordial importance. Several points are to be considered. The first is the working mode—internal or external reflection. Then there is the choice of materials: they must be as stable as possible in contact with the solution in order to prevent any risk of pollution of the surface.

3.3.1. Electrochemical Cells for External Reflection

The cell design is particularly difficult, due to the fact that the solvent strongly absorbs the infrared radiation: in aqueous electrolyte, for instance, it is hardly possible to work with a solution thickness greater than a few microns, which practically imposes the use of flat, smooth (perfectly polished) electrodes.[36] Bewick has examined in detail the problem of the absorption of the infrared radiation at the electrode/aqueous solution interface.[36] Two types of estimations were made for wavelengths close to the O—H stretching vibration bands—firstly, by the direct application of Beer's law and, secondly, by a rigorous application of the Fresnel equations. Fortunately, it appears that the absorption of the infrared radiation is much less for the solution and much more for the adsorbed monolayer than that which is predicted, in first approximation, by Beer's law.

Choice of Material for the Window. The material used for the window has to meet certain criteria. First, the wavelength range over which the transmission is satisfactory is important. The stability of the material in contact with the solution is also of importance (particularly for aqueous acid or alkaline electrolyte solutions). However, other parameters such as the refractive index, the hardness, the thermal expansion coefficient, the standard form in which the material is available, and even the price also influence its choice. Practically, for aqueous solutions, the selection is restricted to a few materials, including silicon, calcium fluoride, zinc sulfide, and zinc selenide. The main properties of these materials, together with those of KRS-5, KRS-6, and CdTe, are gathered in Table VI. Various examples of transmission curves are given also in Figure 21.

The solubility is really a critical point due to two factors. Firstly, a more or less long contact of the window with the solution can depolish it. Secondly, the metallic ions formed by dissolution can be electrochemically redeposited on the electrode surface and can thus modify its adsorption properties.

Fabrication of the Electrochemical Cell. The three-electrode cell can be constructed in Pyrex glass or in synthetic materials like Kel-F. In the initial Pyrex cell proposed by Bewick *et al.*,[36] the window was silicon directly sealed to a glass joint (Figure 22). Such an arrangement is convenient, except for the

TABLE VI. Properties of Different Materials Used in IR Spectroscopy[a]

	Material								
	Quartz (crystal) (d = 2 mm)	Ge (d = 1.5 mm)	Si (d = 2 mm)	CaF$_2$ (d = 1 mm)	ZnS (d = 2 mm)	ZnSe (d = 3 mm)	KRS-5 (d = 1 mm)	KRS-6 (d = 1 mm)	CdTe (d = 2 mm)
Transmission maximum, T_{max}	90%	45%	50%	90%	70%	70%	75%	80%	65%
Wavelength range (μm) with $T > 50\%$	0.15-4, >50	—[b]	1.2-15, 22-42	0.12-12	0.14-14	0.5-20	0.45-45	0.4-32	0.9-24
Solubility in water (g/100 g H$_2$O) at 300 K	1 × 10^{-3}	i.	5 × 10^{-3}	1.3 × 10^{-3}, 1.6 × 10^{-3c}	6.9 × 10^{-4}, 6.5 × 10^{-4c}	1 × 10^{-3}	5 × 10^{-2}	0.32	i.
Solubility in other solvents[d]	s. HF, v.sl.s.alk.	s.h. H$_2$SO$_4$, i. alk.	s. HF + HNO$_3$	s.NH$_4$ salts, sl.s.a.	v.s. a.	d. HNO$_3$, s.a.			d. HNO$_3$, i.a.
Thermal expansion coefficient (°C^{-1})	0.8 × 10^{-6}	6 × 10^{-6}, 5.7 × 10^{-6c}	4.1 × 10^{-6}	18 × 10^{-6}	7.9 × 10^{-6}, 7.8 × 10^{-6c}	7.6 × 10^{-6}	50 × 10^{-6}	58 × 10^{-6}	5.9 × 10^{-6}
Hardness (Knoop)	740	692	1150	158	250	150, 130^c	40	35	45
Specific gravity (g cm^{-3})	2.648	5.35	2.33	3.18	4.08, 3.98^c	5.27, 5.42^c	7.371	7.192	5.849
Index of refraction (optical grade)									
3 μm	1.499	4.05	3.432	1.417	2.29	2.43	2.45	2.25	2.7
7 μm	1.167	4.01	3.419	1.368	2.25	2.42	2.376	2.187	2.683
10 μm		4.003	3.418	1.281	2.20	2.41	2.371	2.177	2.676

[a] Data taken from ORIEL, *Eigenschaften optischer Materialen*, except where otherwise noted.
[b] $T > 20\%$ for $\lambda = 1.8$-23 μm.
[c] Data taken from *Handbook of Chemistry and Physics*.
[d] Abbreviations: a. = acid; alk. = alkali; d. = dilute; h. = hot; i. = insoluble; s. = soluble; sl. = slightly; v. = very.

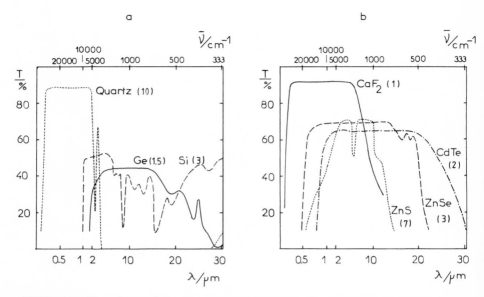

FIGURE 21. Plots of transmission T versus wavelength in the infrared range for various materials. After ORIEL, *Eigenschaften optischer Materialen*, Darmstadt, FRG (1981). Thickness given in millimeters.

fact that silicon is opaque to visible light, which complicates the alignment procedures somewhat. So, when possible, it is preferable to replace silicon by silica or calcium fluoride, or any transparent material. A complication arises with the use of high-refractive index materials, in that they produce high reflection losses. A way to solve this problem is to make beveled edges. For instance, when using calcium fluoride, the window is beveled at 65°, as in Figure 20.

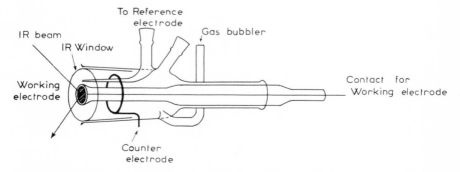

FIGURE 22. Schematic diagram of the electrochemical cell used for external reflection.[36]

The working electrodes are made of 5- to 10-mm-diameter smooth metallic disks, and they are positioned at the extremity of a syringe plunger or a Kel-F sleeve. Whenever possible, it is certainly better to seal the material directly to glass. This is easy for platinum. However, for materials not compatible with Pyrex or soft glasses, certain epoxy glues may be used as long as they are stable in the solvents. In the method used by Pons et al.,[47] the electrodes are attached to a copper rod, glued, and forced into a Kel-F outer sleeve. There is no leakage problem with such electrodes. The parallelism between the window and the electrode surface must be rigorously controlled. Pons et al. proposed a special holder and polishing procedure which allows reduction of the fluctuation down to 0.5 μm.[47] A similar technique is used by Habib and Bockris.[54]

3.3.2. Electrochemical Cells for Internal Reflection

Internal reflection avoids the complication of radiation absorption by the solvent. However, the problem of solubility of the various materials in contact with the solution remains as acute as for external reflection. Furthermore, the material which constitutes the working electrode is generally opaque to the infrared radiation. It is therefore necessary to reduce the thickness to a very thin film deposited on the surface of an infrared-transparent substrate.

Two electrochemical cells have been proposed in the literature. The first one, designed by Neugebauer et al.,[52] had a germanium substrate, from which the beam was multiply reflected, and a Teflon case (Figure 23). Fe or Cd was evaporated to serve as the working electrode. The second one was designed

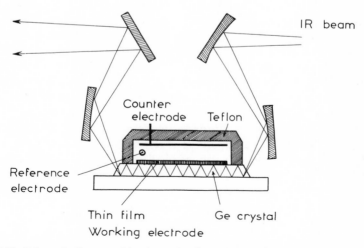

FIGURE 23. Schematic diagram of the electrochemical cell used for internal reflection, according to Ref. 52.

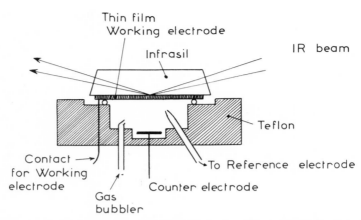

FIGURE 24. Schematic diagram of the electrochemical cell used for internal reflection, according to Ref. 55.

by Neff *et al.*[55] for electromodulated internal reflectance spectroscopy. The substrate was an infrasil prism and the case was again Teflon. The thickness of the gold deposited layer was about 150 Å. Temperature treatment of this film at 470 K formed (111) single crystals[56] (Figure 24).

3.4. Discussion of the Techniques

Following the above review of the various spectroscopic techniques for *in situ* electrode/solution interface investigations, several points should be discussed.

The first one concerns the use of polarization modulation in conjunction with Fourier transform infrared spectroscopy. Although this technique has only been developed recently, the first results are very encouraging. They demonstrate the feasibility of using PM-FTIRS to gain information on monolayers, or even submonolayers, of adsorbed species on metallic surfaces with a very low specific area.

According to Pons *et al.*,[57] the response depends on the characteristics of the spectrometer. If the apparatus is a very sophisticated one, carefully aligned with respect to the electrochemical cell, there does not seem to be much possibility of improving the signal-to-noise ratio by using polarization modulation. However, for a spectrometer with lower performance, the coupling may be advantageous, insomuch as synchronous detection is a very good way to eliminate the usual noise, both electronic and mechanical.

The second question concerns the comparison between IRRAS and PM-FTIRS techniques, i.e., the comparison between dispersive and Fourier transform spectrometers. In a recent interesting report, Golden *et al.*[58] concluded that if resolution is judged to be essential, FTIRS would be the better choice; but if, on the contrary, the highest sensitivity is required, a grating spectrometer might be favored. In particular, when looking for very weak absorption bands, the EMIRS technique is likely the most appropriate.

4. Applications to Selected Examples

4.1. General Survey

The various recent infrared spectroscopic techniques which are appropriate for *in situ* reflectance investigations at the electrode/electrolyte interface have been reviewed in Section 3. It may seem surprising that so many different approaches have been developed in such a short time (less than eight years for most of the techniques). This simply demonstrates how imperative was the necessity to carry out *in situ* infrared vibrational investigations of adsorbates, for which the lack of information was particularly acute. The obstacle to progress, namely, that infrared spectroscopic investigations would not be possible because of too strong an absorption of the IR beam by the solvent, has now been overcome. Only the future will tell which of the various techniques are the most appropriate for solving given problems at the electrode/electrolyte interface. But it is necessary to point out once more that the success of reflectance spectroscopy is above all due to the fruitful contribution of electrochemistry. It is a good demonstration of the efficiency of coupled methods, when compared to conventional techniques.

Since 1980, several dozen important papers have been published concerning investigations on the electrode/electrolyte interface. It is thus possible to give a first overview of the various applications of the techniques. The subjects of various investigations are collected in Table VII, together with the infrared technique which was used and the corresponding references. It can be seen that there is now a wide range of applications, from aqueous to nonaqueous solvents and from adsorbed species on the electrode to species formed in the vicinity of the electrode. It is therefore relevant to select a few examples to illustrate, as well as possible, the appropriateness of each technique.

The first example concerns the adsorption of hydrogen on platinum in acid media. The second deals with the comparative adsorption of carbon monoxide on noble metals in aqueous medium, with a special emphasis on platinum and palladium. The third example is a discussion of the contribution of *in situ* reflectance spectroscopy to solving certain problems encountered in electrocatalysis. Finally, the last example concerns nonaqueous solvents in particular and the detection of species in the double layer.

Other applications of the techniques are discussed in a recent review by Bewick and Pons.[59]

TABLE VII. Survey of Applications of Infrared Spectroscopic Techniques at the Electrode/Electrolyte Interface[a]

| | | Dispersive spectrometers | | | | Fourier transform spectrometers | | |
| | | Internal reflection | External reflection | | | Internal reflection | External reflection | |
Investigated process/species	Electrode	IR + ATR (55)	EMIRS (36, 80, 82)	IRRAS (38)	LPSIRS (37)	FTIRS + ATR (53, 83)	SNIFTIRS (47, 48, 49)	PM-FTIRS (51, 84)
Studies in aqueous media								
Structure of the double layer								
Adsorption of H_2O and D_2O	Pt	56	34, 18, 61, 62, 63, 36					50, 51, 84,
	Au	55, 85	62					
Adsorption of hydrogen	Ag		86					
	Pt		61, 80					
	Rh		87, 80					
Adsorption of CO in aqueous electrolyte	Pt		41, 67, 46, 107	43, 88, 45, 46				72, 71
	Rh		67					
	Au		67					
	Pd		80, 39		39			
Isotopic exchange, ^{12}CO–^{13}CO			80, 89					
Adsorption of CN^-	Pt							71, 90
	Au		82					51, 71, 90
	Ag		82					91, 51, 71, 90
	Cu							71, 90, 92
Adsorption of SCN^-	Pt						48	
	Au		59					
Adsorption of NO	Pt,Rh,Au			46				
Dissociative adsorption of CH_3OH	Pt		40, 41, 93, 77, 80, 108, 109		37, 93, 80			
	Pt		112, 113					

	Metal			
Dissociative adsorption of HCOOH	Pt	41, 42, 70, 94, 80		
Dissociative adsorption of HCHO	Rh	110, 111		
Molecular adsorption of HCOOH	Pt,Rh	95		
Molecular adsorption of H_2SO_4	Au	95		
Reduction of CO_2	Au	59		
	Pt	96		
Anion adsorption of H_3PO_4	Pt			97
	Au			54
Anion adsorption of $CF_3SO_3^-$	Pt			98
Adsorption of acetonitrile	Au	36, 80, 82, 86, 99		
Adsorption of difluorobenzene	Pt	100		
Adsorption of benzonitrile	Au	59		
Redox reaction $Fe(CN)_6^{4-}/Fe(CN)_6^{3-}$	Pt		35, 52, 53	59, 101
Corrosion behavior of metals	Fe		83	
	Cd			
Studies in nonaqueous solvents				
Adsorption of CH_3CN				
In $LiClO_4$ supporting electrolyte	Pt			102, 103, 48, 59
In TBAF supporting electrolyte	Pt			102, 103, 46, 59
Poly(methyl thiophene) in CH_3CN	Pt		104, 105	
Polyphenylene in CH_3OH + NaOH	Pt, Fe		114	
Organic species in the diffuse layer				
Reduction of tetracyanoethylene in CH_3CN	Pt			79, 48, 106, 59
Reduction of benzophenone in CH_3CN	Pt			47, 59
Reduction of anthracene in CH_3CN	Pt			59
Oxidation of 2,6-di-*t*-butyl-4-phenylaniline	Pt			48, 59

[a] For general papers on this topic, see Refs. 46, 47, 49, 57–59, 80, and 81.
[b] For a description of the equipment, see references given in parentheses with each technique.

4.2. Adsorption of Hydrogen on Platinum in Acid Media

4.2.1. Why This Example?

This example was chosen for several reasons. First, the adsorption of hydrogen on smooth, noble metal electrodes is of fundamental interest for electrochemists. In fact, it has been so extensively studied by conventional electrochemical methods that it can be almost said that the measure of the quantity of adsorbed hydrogen is the best way to characterize such surfaces in contact with aqueous solutions. This is particularly true for platinum, for which not only the polycrystalline surface but also the low-index (100), (110), and (111) planes have already been studied in detail.[60] Secondly, hydrogen adsorption is also a well-known process at the solid/gas interface (for which a lot of infrared investigations have been done), which gives *a priori* a solid basis for investigations at the electrode/electrolyte interface. Furthermore, the discussion of the IR reflectance spectroscopic results necessitates taking into account adsorption of water molecules and requires the use of the surface selection rules. Finally, it is an example which illustrates the fact that electrochemical concepts are indispensable for a good interpretation of the information obtained by infrared reflectance spectroscopy.

4.2.2. Experimental Conditions and Data Acquisition

It is beyond the scope of this work to describe in detail an electrochemical technique such as cyclic voltammetry; such a description can be found in classical textbooks such as Ref. 3. The voltammogram [i.e., the $i(E)$ curve, where i is the current density and E the potential applied to the electrode, which is a triangular function of time] of platinum in acid media is well known. Under dynamical conditions (sweep rate $v = 50 \text{ mV s}^{-1}$), two distinct regions are clearly observed, well separated by the so-called "double layer" region where no faradaic process occurs (Figure 25). The most negative region (i.e., the left part of the voltammogram) corresponds to the adsorption–desorption process of hydrogen:

$$H_{aq}^+ + e^- \rightleftarrows H_{ads}$$

It is the investigation of this region by infrared spectroscopy that will be described in this section. The most positive region (i.e., on the right) is associated with the formation of superficial oxides. When the potential limits are those given in Figure 25 for polycrystalline platinum in 0.5 M H_2SO_4 at 25°C, it is observed that a saturation of the adsorption sites results; namely, one hydrogen atom per site or nearly two oxygen atoms per site at the onset of oxygen evolution. A monolayer, and not more, is thus formed. The calculations, especially in the case of hydrogen, are particularly accurate: knowing the superficial crystallographic density of sites N_s, estimated on a face-centered

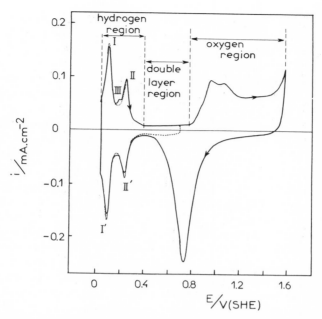

FIGURE 25. Voltammogram of a Pt electrode in the supporting electrolyte (0.5 M H_2SO_4) at a sweep rate of 50 mV s^{-1}.

cubic model and introducing a weighted contribution of the preferential orientations (100), (110), and (111), it is possible to calculate the charge Q_H° associated with the oxidation process of a monolayer of adsorbed hydrogen by:

$$Q_H^\circ = [\text{number of sites per cm}^2] \times [\text{elementary charge}]$$
$$= N_s \times e_0 = 1.3 \times 10^{15} \times 1.6 \times 10^{-19} = 208 \ \mu C \ cm^{-2}$$

As the experiment gives a value within 1% of Q_H°, it is concluded that the model is correct and that hydrogen adsorbs on platinum at one atom per site.

However, as seen in Figure 25, the situation may be more complex. The voltammogram of the hydrogen shows clearly several peaks or shoulders. Two contradictory theories have been proposed: the first one, by Clavilier *et al.*,[60] assumes that peaks I and II (and the homologous peaks I' and II') are related to hydrogen adsorbed on two different preferential orientation planes—(110) for peaks I and I' and (100) for peaks II and II'. [The lack of a peak due to hydrogen adsorbed on the (111) plane is explained by the reconstruction of this plane when in contact with aqueous solution, after cycling in the oxygen region.] The second theory by Bewick *et al.*[61,62] postulates the existence of two different varieties of bound hydrogen, having different binding energies— the "weakly adsorbed" hydrogen for peaks I and I' and the "strongly adsorbed" hydrogen for peaks II and II'. These two theories could be reconciled by assuming that hydrogen is weakly bonded to the (110) plane and strongly bonded to the (100) plane. However, this simple view does not fit the results obtained by UV–visible reflectance spectroscopy,[61] which have shown that the hydrogen giving rise to peaks I and I' behaves more as an adsorbed species, while the hydrogen giving rise to peaks II and II' imports to the surface a character close to that expected for a metallic hydride (which would demonstrate that the hydrogen is incorporated in the lattice).

With these preliminary remarks, it can be seen that the platinum–aqueous solution system was a very good candidate for an *in situ* infrared spectroscopic investigation. The first experiments were done by Bewick and coworkers,[18,34,63] using EMIRS at different modulation amplitudes. Some experiments were made totally in the double-layer region, others totally in the hydrogen region, and others with one limit in the double layer and the other in the hydrogen region.

The most interesting fact to note is that none of these experiments have shown any absorption corresponding to the Pt—H vibration (which was expected around 2100 cm^{-1}, as in the gas-phase experiments). A second fact is that those modulations with one limit in the potential range of peaks II and II' always produce a considerable enhancement of the reflectivity but without any definite absorption peak. A third fact is that those modulations with one limit in the potential range of peaks I and I' are all accompanied by vibrational bands which can be attributed to the vibrational modes of water. All of them are single bands, which, recalling the discussion of Figure 12 (Section 3.1.2), denotes that the absorbing species preponderates at the negative limit. Figure 26 gives some examples of the EMIRS spectra obtained.

The first conclusion is thus that there are strong interactions between hydrogen and water and that the adsorption of hydrogen is accompanied by an increased quantity of water bonded to the surface.

4.2.3. Interpretation of the Results

Interpretation of the results requires a knowledge of the orientation of the water molecules on the platinum surface and the mechanism of infrared radiation absorption.

The vibrational modes of water molecules are given in Figure 27a, where ν_1 and ν_2 are the symmetric modes and ν_3 the antisymmetric mode.

Four models for water adsorption on a metallic surface can be distinguished (Figure 27b). Due to the surface selection rules and remembering that only the vibrational modes which induce a change in the dipolar component perpendicular to the surface are able to absorb the infrared radiation, it is possible to predict the configurations which are active for infrared absorption (Table VIIIa). It is seen that configurations A and B absorb only in the symmetric modes, while configuration D absorbs in the antisymmetric mode and a little in the symmetric modes.

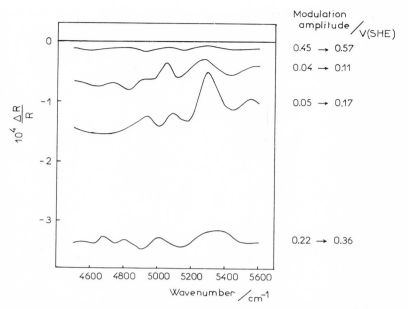

FIGURE 26. EMIRS spectra obtained for the vibrational modes of H_2O with 1 M H_2SO_4 using different modulation limits. After Bewick and Russell.[63]

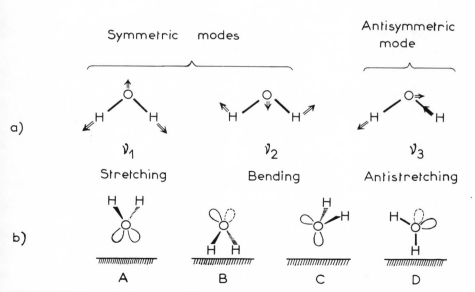

FIGURE 27. Schematic diagrams of (a) the vibrational modes of water and (b) different configurations of adsorbed water molecules.

TABLE VIIIa. Infrared-Active Configurations of Adsorbed Water Molecules

| Configuration | Absorption of infrared radiation | | |
	Symmetric modes, ν_1, ν_2		Antisymmetric mode, ν_3
A	Yes	Yes	No
B	Yes	Yes	No
C	A little	A little	No
D	A little	A little	Yes

TABLE VIIIb. Transitions between the Configurations of Adsorbed Water Molecules

| Transition | Sign of $\Delta R/R$ | |
	Symmetric modes	Antisymmetric mode
C → A	Negative	—
C → B	Negative	—
C → D	—	Strongly negative

To further discuss this model, it is necessary to make a choice of a reference state for the water molecule adsorbed in the double layer. According to Bewick and Russell,[63] the most probable configuration is C, the water molecule having an average orientation with one lobe of the oxygen orbital bonded perpendicular to the surface. Thus, the sign of the relative reflectivity change $\Delta R/R$ depends on the transition between C and the other configurations (Table VIIIb). $\Delta R/R$ is negative for the transition between C and A or C and B when observed in the symmetric modes, and strongly negative in the antisymmetric mode for the transition between configurations C and D. Of course, configurations A, B, and D were also considered as possible reference states by Bewick and Russell. All the results, including the systematic investigation of all types of bands, such as the combination bands $\nu_2 + \nu_3$, and of the shifts observed in the presence of H_2O/D_2O isotopic mixtures, led these authors to decide in favor of a water dimer model in which the hydrogen giving rise to peaks I and I' would be associated (Figure 28).

4.3. Adsorption of Carbon Monoxide on Noble Metals in Aqueous Media

4.3.1. Choice of This Example

Generally speaking, the adsorption of organic molecules affects the active sites of an electrode surface. In order to understand the true activity of a given

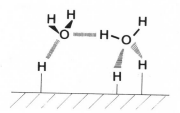

FIGURE 28. The most probable model for hydrogen adsorption on Pt in $1 M$ H_2SO_4, involving a water dimer.[63]

electrode, it is necessary to measure the degree of coverage by the adsorbed organic species and to make a correlation with the geometric area. From that point of view, CO is often considered as a model molecule (or a "test" molecule), already widely studied with regard to its adsorption at the solid/gas interface, at which CO is known to give exceptionally intense infrared absorption bands. Therefore, one can expect it to be the easiest molecule to detect at the electrode/electrolyte interface. Furthermore, its electrochemical behavior is well known. Its electrosorption has been studied in detail on most of the noble metals and, even in the case of platinum, on low-index—(100), (110), (111)—single-crystal planes.[64]

Under these conditions, what information can one expect to gain from an infrared reflectance spectroscopic study? Apart from proving that CO_{ads} itself is detectable, there are at least two main questions:

1. Does CO form multibonded species at the metal/solution interface, and if so, is there any relation with the degree of coverage?
2. Does CO interact with other coadsorbed species, such as adsorbed hydrogen or adsorbed anions?

To answer these questions, two examples are considered: the adsorption of CO on platinum and the adsorption of CO on palladium electrodes in acid media.

4.3.2. Adsorption of CO on Platinum Electrodes

Adsorption and oxidation of CO on platinum electrodes have been the subject of numerous investigations, due to the importance of this molecule in the electrochemical processes that occur in fuel cells.[65,66] The potential at which CO is oxidized depends on the potential at which it was previously adsorbed. It depends also on the coverage of the surface and is sensitive to the structure (i.e., the process is different according to the superficial crystallographic orientation). Furthermore, various authors have suggested the possibility of CO being reduced, at least partially, in the presence of coadsorbed hydrogen.

The first investigations by EMIRS have already shed some light on these various problems.[41,67] Figure 29 contains the voltammograms at 25°C and

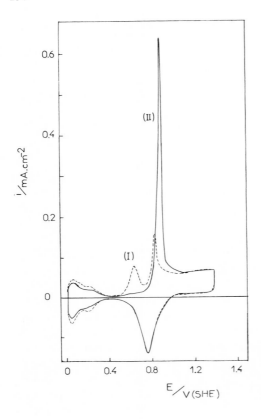

FIGURE 29. Voltammograms of CO adsorbed on a Pt electrode in 0.25 M HClO$_4$ at 25°C and 50 mV s^{-1}: at low CO coverage (— — —); at high CO coverage (——).[68]

$v = 50$ mV s^{-1} of the platinum electrode in 0.25 M HClO$_4$ for a solution either saturated in CO (solid line) or with a low CO concentration (dashed line).[68] In this latter case, the presence of two adsorbed CO species, labeled (I) and (II), is clearly seen. The corresponding electromodulated infrared reflectance spectra in the range 1800–2400 cm^{-1} are given at a constant modulation amplitude of 350 mV, but with the negative limit progressively shifted towards more positive potentials (Figure 30). A strong bipolar EMIRS band, centered at around 2070 cm^{-1}, is seen, with a peak-to-peak intensity first constant, then decreasing when the limits of the modulation become more positive. A perturbation also is seen near 2350 cm^{-1}; its appearance more or less coinciding with the disappearance of the main band. Finally, a weak band also is detectable around 1830 cm^{-1}.

The first conclusion of these preliminary studies was that adsorbed CO was unambiguously detectable at the electrode/electrolyte interface. The discussion which followed led to the conclusion that the main bipolar EMIRS band was due to linearly adsorbed CO. By looking at Figure 12 (Section 3.1.2), it is clear that such a band shape is to be expected when the species which

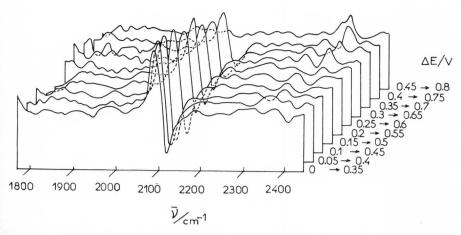

$\Delta E/V$

0.45 → 0.8
0.4 → 0.75
0.35 → 0.7
0.3 → 0.65
0.25 → 0.6
0.2 → 0.55
0.15 → 0.5
0.1 → 0.45
0.05 → 0.4
0 → 0.35

1800   1900   2000   2100   2200   2300   2400

$\bar{\nu}/_{cm^{-1}}$

FIGURE 30. EMIRS spectra of CO adsorbed on Pt in 1 M HClO$_4$ at 25°C. Modulation at 8.5 Hz with the potential limits given in the figure. After Beden *et al.*[67]

absorbs the infrared radiation is present on the surface at both limits of potential, but slightly shifted in frequency, according to a mechanism already proposed by Blyholder.[69]

The second conclusion arose from the detection of a weak band at around 1830 cm^{-1}, appearing either as a single band or a bipolar band according to the bulk concentration of the electroactive species, and of the potential modulation. By analogy with CO adsorption at the solid/gas interface, this weak band was attributed to multibonded CO, i.e., a CO species engaged in bonding at more than one adsorption site. This band is particularly clear in Figure 31a, for CO produced by decomposition, i.e., the dissociative chemisorption of formic acid on a platinum electrode in an acid medium. It is striking how well this spectrum corresponds to that obtained at the solid/gas interface [Figure 31b, for CO adsorbed on Pt(111)[38]], except for a small negative shift in wavenumber. The band at 2350 cm^{-1} is without any doubt due to CO$_2$.

Further detailed studies were made by Kunimatsu *et al.*,[51] using PM-FTIRS, which possesses the advantage of being able to detect real absorption bands. Two series of experiments were considered, one for CO adsorbed at 0.4 V versus SHE (i.e., the double-layer region of platinum), the other at 0.05 V versus SHE (i.e., in the presence of adsorbed hydrogen). The results are given in Figures 32a and 32b, in the wavenumber range of the linearly adsorbed CO species. Some differences in the behavior become obvious. For CO adsorbed at 0.4 V (Figure 32a), the integrated band intensity remains constant up to 0.6 V, just at the potential at which the oxidation process starts leading to CO$_2$ evolution (Figure 33a). There is thus a strong correlation between these two processes. For CO adsorbed at 0.05 V (Figure 32b), the oxidation of CO starts at much lower potentials, around 0.2 V, also as indicated by CO$_2$ evolution.

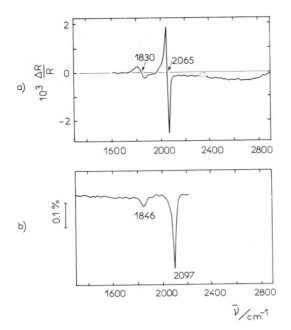

FIGURE 31. Comparison of the IR absorption spectra of CO (a) from the dissociative chemisorption of 0.25 M HCOOH on Pt in 0.25 M H_2SO_4 (EMIRS spectrum)[42] and (b) at the Pt(111)–gaseous CO interface (IRRAS spectrum).[38]

Simultaneously, the integrated band intensity decreases and falls abruptly near 0.4 V (Figure 33b). Interpretation of these results by the authors has led to two different models for adsorption, depending on the potential. When adsorbed at 0.4 V, CO is thought to form islands, at the edge of which oxidation takes place by a process similar to that at the solid/gas interface. However, when CO is adsorbed at 0.05 V, the mechanism of oxidation is different and takes place randomly, which implies that the CO_{ads} species of this type are not bonded together, in contrast to those adsorbed at 0.4 V. Furthermore, the fact that CO_2 evolution occurs at potentials as low as 0.2 V probably implies the existence of at least another CO species, either bridged or multibonded to the surface, thus confirming the electrochemical measurements. In this latter case, the oxidation might proceed by making holes in the CO layer, thus favoring the approach of the next water molecules necessary for the oxidation process to be sustained.

These experiments by PM-FTIRS show clearly the shift in frequency of the linearly adsorbed CO band with respect to potential and thus account for the shape of the absorption band previously obtained by EMIRS. The magnitude of this linear shift, which is about $30\ cm^{-1}\ V^{-1}$, is not affected by the specific adsorption of anions, as shown in Figure 34.[51] A similar shift value was measured by EMIRS for the CO species formed by the dissociative chemisorption of formic acid on platinum,[70] and also for the isoelectronic species CN^-.[71]

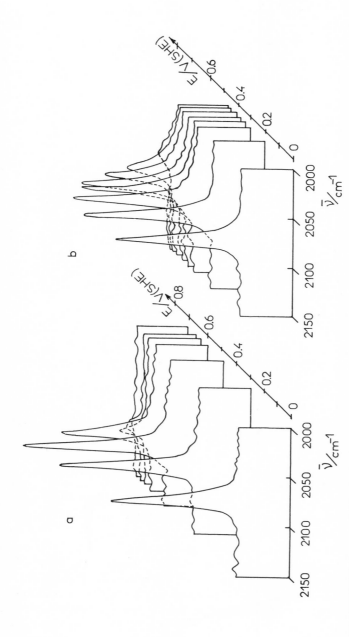

FIGURE 32. PM-FTIRS spectra of CO adsorbed on a Pt electrode in 0.5 *M* H$_2$SO$_4$: (a) At 0.4 V versus SHE; (b) at 0.05 V versus SHE. After Kunimatsu *et al.*[51]

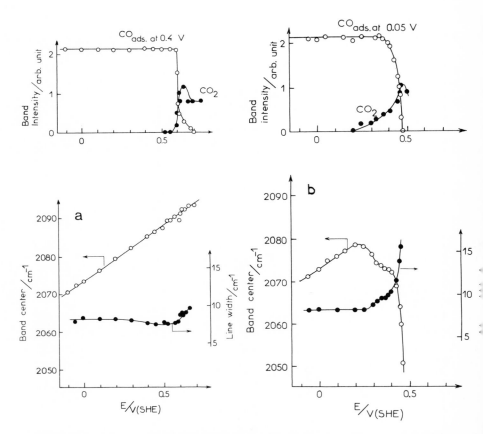

FIGURE 33. Plots of the integrated IR absorption intensity, the band center wavenumber, and the linewidth versus the electrode potential for CO adsorbed on a Pt electrode in 0.5 M H$_2$SO$_4$: (a) At 0.4 V versus SHE; (b) at 0.05 V versus SHE.[51]

Finally, it is important to note that the whole range of wavenumbers, from 4000 to 500 cm^{-1}, was investigated without detecting any species which would correspond to $-$CHO or C$-$OH reduced species.[72]

On the other hand, the isotopic exchange ^{13}CO-^{12}CO was studied by Bewick et al. for various mixture compositions.[89] A shift of around 50 cm^{-1} (towards lower wavenumbers) was observed when ^{12}C is replaced by ^{13}C (Figure 35). This magnitude corresponds to theoretical calculations. It is surprising that in none of the mixtures can the coexistence of the two bands due to ^{12}CO and ^{13}CO be observed. An interpretation is that there is a very strong interaction between the adsorbed molecules (at high coverage, of course).

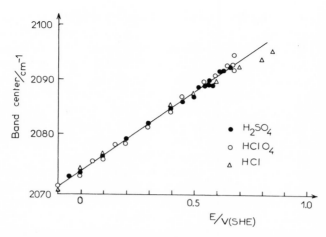

FIGURE 34. Potential shift of the center wavenumber of the band associated with linearly bonded CO adsorbed at 0.4 V versus SHE on Pt in various electrolytes.[51]

4.3.3. Adsorption of CO on Palladium

Adsorption of CO on a palladium electrode is interesting because this metal is known to behave very differently from platinum. Breiter could conclude from electrochemical measurements that most of the adsorbed CO species were occupying more than one site.[73] Similarly, spectroscopic investigations

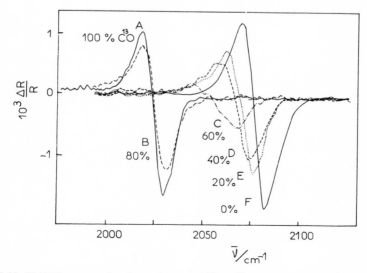

FIGURE 35. EMIRS spectra of various mixtures of $^{12}CO/^{13}CO$ adsorbed on Pt in 1 M H_2SO_4 as a function of % ^{13}CO. After Bewick *et al.*[89]

at the solid/gas interface led to the conclusion that bridged CO species were dominant on palladium.[29]

A detailed study of this system was made by Kunimatsu, using LPSIRS, i.e., by recording reflectograms at various wavelengths and then reconstructing the spectrum.[39] At saturation coverage, two infrared bands are seen (Figure 36), one in the range 2020–2060 cm^{-1} and the second in the range 1920–1980 cm^{-1}. Both shift with the potential, but only the first one shifts linearly. By analogy with the gas phase, the less intense band at 2020–2060 cm^{-1} is attributed to linearly adsorbed species, while the band at 1920–1980 cm^{-1} is associated with the dominant bridge-bonded CO species.

In addition, the CO/Pd system presents a very interesting feature, observed by Bewick *et al.* using EMIRS.[59] At a low coverage θ, a third variety of adsorbed CO appears near 1850 cm^{-1} (Figure 37). Then, there is a steep transition, at $\theta > 0.4$, which leads to the configuration described by Kunimatsu. This system should thus exhibit a phase transition between low coverages (with predominant multibonded species) and high coverages (with the presence of linearly and bridge-bonded species).

4.3.4. Infrared Bands of Adsorbed CO

Table IX gives all the information available at the present time for CO adsorption on noble metals at the electrode/electrolyte interface. By comparing this with the data for the solid/gas interface,[29] it is seen that there is a great similarity in behavior at these two interfaces. However, the spectra in solution are systematically shifted towards lower wavenumbers, by several tens of cm^{-1}.

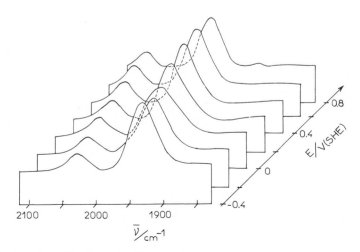

FIGURE 36. LPSIRS spectra of CO adsorbed on a palladium electrode in 1 M HClO$_4$ at different constant potentials. After Kunimatsu.[39]

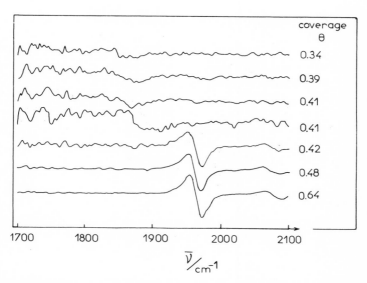

FIGURE 37. EMIRS spectra of CO adsorbed on Pd in 1 M H_2SO_4 for different degrees of coverage θ by adsorbed CO.[80]

No doubt this results from the influence of solvation and of the electrode potential.

The data given in this table are a very good example of the success of infrared reflectance spectroscopy in investigations of the solid/solution interface.

4.4. Adsorbed Intermediates in Electrocatalysis

The electrocatalytic oxidation of small organic molecules, such as methanol or formic acid, involves various adsorbed intermediates which play a major role in the reaction mechanism. Some of them are main intermediates in the reaction path, and some others behave as catalytic poisons, blocking the electroactive sites. The elucidation of the nature of these intermediates, which is still a subject of controversy,[4] is of primary importance, not only to understand the reaction mechanisms, but also to increase the catalytic activity of the electrode and to decrease the poisoning, in order to improve the electrode's long-term stability.

4.4.1. Chemisorption of Methanol at a Platinum Electrode

As a typical example, the mechanism of electrooxidation of CH_3OH dissolved in an acid solution can proceed through different parallel paths.

TABLE IX. Infrared Absorption Bands for CO Adsorbed on Noble Metals

Metal	Solid/gas interface[a]			Electrode/electrolyte interface		
	$\bar{\nu}$ (cm^{-1})	Intensity	CO species	$\bar{\nu}$ (cm^{-1})	Intensity	Comments
Pt	2060–2100	Strong	Linearly bonded on smooth surfaces	2040–2090[b]	Strong	On smooth-surface electrodes
	1800–1875	Very weak	Bridge-bonded or multibonded	1830–1860[b]	Very weak	Intensity varies with coverage; increases on rough surfaces
Rh	2015–2170	Strong	Linearly bonded	2030[c]	Strong	
	1925	Medium strong	Bridge-bonded	1900[c]	Medium strong	
Au	2080–2100	Weak	Linearly bonded	2120[c]	Very weak	⎱ Strongly influenced by the
				1930[c]	Very weak	⎰ potential applied to the electrode
Pd	2080–2100	Weak	Linearly bonded	2030–2070[d]	Weak	
	1925–1985[e]	Very strong	Bridge-bonded and multibonded	1930–1980[d]	Very strong[f]	
				1850	Weak	

[a] Ref. 29.
[b] Refs. 67, 88, 45, 50, 84, and 72.
[c] Ref. 67.
[d] Refs. 39 and 59.
[e] Shoulders at 1950, 1985, and 1840 cm^{-1}.
[f] Two maxima.

According to Soviet electrochemists, the main intermediate is the species $\equiv$COH adsorbed on three adjacent Pt sites, and the reaction mechanism may be formally written[74]:

$$CH_3OH \rightarrow \equiv COH + 3\,H_{ads}$$

$$3\,H_{ads} \rightarrow 3\,H_{aq}^+ + 3\,e^- \quad \text{at } E > 0.4\,V \text{ versus SHE}$$

$$\equiv COH + H_2O \rightarrow CO_2 + 3\,H_{aq}^+ + 3\,e^-$$

According to the Western school, the reaction mechanism involves adsorbed CO as a main intermediate, and thus may be written[75,76]:

$$CH_3OH \rightarrow (CO)_{ads} + 4\,H_{ads}$$

$$4\,H_{ads} \rightarrow 4\,H_{aq}^+ + 4\,e^- \quad \text{at } E > 0.4\,V \text{ versus SHE}$$

$$(CO)_{ads} + H_2O \rightarrow CO_2 + 2\,H_{aq}^+ + 2\,e^-$$

The exact nature of the adsorbed intermediate ($\equiv$COH or CO_{ads}) can only be ascertained unequivocally using an *in situ* spectroscopic technique, because the coulometric methods used until now are much too inaccurate to yield the exact number of electrons (three or two, respectively) needed to oxidize the chemisorption residue of CH_3OH to CO_2.

Using first the EMIRS technique, it was possible to identify without any ambiguity the presence of $(CO)_{ads}$ as the main adsorbed species produced by chemisorption of CH_3OH.[40] Not only was the nature of the species ascertained, but its structure was also precisely established. In fact, two or three types of adsorbed CO may be distinguished on the platinum surface: a linearly bonded —CO linked to one metal atom at around 2080 cm^{-1}, a bridge-bonded $>$CO interacting with two surface atoms at around 1860 cm^{-1}, and, in some cases, a multibonded $\equiv$CO, in interaction with at least three metal adatoms. In the experimental conditions used for this first experiment, the main species is linearly bonded CO, with a very small amount of bridge-bonded CO (Figure 38a).

These results were further confirmed using the LPSIRS technique developed by Kunimatsu.[77] In this technique, the change of reflectance is recorded at various fixed wavenumbers, while the electrode potential is varied linearly. It is then possible to reconstruct the absorbance spectrum at a particular potential by taking the difference between the reflectance at this potential and that at a reference potential where no adsorbed species exists. This gives normal absorption spectra, displaying the same type of bands as in EMIRS measurements.

Since this pioneering work, a systematic investigation of the nature and distribution of chemisorbed methanol species at platinum was carried out by EMIRS. Several intermediates adsorbed at a polycrystalline surface were detected, depending on the methanol concentration in the bulk.[108] Above 10^{-2} *M*, only adsorbed CO bands are detected. Among these linearly-bonded

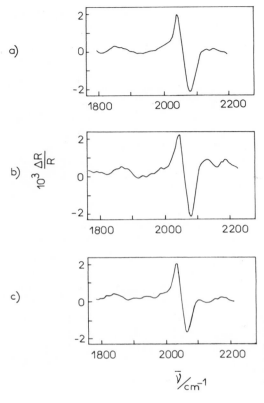

FIGURE 38. EMIRS spectra of CO adsorbed on a Pt electrode in 0.5 M HClO$_4$: (a) CO produced from the chemisorption of 0.25 M CH$_3$OH;[41] (b) CO produced from gaseous CO dissolved in the solution;[67] (c) CO produced from the chemisorption of 0.25 M HCOOH.[42]

CO is predominant, leading to a nearly full monolayer of adsorbed species. This is proved by the magnitude of the IR band compared to that of the EMIRS band of a monolayer of CO adsorbed from gaseous CO dissolved in solution (Figure 38b). At lower concentrations (below $5 \times 10^{-3} M$ CH$_3$OH) almost equal amounts of linearly and bridge-bonded CO species are detected, together with a new intense band, single-sided, around 1700 cm^{-1}. This latter band is associated with the stretching vibration of a carbonyl functional group, possibly a ·CHO-like adsorbed species. Attempts to detect other adsorbed species, such as $\equiv$C—OH, have been unsuccessful.

Furthermore, the surface distribution of adsorbed intermediates depends strongly on the electrode structure, as recently demonstrated using single crystal platinum electrodes.[109] The same intermediates were detected: linearly-bonded CO, bridge-bonded CO, and ·CHO-like species (Figure 39). But conversely to the polycrystalline platinum case, the linearly-adsorbed CO is not the main adsorbed species, even at high methanol concentration, except for Pt(110). The behavior of this crystal is close to that of polycrystalline Pt. Thus, blocking and poisoning of the platinum catalytic surface, which occurs

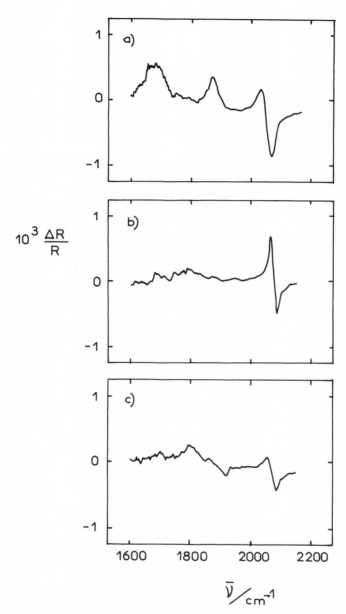

FIGURE 39. EMIRS spectra of the adsorbed species resulting from the chemisorption of 0.1 M CH_3OH in 0.5 M $HClO_4$ at a Pt single-crystal electrode. ($\Delta E = 400$ mV, $\bar{E} = 350$ mV versus RHE, f = 13.6 Hz). Exposed face (a) Pt(100); (b) Pt(110); (c) Pt(111).

during the electrooxidation of methanol, can be interpreted in terms of attractive lateral interactions between the different adsorbed CO species.

4.4.2. Chemisorption of Formic Acid at Platinum, Rhodium, and Gold Electrodes

The electrooxidation of HCOOH at different electrodes has been thoroughly investigated due to the interest in this molecule as a model for an organic fuel.

At platinum, the reaction seems rather simple at first sight, involving in the main path, as a weakly adsorbed intermediate, the species $-COOH_{ads}$. But in a parallel path, a strongly bonded species is formed, which poisons the electrode sites. An investigation by EMIRS showed clearly the occurrence of the same adsorbed $(CO)_{ads}$ as in the case of methanol chemisorption: a linearly bonded $(-CO)_{ads}$ and a bridge-bonded $(>CO)_{ads}$ (Figure 38c).[70] The linearly bonded species predominates, and its intensity is similar to that corresponding to a surface nearly fully covered by CO. For potentials sufficiently positive (greater than about 600 mV versus RHE), these data show the disappearance of $(CO)_{ads}$ and the simultaneous appearance of CO_2, characterized by a band around 2350 cm^{-1}.[42] The $(CO)_{ads}$ is remarkably stable, even in the presence of adsorbed hydrogen and hydrogen evolution (until -700 mV versus RHE).[42]

As mentioned for the CO species adsorbed from dissolved CO, the center of the band of simply bonded $-CO$ varies linearly with the potential, with a positive rate of about $30 \text{ cm}^{-1} \text{ V}^{-1}$.[70] This effect is certainly associated with a reduction in the extent of back donation from the metal d-orbitals into the π^* antibonding orbital of CO as the metal is made more positive. This weakens the Pt$-$C bond and strengthens the C$-$O bond, thus increasing the force constant and the wavenumber.[69]

The adsorption of formic acid at a rhodium electrode is a very interesting case, since this molecule gives two types of adsorbed species by dissociative chemisorption.[78] First, CO type species are observed by EMIRS in the wavenumber range 1850-2180 cm^{-1}: a linearly bonded $-CO$ around 2050 cm^{-1} and a bridge bonded $>CO$ around 1925 cm^{-1} (Figure 40a). These two bands were assigned by comparing them to the EMIRS spectra obtained for adsorbed CO from carbon monoxide dissolved in the supporting electrolyte (Figure 40b). However, the intensity ratio of the two bands of CO arising from the dissociation of HCOOH is slightly different (1.7 instead of 2.0 for gaseous CO, in favor of the bridge-bonded species). Another band is observed in the wavenumber range 1250-1400 cm^{-1} at around 1320 cm^{-1} (Figure 41) and may be due to the symmetric O$-$C$-$O stretch of adsorbed formate, by comparison with the corresponding band for free formate at 1344 cm^{-1}.

The surface distribution of species adsorbed at rhodium also was investigated as a function of the bulk concentration of formic acid, the electrode

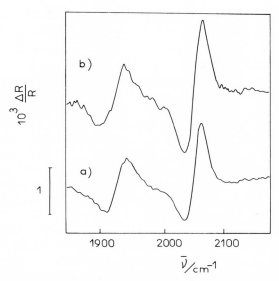

FIGURE 40. EMIRS spectra of CO adsorbed on a Rh electrode in 0.5 M $HClO_4$[78]: (a) CO produced from the chemisorption of 0.1 M HCOOH; (b) CO produced from gaseous CO dissolved in the supporting electrolyte.

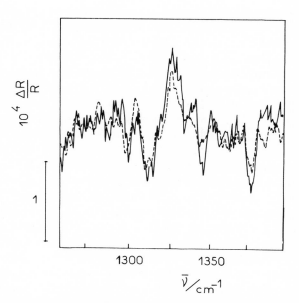

FIGURE 41. EMIRS spectra of adsorbed formate from 0.1 M HCOONa in 0.1 M $NaClO_4$[78]: (————) 6 averaged scans; (– – –) 12 averaged scans.

potential, and the presence of foreign metal adatoms (Pb, Cd, ...)[110,111]. For high HCOOH concentrations ($>10^{-3}$ M), both linearly bonded CO_L and bridge-bonded CO_B species exist with a comparable intensity, whereas for low concentrations ($<10^{-3}$ M) linearly bonded CO_L disappears. Together with CO_B a multibonded CO_m species appears and predominates at very low concentrations ($<10^{-5}$ M). Lead adatoms, which are known to enhance greatly the electrocatalytic activity of rhodium, have a very specific behavior, since they affect only bridge-bonded CO_B, without affecting the other CO_{ads} species (CO_L and CO_m). This leads to a one-by-one replacement of the CO_B by Pb, suggesting that the lead adatoms occupy two adjacent adsorption sites, which certainly may explain the decreasing of poisoning in the presence of Pb adatoms. Cd adatoms, conversely, do not affect the intensity of any adsorbed CO, which means that they do not coadsorb with CO_{ads} species, thus explaining the absence of any catalytic effect.

On the other hand, the adsorption of formic acid at a gold electrode gives two bands, one around 1325 cm^{-1} and the other one around 1720 cm^{-1}.[59] The first one can be attributed, for the same reasons as above, to a formate adsorbed species $(HCOO)_{ads}$, and the second one to adsorption of molecular formic acid, because 1720 cm^{-1} is the wavenumber of the carbonyl stretching mode of formic acid.

4.4.3. Chemisorption of Ethanol at a Platinum Electrode

The study of the adsorption and electrooxidation of ethanol on catalytic surfaces, such as platinum, is of considerable interest, both because of its possible use in fuel cells and because it is the first member of the series of aliphatic alcohols with a C—C bond.

Using EMIR spectroscopy, it was possible to detect numerous absorption bands in the wavenumber range 700–3000 cm^{-1}, depending on the electrode potential and on the bulk concentration of ethanol.[112,113] For a low ethanol concentration (e.g., 10^{-3} M), the EMIR spectrum recorded in the range 1500–3000 cm^{-1} (Figure 42b) displays a main bipolar band at 2070 cm^{-1} due to linearly bonded CO_{ads}, and a weaker band at around 1850 cm^{-1} due to multibonded adsorbed CO_{ads}. This definitively proves that the chemisorption of ethanol leads to a breaking of the C—C bond, even at room temperature. Besides the CO_{ads} species, a unipolar band at ca. 2350 cm^{-1} arises from presumably weakly adsorbed carbon dioxide, CO_2, that is produced by the oxidation of CO_{ads}. Other bands are also seen, around 2880–2980 cm^{-1}, for C—H stretchings of CH_3 groups, and in the range 2600–2800 cm^{-1} due to softened C—H stretching modes, indicative of strong hydrogen bonding. At lower wavenumbers, between 1680 and 1800 cm^{-1}, several bands due to $>C=O$ stretchings of carbonyl functional groups are detected. These may be attributed to either acetaldehyde-like or acetyl-like (or formyl-like) species. A weaker fluctuating band around 1650 cm^{-1} may be associated with the H—O—H

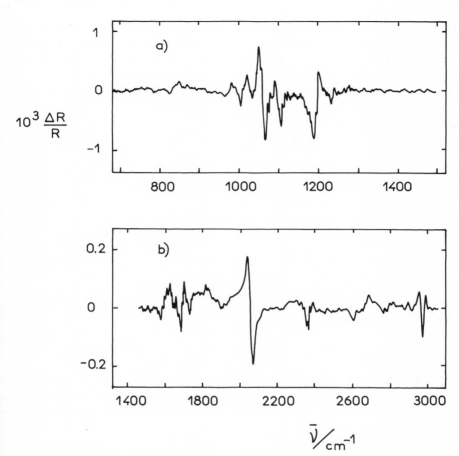

FIGURE 42. EMIRS spectra of the species resulting from the chemisorption of x M C_2H_5OH in 0.5 M $HClO_4$ (ΔE = 400 mV, $\bar{E}$ = 200 mV versus RHE, f = 13.6 Hz, room temperature), at a Pt polycrystalline electrode. (a) x = 0.10^{-1}; (b) x = 10^{-3}.

bending mode of water that is adsorbed at the electrode surface. At 0.1 M CH_3OH, the intensity of the linearly bonded CO_{ads} band becomes much greater and corresponds to coverage of more than one half monolayer of CO_{ads}. In experimental conditions for which the surface is not completely covered by CO_{ads} (Figure 42a), other adsorbed species coexist on the electrode surface as evidenced by the bands observed in the EMIR spectrum recorded between 700 and 1500 cm^{-1}. The most intense band, at 1055 cm^{-1}, would correspond to C—O stretches of a methoxy adsorbed species, while weaker bands at 975 and 1015 cm^{-1} may be attributed to ν C—O of adsorbed ethanol molecules. Another band, relatively intense, at 1200 cm^{-1}, is more difficult to assign, but might correspond to rotation of the CH_3 group. Different models of adsorbed species, that have been discussed in light of the spectral data, include the molecular adsorption of ethanol, and the adsorption of ethoxy, acetaldehyde, and acetyl groups.[112]

4.5. Investigations in Nonaqueous Solvents and Detection of the Intermediates Formed in the Vicinity of the Electrode Surface

4.5.1. Choice of Examples

The examples selected here are chosen for two purposes: on the one hand, to show that reflectance infrared spectroscopy is perfectly appropriate for investigations using organic electrolytes and, on the other, to demonstrate that some of the techniques (such as SNIFTIRS) permit the detection of species formed in the vicinity of the electrode surface by electrolysis.

4.5.2. Spectra of Adsorbed Species in Nonaqueous Media

The spectra of adsorbed species are much easier to obtain in nonaqueous media, as organic solvents generally absorb infrared radiation far less than water. It is possible, with judicious selection of the solvent, to work with solution thicknesses greater than 50 μm. Most of the work in aprotic solvents has been done by Pons et al., using SNIFTIRS (Table VIIb).

The SNIFTIRS spectra for a platinum electrode in a solution of tetra-butylammonium tetrafluoroborate (TBAF) in anhydrous CH_3CN are given in Figure 43.[47] The potentials are given versus the 0.01 M Ag^+/Ag reference electrode in the same solution. Increasing the positive limit of potential leads to a growth of the peaks near 1060 and 2350 cm^{-1}. These peaks are single, indicating that the species which absorbs the radiation is predominant at the positive limit. Complementary studies in lithium perchlorate, instead of TBAF, show that the band at 2350 cm^{-1} persists (and is thus not related to anion adsorption), while the band at 1060 cm^{-1} due to the tetrafluoroborate anion is replaced by one at 1100 cm^{-1} due to the perchlorate ion. The 2350 cm^{-1} band is assigned to the $C\equiv N$ vibration of adsorbed acetonitrile. This band is significantly blue-shifted by comparison to the usual absorption band of acetonitrile at 2220 cm^{-1}.

The spikes near 3000 cm^{-1} are related to partially compensated $C-H$ bands, as far as they exist at both limits of potential. When a little water is added to the solution, new bands appear near 1630 and 3350 cm^{-1} (Figure 44). The band at 1630 cm^{-1} has a negative sign, which corresponds to less adsorption of water at the more positive limit, whereas the band at 3350 cm^{-1} has a bipolar shape. This latter effect is probably the consequence of a shift of the $O-H$ vibration frequency, as a result of both a dependence on potential and the influence of hydrogen bonds.

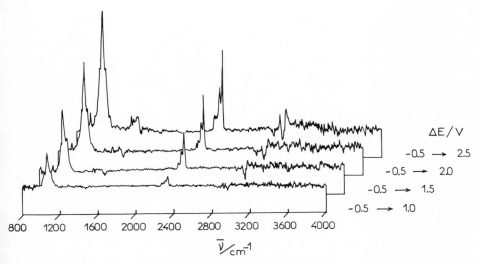

FIGURE 43. SNIFTIRS spectra of 0.1 *M* tetrabutylammonium tetrafluoroborate in anhydrous acetonitrile for different potential pulses (Pt electrode, potential limits of the pulse in volts as indicated), according to Pons *et al.*[47]

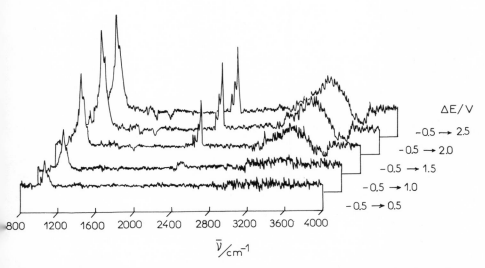

FIGURE 44. SNIFTIRS spectra of 0.1 *M* TBAF + 0.1 *M* H_2O in acetonitrile for different potential pulses (Pt electrode, potential limits in volts as indicated), according to Pons *et al.*[47]

4.5.3. Observation of Anion and Cation Radicals

The SNIFTIRS technique is well suited to the investigation of electrolysis products which accumulate in the vicinity of electrode surfaces. The base potential, E_1, is chosen in a potential range where it is supposed that no faradaic process occurs. The potential is then pulsed to E_2, at which potential the electron transfer takes place. Typical results are presented in Figure 45 for the reduction of tetracyanoethylene (TCNE) on a platinum electrode in acetonitrile[79] and in Figure 46 for the oxidation of 2,6-di-t-butyl-4-phenylaniline.[48] The reduction of TCNE yields an anion radical with characteristic frequencies at 2148 and 2187 cm^{-1} (Figure 45b), i.e., shifted by about 100 cm^{-1} towards lower frequencies when compared to the transmission spectrum of TCNE (Figure 45a). All the bands in the range 800–1200 cm^{-1}, which correspond to the skeletal vibrations, are identical for TCNE and its anion radical. They are thus eliminated by subtraction in the SNIFTIRS spectrum. In contrast, some bands of the cation radical formed during the oxidation of the aniline

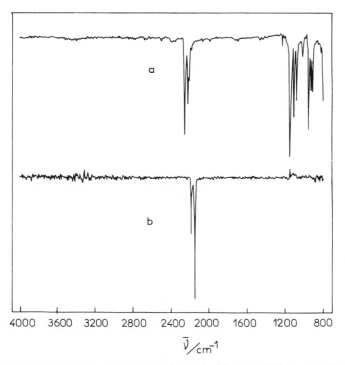

$$\bar{\nu}/\text{cm}^{-1}$$

FIGURE 45. IR spectra of tetracyanoethylene (TCNE): (a) Transmission spectrum of TCNE; (b) SNIFTIRS spectrum recorded during the reduction of TCNE in acetonitrile at a platinum electrode (potential pulses between +0.25 V and −0.25 V versus the 0.01 M Ag$^+$/Ag reference electrode).[79]

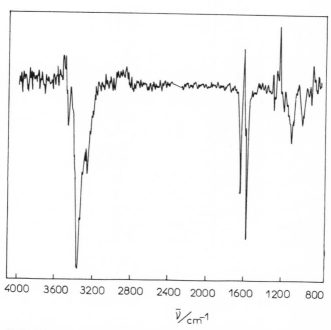

FIGURE 46. SNIFTIRS spectrum recorded during the oxidation of 2,6-di-*t*-butyl-4-phenylaniline in acetonitrile at a Pt electrode (potential pulses between 0.0 and 0.8 V versus Ag$^+$/Ag).[48]

derivative are shifted and thus are not compensated by subtraction. They appear with an inverted sign (Figure 46).

Recently, more detailed information has been obtained by Bewick and Pons[59] to elucidate the behavior of TCNE in solutions of TBAF in acetonitrile. For very low concentrations of TCNE, and using polarized light, it was shown that the molecule was adsorbed flat on the platinum surface.

5. Conclusions

The infrared spectroscopic techniques which are discussed in this chapter and illustrated by a few selected examples, have been developed very recently (within the past five years in most cases). The feasibility of applying *in situ* infrared reflectance spectroscopy to the investigation of the electrode/electrolyte interface is now firmly established.

Since the pioneering work of Bewick *et al.*,[18,36] rapid progress has been made, leading to a considerably enlarged range of applications, from aqueous to nonaqueous systems, and from submonolayers of species adsorbed at the electrode surface to species generated in the double layer (refer to Table VII for details). Furthermore, in the case of adsorbed species, it is not only possible to observe the vibrational spectra, but it also appears possible to make interesting deductions concerning the surface orientation.

Undoubtedly, many aspects of the techniques can still be improved. For example, more effort is needed both to increase the sensitivity and to extend the wavenumber range to higher wavenumbers (where combination bands and overtones are expected) or to lower wavenumbers, down to 100 cm^{-1} (in the far-infrared range, in order to reach the metal–substrate vibrations). In the very near future, other exciting applications are expected, such as *in situ* spectroscopic investigations on well-defined single-crystal electrodes, as well as spectroscopic investigations of short-lived intermediates at catalytic metal surfaces. Such information, which should help to identify the various intermediates and poisons formed on catalytic surfaces, is of major interest in understanding the mechanisms of electrocatalytic reactions.

This optimistic point of view has to be tempered somewhat, however. Even with the latest improvements of the techniques, making them highly sensitive, *in situ* infrared spectroelectrochemical investigations of the electrode/electrolyte interface remain difficult. Each of the techniques described (using dispersive or Fourier transform spectrometers, internal or external reflection) possesses its own advantages and limits. The most appropriate technique to employ depends on the subject under investigation.

If there is one more point to be underlined, it is certainly the absolute necessity, in all cases, to work on very "clean" systems (very clean cells and solutions, well-defined and reproducible surfaces, stable interface during data acquisition, etc.). In practice, this means that for a good interpretation of the *in situ* infrared reflectance spectra, an extensive understanding of the electrochemical behavior of the systems under investigation is a prerequisite.

Acknowledgments

The authors are very grateful to Dr. A. Bewick (University of Southampton, U.K.) for initiating them to the EMIRS technique and for the facilities he has provided during several stays and visits to his laboratory. The fruitful collaboration which ensued led to the first applications in the electrocatalysis field and, more recently, to the establishment of the EMIRS equipment in Poitiers.

Many thanks are also due to Dr. S. Pons (University of Utah, Salt Lake City) and to Dr. K. Kunimatsu (Research Institute for Catalysis, Hokkaido University, Sapporo, Japan) for their continued interest and useful discussions.

Financial support from the CNRS-PIRSEM (grants no. 332 and no. 2004 of the ATP "Applications de l'Electricité à la Chimie, Générateurs électrochimiques") and from the DRET (grant no. 83/1060) for the equipment in Poitiers are gratefully acknowledged.

References

1. J. O'M. Bockris, B. E. Conway, and E. Yeager (eds.), *Comprehensive Treatise of Electrochemisty*, Vol. 1, Plenum Press, New York (1980).
2. A. J. Appleby, in: *Comprehensive Treatise of Electrochemistry*, Vol. 7 (B. E. Conway, J. O'M. Bockris, E. Yeager, S. U. M. Khan, and R. E. White, eds.), pp. 173–239, Plenum Press, New York (1983).
3. A. J. Bard and L. R. Faulkner, *Electrochemical Methods. Fundamentals and Applications*, John Wiley and Sons, New York (1980).
4. D. Pletcher and V. Solis, *Electrochim. Acta* **27**(6), 775 (1982).
5. E. Yeager, *J. Electroanal. Chem.* **150**, 679 (1983).
6. C. Lamy, *J. Electroanal. Chem.* **150**, 545 (1983).
7. D. A. Scherson, S. B. Yao, E. Yeager, J. Eldridge, M. E. Kordesch, and R. W. Hoffman, *J. Electroanal. Chem.* **150**, 535 (1983).
8. J. D. E. McIntyre, in: *Optical Properties of Solids. New Developments* (B. O. Seraphin, ed.), pp. 559–630, North-Holland Publishing Co., Amsterdam (1976).
9. H. B. Mark and B. S. Pons, *Anal. Chem.* **38**, 119 (1966).
10. A. H. Reed and E. Yeager, *Electrochim. Acta* **15**(8), 1345 (1970).
11. M. Cardona, *Modulation Spectroscopy*, Academic Press, New York (1969).
12. J. D. E. McIntyre and D. M. Kolb, *Symp. Faraday Soc.* **4**, 99 (1970).
13. J. D. E. McIntyre, *Surface Sci.* **37**, 658 (1973).
14. J. Pritchard, T. Catterick, and R. K. Gupta, *Surface Sci.* **53**, 1 (1975).
15. F. Hoffman and A. M. Bradshaw, *Surface Sci.* **72**, 513 (1977).
16. J. Corset, *Le Vide, Les Couches Minces* **216**, 247 (1983).
17. J. S. Clarke, A. T. Kuhn, W. J. Orville-Thomas, and M. Stedman, *J. Electroanal. Chem.* **49**, 199 (1974).
18. A. Bewick, K. Kunimatsu, and B. S. Pons, *Electrochim. Acta* **25**(4), 465 (1980).
19. M. Cardona, K. L. Shaklee, and F. H. Pollak, *Phys. Lett.* **23**, 37 (1966).
20. B. O. Seraphin, in: *Optical Properties of Solids* (F. Abélés, ed.), p. 163, North-Holland Publishing Co., Amsterdam (1972).
21. M. Fleischmann, P. J. Hendra, and A. J. McQuillan, *Chem. Phys. Lett.* **26**, 163 (1974).
22. A. Otto, *Appl. Surface Sci.* **6**, 309 (1980).
23. R. K. Chang and T. E. Furtak (eds.), *Electrochemical Effects in Surface-Enhanced Raman Scattering*, Plenum Press, New York (1982).
24. M. Born and E. Wolf, *Principles of Optics*, Pergamon Press, London (1964).
25. W. J. Moore, *Physical Chemistry*, Longman, London (1978), Chapter 17, pp. 747–827.
26. P. W. Atkins, *Physical Chemistry*, Oxford University Press, London (1984), Chapter 17, pp. 563–601.
27. G. Kortüm, *Reflectance Spectroscopy*, Springer-Verlag, Heidelberg (1969).
28. M. J. Dignam and J. Fedyk, *Appl. Spectrosc. Rev.* **14**(2), 249 (1978).
29. J. Pritchard, in: *Chemical Physics of Solids and Their Surfaces*, Specialist Periodical Reports, Vol. 7, pp. 157–179, The Chemical Society, London (1978).
30. W. R. Hunter, *J. Opt. Soc. Am.* **55**, 1197 (1965).
31. W. N. Hansen, *J. Opt. Soc. Am.* **58**, 380 (1968).
32. J. D. E. McIntyre and D. E. Aspnes, *Surface Sci.* **24**, 417 (1971).
33. L. H. Little, *Infrared Spectra of Adsorbed Species*, Academic Press, London (1966), Chapter 15, pp. 382–402.
34. A. Bewick and K. Kunimatsu, *Surface Sci.* **101**, 131 (1981).
35. H. Neugebauer, G. Nauer, and N. Brinda-Konopik, 32nd I.S.E. Meeting, Dubrovnik/Cavtat, Yugoslavia (1981).
36. A. Bewick, K. Kunimatsu, B. S. Pons, and J. W. Russell, *J. Electroanal. Chem.* **160**, 47 (1984).
37. K. Kunimatsu, *J. Electroanal. Chem.* **140**, 205 (1982).

38. W. G. Golden, D. S. Dunn, and J. Overend, *J. Catal.* **71,** 395 (1981).
39. K. Kunimatsu, *J. Phys. Chem.* **88,** 2195 (1984).
40. B. Beden, A. Bewick, K. Kunimatsu, and C. Lamy, *J. Electroanal. Chem.* **121,** 343 (1981).
41. A. Bewick, K. Kunimatsu, B. Beden, and C. Lamy, 32nd I.S.E. Meeting, Dubrovnik/Cavtat, Yugoslavia (1981), Extended Abstract A 28, p. 92.
42. B. Beden, A. Bewick, and C. Lamy, *J. Electroanal. Chem.* **148,** 147 (1983).
43. J. W. Russell, J. Overend, K. Scanlon, M. W. Severson, and A. Bewick, *J. Phys. Chem.* **86,** 3066 (1982).
44. A. Bewick and J. Robinson, *J. Electroanal. Chem.* **60,** 163 (1975); **71,** 131 (1976).
45. J. W. Russell, M. Severson, K. Scanlon, J. Overend, and A. Bewick, *J. Phys. Chem.* **87,** 293 (1983).
46. J. W. Russell and M. Severson, 35th I.S.E. Meeting, Berkeley, California (1984), Extended Abstract A 8-25, p. 512.
47. S. Pons, T. Davidson, and A. Bewick, *J. Electroanal. Chem.* **160,** 63 (1984).
48. S. Pons, *J. Electroanal. Chem.* **150,** 495 (1983).
49. M. A. Habib and J. O'M. Bockris, *J. Electroanal. Chem.* **180,** 287 (1984).
50. W. G. Golden, K. Kunimatsu, and H. Seki, *J. Phys. Chem.* **88,** 1275 (1984).
51. K. Kunimatsu, H. Seki, W. G. Golden, J. G. Gordon II, and M. R. Philpott, *Surface Sci.* **158,** 596 (1985).
52. H. Neugebauer, G. Nauer, N. Brinda-Konopik, and G. Gidaly, *J. Electroanal. Chem.* **122,** 381 (1981).
53. N. Brinda-Konopik, H. Neugebauer, G. Gidaly, and G. Nauer, *Mikrochim. Acta,* Suppl. 9, 329 (1981).
54. M. A. Habib and J. O'M. Bockris, *J. Electrochem. Soc.* **132,** 108 (1985).
55. H. Neff, E. Piltz, K. Bange, and J. K. Sass, *Vacuum,* in press.
56. H. Neff, P. Lange, D. K. Roe, and J. K. Sass, *J. Electroanal. Chem.* **150,** 513 (1983).
57. S. Pons, J. Foley, and C. Marcott, 35th I.S.E. Meeting, Berkeley, California (1984), Extended Abstract A 8-2, p. 448.
58. W. G. Golden, D. D. Saperstein, M. W. Severson, and J. Overend, *J. Phys. Chem.* **88,** 574 (1984).
59. A. Bewick and S. Pons, in: *Advances in Infrared and Raman Spectroscopy* (R. J. H. Clark and R. E. Hester, eds.), Vol. 12, pp. 1-63, Wiley Heyden, London (1985).
60. J. Clavilier, D. Armand, and B. L. Wu, *J. Electroanal. Chem.* **135,** 159 (1982).
61. A. Bewick, K. Kunimatsu, J. Robinson, and J. W. Russell, *J. Electroanal. Chem.* **119,** 175 (1981).
62. A. Bewick, M. Fleischmann, and J. Robinson, *Dechema Monographie* **90,** 1851; *Elektrochem. Elektron.,* 87 (1981).
63. A. Bewick and J. W. Russell, *J. Electroanal. Chem.* **132,** 329 (1982).
64. J. M. Leger, B. Beden, C. Lamy, and S. Bilmes, *J. Electroanal. Chem.* **170,** 305 (1984).
65. M. W. Breiter, *J. Electroanal. Chem.* **101,** 329 (1979).
66. J. Sobkowski and A. Czerwinski, *J. Phys. Chem.* **89,** 365 (1985).
67. B. Beden, A. Bewick, K. Kunimatsu, and C. Lamy, *J. Electroanal. Chem.* **142,** 345 (1982).
68. N. Collas, B. Beden, J. M. Leger, and C. Lamy, *J. Electroanal. Chem.* **186,** 287 (1985).
69. G. Blyholder, *J. Phys. Chem.* **68,** 2772 (1964).
70. B. Beden, C. Lamy, and A. Bewick, *J. Electroanal. Chem.* **150,** 505 (1983).
71. K. Kunimatsu, H. Seki, W. G. Golden, and J. G. Gordon II, 35th I.S.E. Meeting, Berkeley, California (1984), Extended Abstract A 8-6, p. 457.
72. K. Kunimatsu, H. Seki, W. G. Golden, J. Gordon II, and M. Philpott, *Langmuir* **2,** 464 (1986).
73. M. W. Breiter, *J. Electroanal. Chem.* **109,** 243 (1980).
74. V. S. Bagotzky and Yu. B. Vassiliev, *Electrochim. Acta* **12,** 1323 (1967).
75. S. Gilman, *J. Phys. Chem.* **68,** 70 (1964).
76. T. Biegler, *J. Phys. Chem.* **72,** 1571 (1968).
77. K. Kunimatsu, *J. Electroanal. Chem.* **145,** 219 (1983).

78. F. Hahn, B. Beden, and C. Lamy, *J. Electroanal. Chem.* **204**, 315 (1986).
79. S. Pons, T. Davidson, and A. Bewick, *J. Am. Chem. Soc.* **105**, 1802 (1983).
80. A. Bewick, *J. Electroanal. Chem.* **150**, 481 (1983).
81. B. Beden, *Spectra 2000* **13**(95), 19 (1984); **13**(96), 31 (1984).
82. C. Gibilaro, Ph.D. thesis, University of Southampton (1982).
83. H. Neugebauer, G. Nauer, N. Brinda-Konopik, and R. Kellner, *Fresenius Z. Anal. Chem.* **314**, 266 (1983).
84. K. Kunimatsu, W. G. Golden, H. Seki, and M. R. Philpott, *Langmuir* **1**, 245 (1985).
85. P. Lange, V. Glaw, H. Neff, E. Piltz, and J. K. Sass, *Vacuum* **33**(10–12), 763 (1983).
86. A. Bewick, C. Gibilaro, M. Razak, and J. W. Russell, *J. Electron Spectrosc. Relat. Phenom.* **30**, 191 (1983).
87. A. Bewick and J. W. Russell, *J. Electroanal. Chem.* **142**, 337 (1982).
88. J. W. Russell, J. Overend, K. Scanlon, M. Severson, and A. Bewick, *J. Phys. Chem.* **150**, 495 (1983).
89. A. Bewick, M. Razaq, and J. W. Russell, *Surface Sci.*, in press.
90. H. Seki, K. Kunimatsu, and W. G. Golden, 35th I.S.E. Meeting, Berkeley, California (1984), Extended Abstract A 9–29, p. 522.
91. K. Kunimatsu, H. Seki, and W. G. Golden, *Chem. Phys. Lett.* **108**, 195 (1984).
92. K. A. Bunding, K. Kunimatsu, J. G. Gordon, W. G. Golden, and H. Seki, Electrodynamics and Quantum Phenomena at Interfaces Meeting, Telavi (USSR) (1984).
93. K. Kunimatsu, *J. Electron. Spectroscopy* **30**, 215 (1983).
94. B. Beden, C. Lamy, and A. Bewick, 33rd I.S.E. Meeting, Lyon, France (1982), Extended Abstract IA 17, p. 49.
95. A. Bewick in: *The Chemistry and Physics of Electrocatalysis* (J. D. E. McIntyre, M. J. Weaver, and E. B. Yeager, eds.), *Proc. Electrochem. Soc.* **84**(12), 301 (1984).
96. B. Beden, A. Bewick, M. Razak, and J. Weber, *J. Electroanal. Chem.* **139**, 203 (1982).
97. M. A. Habib and J. O'M. Bockris, *J. Electrochem. Soc.* **130**(12), 2510 (1983).
98. P. Zelenay, M. A. Habib, and J. O'M. Bockris, *J. Electrochem. Soc.* **131**, 2464 (1984).
99. A. Bewick, C. Gibilaro, and S. Pons, Report 1984 TR-39 Order no. AD A14866: 4/6 GAR. Avail. NTIS, From Gov. Rep. Announce Index US, 85-6 (1985) 58.
100. S. Pons and A. Bewick, *Langmuir* **1**, 141 (1985).
101. S. Pons, M. Datta, J. F. McAleer, and A. S. Hinman, *J. Electroanal. Chem.* **160**, 369 (1984).
102. T. Davidson, S. Pons, A. Bewick, and P. Schmidt, *J. Electroanal. Chem.* **125**, 237 (1981).
103. S. Pons, T. Davidson, and A. Bewick, *J. Electroanal. Chem.* **140**, 211 (1982).
104. H. Neugebauer, G. Nauer, N. Brinda-Konopik, A. Neckel, G. Tourillon, and F. Garnier, 34th I.S.E. Meeting, Erlangen, FRG (1983), Extended Abstract 0826.
105. H. Neugebauer, A. Neckel, G. Nauer, N. Brinda-Konopik, F. Garnier, and G. Tourillon, *J. Phys.* (*Paris*) *C10*, **44**(12), 517 (1983).
106. S. Pons, S. B. Khoo, A. Bewick, M. Datta, J. J. Smith, A. C. Hinman, and G. Zachmann, *J. Phys. Chem.* **88**, 3575 (1984).
107. H. Nakajima, H. Kita, K. Kunimatsu, and A. Aramata, *J. Electroanal. Chem.* **201**, 175 (1986).
108. B. Beden, F. Hahn, S. Juanto, C. Lamy and J-M. Leger, *J. Electroanal. Chem.* **225**, 215 (1987).
109. B. Beden, S. Juanto, J-M. Leger and C. Lamy, *J. Electroanal. Chem.* **238**, 323 (1987).
110. M. Choy de Martinez, B. Beden and C. Lamy, 38th I.S.E. Meeting, Maastricht, The Netherlands (September 1987), Extended Abstract 4.66, Volume I, p. 404.
111. M. Choy de Martinez, B. Beden, F. Hahn and C. Lamy, *J. Electroanal. Chem.*, submitted.
112. B. Beden, M-C. Morin, F. Hahn and C. Lamy, *J. Electroanal. Chem.* **229**, 353 (1987).
113. J. M. Perez, A. Aldaz, B. Beden and C. Lamy, *J. Electroanal. Chem.*, submitted.
114. M. C. Pham, F. Adami, P. C. Lacaze, J. P. Doucet, and J. E. Dubois, *J. Electroanal. Chem.* **201**, 413 (1986); **210**, 295 (1986).

6.

Surface-Enhanced Raman Scattering

Ronald L. Birke and John R. Lombardi

1. Overview

1.1. Introduction

Surface-enhanced Raman scattering (SERS) has been observed at solid/solution, solid/gas, solid/vacuum, and solid/solid interfaces, and it is possibly the most sensitive surface high-resolution vibrational spectrosopic technique available as an analytical probe. The great interest in SERS is evidenced by the virtual explosion in the number of publications dealing with this technique since 1977 from the physics, chemistry, and materials science research communities. One reason for this interest in SERS is the generality of the technique with regard to the nature of the material phase in contact with the solid surface. Another reason is that there are several mechanisms that contribute to SERS, and a considerable scientific challenge exists in determining the extent of the contributions of various mechanisms to the overall enhancement.

This chapter is mainly concerned with the metal/solution interface in an electrochemical cell. SERS and resonance Raman scattering (RRS) are particularly important for this system, for which there are relatively few *in situ* surface analytical probes available, in comparison to the many varieties of such probes available for solid/ultrahigh vacuum (UHV) interfacial systems. The organization of this chapter is as follows. Section 1 contains an overview of the SERS technique, including some experimental observations and theoretical considerations. In Section 2, experimental details are given for obtaining SERS in electrochemical cells; in Sections 3 and 4, a theoretical treatment of different SERS enhancement mechanisms is developed; and in Section 5, overall enhancement equations are discussed. Symmetry considerations for

Ronald L. Birke and John R. Lombardi • Department of Chemistry, City College, City University of New York, New York, New York 10031.

SERS are given in Section 6; in Section 7, the effect of electrode potential on SERS is considered; and in Section 8, chemical systems which have been investigated with SERS are discussed.

1.2. Light Scattering by Molecules

When the oscillating electric field associated with light interacts with a molecule, a small dipole moment is induced through the polarizability of the molecule. This induced dipole oscillates at the same frequency as the exciting light. However, it is known that an oscillating dipole radiates light at the same frequency at which it oscillates. This radiated light is not necessarily in the same direction as the exciting light and is thus scattered. This phenomenon is termed Rayleigh scattering. The intensity of scattering is proportional to the fourth power of the frequency of the exciting light so light of higher frequencies is considerably more likely to be scattered.

An additional type of scattering is obtained due to the fact that molecules themselves are vibrating at frequencies corresponding to various normal modes of motion. These characteristic vibrational frequencies can mix with the exciting light to form sum and difference frequencies in the scattered radiation. Although weak, these frequencies may be detected as shifts from the Rayleigh frequency and are called normal Raman spectra. The measure of these shifts reflects the characteristic vibrations of the molecule and may be utilized as a complement to infrared spectroscopy.

In the above, we have assumed that the exciting light is not oscillating at an absorption frequency of the molecule so that the process does not involve an excited electronic state of the molecule. However, if the exciting light is chosen to coincide with an electronic absorption, the Raman spectra are greatly intensified. This phenomenon is very useful in obtaining spectra when the normal (nonresonance) Raman (NR) spectrum is too weak to be detected and is termed resonance Raman spectroscopy (RRS).

When a molecule which can scatter light is either adsorbed on a *rough* metal surface with the right dielectric properties or within a distance of molecular dimensions to this surface, the exciting light and the light scattered by the molecule can both interact with the surface. This is the situation that is found in SERS, and the scattered light, which is detected by the Raman spectrometer, shows some of the features of both normal Raman scattering and resonance Raman scattering in addition to new features. It is, in fact, a new type of molecular light scattering in which the proximity of the molecule to the rough metal surface leads to an amplification of the molecularly scattered light and to a unique type of light scattering!

1.3. Characteristics of Surface Raman Scattering

At an electrode/solution interface, it is possible to obtain all three types of Raman scattering. If a thick surface film is formed by electrode reaction or multilayer adsorption, the concentration may be large enough in the film to give normal Raman scattering. This situation is easy to detect since the NR spectrum obtained is identical in relative intensities and Raman-shifted frequencies to a bulk sample of the film. On the other hand, an optically absorbing molecular species within the diffusion layer of an electrode system, when excited by light within its absorption band, has enough scattering intensity even at very low bulk concentrations to give a resonance Raman spectrum. This type of spectrum will be potential dependent if the RR-active species is involved in an electrode process.

A RRS process can be distinguished from a SERS process by several characteristics. Among these are: (1) a RR spectrum can be obtained on a non-SERS-active metal such as Pt or Hg or on a smooth SERS-active metal, while SERS can only be obtained on a roughened SERS-active metal such as Ag, Cu, or Au, (2) a RR spectrum may show strong overtones or combination bands in contrast to a SERS spectrum, which does not show such bands with intensities comparable to those of fundamental modes, and (3) the influence of parallel and perpendicularly polarized light on band intensities may be different in the two light scattering processes (see below).

It should be mentioned that RR scattering also can be enhanced by a SERS-active metal, i.e., a surface-enhanced RRS process. In the case of an adsorbed dye molecule, for example, excited in its absorption band, Raman spectra can be obtained at dye concentrations below those necessary for a solution RR spectrum. However, the surface enhancement mechanism, in this case, will not contain the same type of resonance enhancement proposed for the SERS of nonabsorbing molecules. It is also expected for molecules near a metal surface that the molecular RRS process may be damped by energy transfer to the metal leading to lower excited-state lifetimes and smaller resonance enhancements. On the other hand, fluorescence will be even more strongly damped at metal surfaces, making it possible to obtain surface RR spectra in the absence of an interfering fluorescence background.

A SERS spectrum can be obtained for nonoptically absorbing species and will be localized to molecules directly at the electrode surface or within the electric double-layer region. Thus submonolayer, monolayer, or multilayer systems a few layers thick, from which it is impossible to obtain a NR spectrum, will give SERS on an activated (roughened) SERS electrode. The majority of SERS band positions are found to be slightly shifted from NR and RR spectra of the same molecule, and relative intensities will be different for the three types of light scattering processes. However, the NR and SERS spectra are similar enough so that molecular identification is easily achieved. Also, SERS intensities are found to be a function of electrode potential even in a region where only nonfaradaic electrode processes occur. A very characteristic and distinguishing feature of SERS is the insensitivity of the band intensities of molecular vibrations of different symmetry types to the polarization direction of the exciting light, i.e., SERS bands are said to be depolarized. The fact that

SERS is localized to the vicinity of a metal surface means that it is a highly specific surface probe which can be used to study both static and dynamic molecular processes at activated metal surfaces.

1.4. The SERS Experiment

The first report of surface Raman spectroscopy of a molecule adsorbed at the metal/solution interface under potentiostatic electrochemical control was the paper of Fleischmann et al.[1] The Raman spectrum of pyridine adsorbed on an Ag electrode from KCl electrolyte medium was observed after 15 min of an oxidation–reduction cycle (ORC) applied by a repeating triangular potential waveform. To obtain the surface Raman spectrum, the electrode was exposed to a beam of laser light and the scattered light from the electrode surface was collected with a Raman spectrometer. These authors initially thought that the OR cycling had significantly increased the surface area of the electrode so that a well-defined Raman spectrum of pyridine could be obtained. It was soon pointed out independently by Jeanmaire and Van Duyne[2] and Albrecht and Creighton[3] that only a single-cycle ORC was necessary to produce an intense Raman spectrum and that the scattered light from pyridine was anomalously enhanced by at least five to six orders of magnitude over that of an isolated molecule in the bulk of the solution. These reports opened a new field of scientific inquiry—the study of enhanced optical scattering from surfaces—which has grown rapidly in the last few years.[4-8] In this chapter, we only will treat linear optical scattering; methods based on nonlinear optical phenomena have been reviewed elsewhere.[8]

For the pretreated electrochemical interface, it has been postulated[4-8] that two main mechanisms are operating: (a) an electromagnetic enhancement related to the enhancement of the oscillating electric field of the exciting light by large-scale metal surface structures and (b) a charge transfer resonance Raman type enhancement possibly related to atomic-scale structures. Both of these types of surface roughness are created during the ORC pretreatment of the SERS-active electrode. Thus, electrochemical pretreatment produces a type of "modified" electrode with new optical properties and possibly new electrochemical properties due to active atomic-scale sites, probably adatom sites. Most of the earlier SERS work was accomplished with a Ag electrode but it was soon discovered that Cu and Au surfaces[9-16] show SERS. The first observation of SERS on Cu with blue laser exciting light was very weak;[9] however, it was found later that red laser excitation allowed well-defined SERS spectra.[10-16] Other metals such as the alkali metals[17,18] and possibly Ni, Ti, Pd,[19] and Al[20] show SERS, and SERS of pyridine on a roughened beta palladium hydride electrode has also been reported.[21] Alkali metals, Ni, Ti, and Al metal substrates have not been demonstrated to give SERS in the electrochemical environment. Although reports of SERS at Hg have appeared,[22,23] it is unlikely that a Hg surface give observable SERS.[4]

Even if it proves that only Ag, Au, and Cu can be routinely used for SERS in the electrochemical environment, these three electrodes can be used to investigate an *immense* range of electrochemical phenomena. This fact is obvious when one considers that both Ag and Au are general-purpose working electrodes which can be used to study nonfaradaic processes as well as electroreduction and electrooxidation reactions over a wide range of potentials. Ag is an electrode with a high hydrogen overpotential, much like the Hg electrode, and Au is a noble metal with a large, useful potential range, much like Pt. The Cu electrode, on the other hand, has a more limited range as a working electrode; however, it can be used to study (a) cathodic deposition (electrocrystallization), (b) anodic dissolution (corrosion), (c) cathodic hydrogen evolution, and (d) anodic formation of films such as halide and oxide films.

1.5. Active Sites and the Quenching of SERS

A rough metal surface can also be made by coldly evaporating Ag films in an ultrahigh-vacuum (UHV) system. The Raman intensities found from such a system are comparable to those observed with ORC-pretreated electrodes. When these surfaces are annealed by warming to room temperature, the SERS intensity is irreversibly lost.[24-26] Such experiments have been interpreted in terms of the loss of atomic-scale defects during the annealing process[7]; however, it has been argued that a loss of large-scale surface structures is also possible.[27] An analogous process occurs for pyridine on electrodes which have been pretreated by an ORC and then held at potentials negative to −1.2 V versus SCE. When the potential is brought back in a positive direction, short of that required for oxidation of the Ag, the scattering intensity is not reestablished.[28-31] This hysteresis, which depends on scan rate in a cyclic sweep, is illustrated in Figure 1b. A similar result has been obtained for thiourea on Ag.[32] This phenomenon has been interpreted as resulting from the desorption of ligands from a SERS-active silver adatom–ligand complex, allowing the adatoms to be incorporated into the metal lattice by surface diffusion and destroying SERS atomic-scale active sites.[30,32] Thus, electrode potential quenching is analogous to temperature annealing. Large-scale surface structures which give rise to an electromagnetic enhancement should still exist after potential quenching.

Another possibility for quenching SERS-active atomic-scale sites in the electrochemical environment is by underpotential deposition of metals such as Tl and Pb.[33-35] A very small underpotential coverage of deposited Tl, ca. 3% leads to almost complete quenching of SERS from pyridine on Ag.[33-35] This quenching is irreversible since anodic stripping of Tl does not lead to restoration of the SERS signal. Underpotential deposition of Pb gives much weaker quenching and the quenching uniformly increases as the Pb coverage increases.[33] The Tl experimental results are interpreted in terms of the destruction of atomic-scale active sites with an Ag atom being replaced by a Tl atom

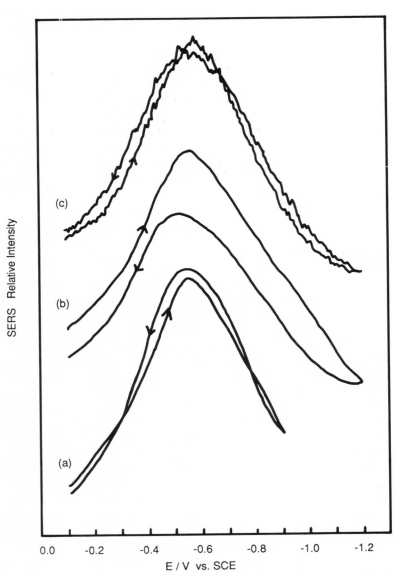

FIGURE 1. SERS intensity versus a triangular sweep potential for the 1006-cm^{-1} band of pyridine as a function of switching potential and scan rate. Solution: 0.05 M pyridine and 0.1 M KCl. (a) Intensity versus potential (I-V) curves shows no hysteresis for a switching potential of -0.9 V versus SCE at a scan rate of 100 mV s^{-1}. (b) I-V curve shows hysteresis is observed for a switching potential of -1.2 V versus SCE at a scan rate of 5 mV s^{-1}. At slower scan rates, still more hysteresis develops.$^{(122)}$ (c) I-V curve shows no hysteresis for a switching potential of -1.2 V versus SCE at a scan rate of 500 mV s^{-1}.

at a SERS active site.[33,34] Cu deposition on Ag at only 0.1 monolayer leads to a shift in the pyridine spectrum typical of an Ag surface to that typical of a Cu surface.[35] All of these results indicate very high enhancements for molecules adsorbed at atomic-scale active sites.

Additional evidence for atomic-scale active sites from SERS on Ag are the deposition experiments of Furtak and coworkers[36,37] and those of Marinyuk *et al.*[38] In the former experiment, monolayer deposits of Ag on Au show SERS from pyridine under excitation with blue laser light. An Au electrode is similar to a Cu electrode in that it only gives SERS with red laser excitation; thus with blue light it would be impossible to observe SERS of pyridine on Au. In the latter experiment,[38] SERS is observed after monolayer amounts of Ag are deposited on Pt, a surface which does not normally show SERS in the electrochemical environment.

1.6. Metal–Molecule Complex

The existence of an adatom–molecule complex formed after an ORC pretreatment of an Ag electrode has been deduced from the examination of spectral data for the SERS of cyanide[39] and 2,6-dimethylpyridine[40] (lutidine) on Ag. Additional evidence for formation of an Ag adatom–molecule complex was provided by spectral studies of SERS from interfacial H_2O. A SERS spectrum of adsorbed water apparently is only possible if a coadsorbed halide is present.[41-44] The concentration of the halide must be above 1 M for Cl^- and Br^-, whereas I^- solutions show the enhanced H_2O spectrum at concentrations as low as 0.1 M.[44] The halide ion is viewed as specifically adsorbed and acting to break up hydrogen bonding in H_2O through its structure-breaking capacity. Thus, the halide promotes the formation of water monomers bound to silver adatoms by disrupting hydrogen bonding and stabilizing the active site by its involvment in the surface complex. A similar phenomenon has been observed for ammonia on Ag.[45] When the supporting electrolyte (KCl) was below 0.5 M in concentration, a SERS spectrum of NH_3 could not be obtained, and the most intense spectra were obtained at the highest concentration used, 4.0 M KCl. Again the halide effect can be interpreted as (a) causing the disruption of hydrogen bonding, allowing the formation of the silver–molecule species and (b) stabilizing the surface complex by participation of specifically adsorbed halide.

Several experiments have demonstrated the necessity for sites in the molecule that can coordinate chemically to the surface in order to achieve intense SERS. Comparison of 4-pyridine carboxylic acid (isonicotinic acid) with benzoic acid on a silver island film SERS surface (Figure 2) showed that only when the carboxylate end of the benzoic acid was chemisorbed could SERS be obtained;[46] whereas the isonicotinic acid, which has two coordinating sites, the carboxylate and the pyridine nitrogen, gave SERS in either configuration.[47] Similar experiments[48] indicated that the isonicotinic acid,

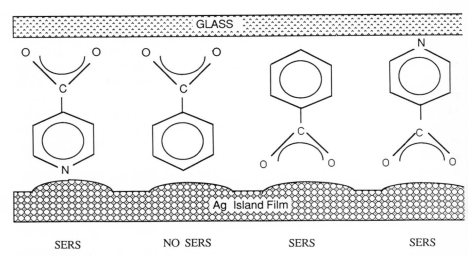

SERS   NO SERS   SERS   SERS

FIGURE 2. Schematic of the effect of coordination at a Ag island film surface–glass sandwich
on the SERS of benzoic acid and isonicotinic acid.

with its lone pair on the nitrogen, was more effective in giving SERS than
benzoic acid. It has also been shown that acridine gives SERS on Ag in an
electrochemical cell; whereas anthracene, without the coordinating nitrogen,
does not exhibit SERS on Ag.[49] Analogous results are obtained from ethylene
and ethane[50] with another type of SERS substrate, i.e., a colloidal silver/argon
matrix. Here, while ethylene showed SERS, none was observed for ethane.
Indeed, olefinic and aromatic molecules can bind to the SERS metal surfaces
through their π-orbital system. Apparently, what is necessary for intense
enhancements is a chemical bond of moderate strength between the SERS-
active molecule and the SERS-active metal substrate. This requirement is
consistent with a charge transfer type mechanism.

A wide variety of molecules and ions have shown SERS (more than 80
different species have been observed to give SERS in a electrochemical cell[51]),
and nearly all of these species can enter into a bonding relationship with the
metal surface. It would appear, then, that for the 10^5 to 10^6 enhancements in
the electrochemical environment, the formation of a weak chemical bond with
the active site is a necessity. The total enhancement can be attributed to both
a "classical" electromagnetic enhancement, which does not require a surface
bond, and a chemical enhancement, most likely a charge transfer resonance
Raman enhancement, which would require the surface–molecule interaction.

1.7. Theoretical Considerations

The main electromagnetic field enhancement is now considered to come
from a geometrically defined localized plasmon resonance within metal particles such as produced by an ORC pretreatment in an electrochemical system.

The theory of the localized plasmon resonance will be treated in detail in Section 3. Other SERS metal substrates such as colloidal systems, metal island films, metal surfaces in the form of a grating, and lithographically produced metal-coated SiO_4 posts allow geometrical microstructures that can give rise to enhanced electromagnetic fields.[8] When the incident light is at the right frequency and direction to satisfy a resonance condition for a localized surface plasmon, a large field is generated at the metal interface. The resonance condition depends on the dielectric constants of the metal and the surrounding medium and a factor defined by the geometry of the metal particle. The electromagnetic (EM) theory for localized plasmons has been developed for various isolated microstructures such as spherical particles,[52] spheroids,[53] prolate spheroids,[54,55] and various other geometries.[56] On an electrode surface, the microstructures are of random size and shape and interconnected by a conducting plane. The technique of scanning electron microscopy (SEM) shows various bumps[57,58] on the metal surface after pretreatment which can be modeled as spheres or prolate spheroids, or hemisolids of these shapes. Our treatment of the localized surface plasmon will be limited to the sphere and the prolate spheroid.

The possibility of a photon-assisted charge transfer SERS mechanism involving an electronic resonance between states of the metal and the molecule was first suggested by a quantum mechanical treatment from our laboratory[59] and, independently, by other groups for several types of charge transfer mechanisms.[60,61] This theory, which was further developed,[62] predicts a background continuum, which is observed experimentally.[29,63,64] Subsequent theoretical developments of the photon-assisted charge transfer resonance Raman model for SERS have been given in several investigations.[65-68] In addition to the background continuum, the main experimental evidence for the charge transfer mechanism is the shift of the maximum in the SERS intensity versus voltage when the laser excitation energy is changed.[34,69-76] This effect is illustrated in Figure 3. Such an effect would be predicted since the electrode potential controls the metal energy state (Fermi level) which would be involved in a resonance Raman process. When the laser energy is changed, a compensating change in the electrode potential (Fermi level) could reestablish the resonance condition.

A second type of charge transfer enhancement between a metal and molecular adsorbate also has been suggested.[77-80] These theories involve a non-photon-assisted ground-state charge transfer in which the molecular vibrations of the adsorbate molecule transfer charge back and forth between coupled molecular states of the metal (adatom) and adsorbate. A theoretical treatment of this model by Lippitsch[78] is based on the vibronic coupling theory of Raman scattering as derived by Albrecht.[81,82] Such an effect has also been invoked in conjunction with a SERS model that involved the formation of surface Ag^+ complexes.[83] Although the ground-state charge transfer model includes an effect of electrode potential on SERS intensities, its major

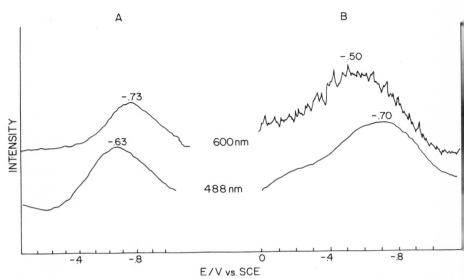

FIGURE 3. Relative Raman intensity versus potential. (A) 0.05 M 2-methylpyridine in 0.1 M KCl for the 1008 cm^{-1} band. (B) 0.05 M imidazole in 0.1 M KCl for the 1163 cm^{-1} band.[73] Scan rate: 20 mV s^{-1}. Lower curves have excitation at 488 nm and upper curves have excitation at 600 nm.

deficiency is that it does not account for the shift of the SERS intensity versus voltage maximum with laser excitation. The charge transfer mechanism is consistent with most of the experimentally observed characteristics of SERS. A theoretical treatment[84] of this mechanism based on vibronic coupling will be given in Section 4. First, however, the experimental details of the SERS method will be discussed.

2. Experimental Methods

2.1. Introduction

The basic elements of a SERS experimental setup are identical to those used in a normal Raman or resonance Raman instrument. A line diagram of such a Raman spectrometer follows:

Monochromatic Radiation Source → Raman Cell

↓

Monochromator → Detector

↓

Amplifier/Signal Processing

↓

Readout

In modern Raman instruments, the monochromatic radiation source is usually a continuous-wave (CW) laser. For a resonance Raman measurement, a tunable dye laser is necessary for use as a source, in order to select an appropriate frequency to satisfy the resonance condition. The use of a laser source is indicated because Raman-scattered light is always a small fraction of the incident light and lasers have the requied power to give a detectable Raman signal. In addition, a laser provides an excitation source which is extremely monochromatic and is self-collimated and plane polarized. In the SERS measurement at electrodes, the Raman cell is an electrochemical cell which contains an activated metal surface with adsorbed molecules and/or ions which are the scattering centers for the incident laser light. The detailed characteristics of each element in the line diagram will be discussed further on; however, first we will consider the factors which determine the sensitivity of the detected Raman signal in the SERS measurement.

2.2. Intensity of Detected Scattered Light

The detected, integrated SERS intensity, I_I, in photons per second is given by[8,28]

$$I_I = \Omega(d\sigma/d\Omega)L^2(\omega_L)L^2(\omega_S)NAJ_LQT_MT_O \tag{1}$$

where Ω is the solid angle of the collection optics in steradians (sr), $d\sigma/d\Omega$ is the differential Raman cross section in cm^2/sr^{-1} molecule^{-1}, $L^2(\omega_L)L^2(\omega_S)$ are the electromagnetic enhancement factors at the incident laser angular frequency ω_L and Raman-shifted angular frequency ω_S, N is the number of molecules per cm^2 adsorbed on the surface, A is the area of the laser beam on the metal surface in cm^2, J_L is the incident laser flux in photons s^{-1} cm^{-2}, Q is quantum efficiency of the detector, T_M is the transmission of the mono-chromator, and T_O is the transmission of the collection optics.

The total number of molecules involved in enhanced scattering will be NA if only adsorbed molecules contribute; however, it is possible that some molecules in a volume element, δV, contiguous to the surface will also produce enhanced scattering and then NA in Eq. (1) would have to be expanded to $(NA + C\delta V)$ to include those solutions-soluble molecules which are within the enhancement range.

In practice it is not necessary to include the area of the laser beam on the surface in a calculation of I_I since AJ_L can be replaced by I_L, the total number of photons striking the surface per second. I_L can be calculated by measuring the total power, P_L, in the laser beam which impinges on the metal surface. If P_L, which is in watts, i.e., joules per second, is divided by the energy per photon, E_λ, the result is I_L. Thus,

$$I_L(\text{photons s}^{-1}) = P_L/E_\lambda = AJ_L \tag{2}$$

and

$$E_\lambda(\text{J photon}^{-1}) = 1.986 \times 10^{-16}/\lambda_L = hc/\lambda_L \tag{3}$$

where λ_L is the wavelength of the laser source in nanometers (nm), h is Planck's constant, and c is the velocity of light. As an example, the value of E_λ for the blue line at 488 nm of an argon-ion laser is 4.07×10^{-19} J photon^{-1} and the laser intensity I_L at $P_L = 200$ mW is 4.91×10^{17} photons s^{-1}.

In order to estimate the expected Raman intensity of an unenhanced signal from pyridine adsorbed from solution on a flat metal surface, we can set the electromagnetic enhancement term, $L^2(\omega_L)L^2(\omega_S)$, in Eq. (1) equal to 16 in accord with the Fresnel reflection equations. These show that a local field enhancement of $L(\omega_L) \approx 2$ and an emitted Raman field enhancement at the detector of $L(\omega_S) \approx 2$ will result when the detected light is reflected from a flat surface. We take for the Raman cross section of pyridine its gas-phase value of 18.5×10^{-30} cm^2 sr^{-1} molecule^{-1} for the 991-cm^{-1} band, which is the symmetrical stretching vibration and the most intense band in the spectrum. Using typical values of solid angle $\Omega = 1$ sr, $P_L = 200$ mW (total focused power of a 488 nm laser line) $N = 5 \times 10^{14}$ molecules cm^{-2}, $Q = 0.2$ for a detection system with a photomultiplier (PM) tube, $T_M = 0.01$ for the transmission of the monochromator, and $T_O = 0.8$ for the transmission of the collection optics, and assuming each photomultiplier pulse is counted in the detection system, we obtain a value of 23.4 Hz (counts s^{-1}) for the detected integrated intensity, I_I. Since it is usual to measure the intensity of the Raman band at its peak value, I_I should be converted to the peak intensity I_P which is given by

$$I_P = I_I / \pi \Gamma_R \qquad (4)$$

where Γ_R is the half-width at half-height of the Raman band in cm^{-1} and a Lorentzian lineshape is assumed for the Raman band.[28] Thus, the estimated peak intensity with $\Gamma_R = 4$ cm^{-1} is 1.86 Hz/cm^{-1}, which is significantly below the background (dark current) of the most sensitive PM tube (about 10 Hz).

This type of calculation was not encouraging to early investigators who envisioned using the Raman technique for surface studies. However, when a surface pretreatment is used to roughen a Ag surface, a peak intensity of ca. 20,000 Hz can be observed for pyridine with conditions similar to those in the above calculation except at one-tenth the laser power. This means that a 10^5-fold enhancement has occurred for the SERS peak intensity. This enhancement can be accounted for by large increases in both the differential Raman cross section, $d\sigma/d\Omega$, and the electromagnetic factor $L^2(\omega_L)L^2(\omega_S)$. We will treat the origin of these enhancements in the theoretical sections.

2.3. Laser Radiation Sources

Laser is an acronym for "light amplification by simulated emission of radiation." In SERS, as well as in other types of Raman scattering experiments, a continuous-wave (CW) gaseous ion laser is normally used, e.g., an argon or krypton-ion laser. It is also possible to use a pulsed laser, such as a neodymium, Nd^{3+}, in yttrium–aluminum garnet (YAG) laser; however, a much more complicated detection system is required for a pulsed laser, and we will only discuss CW lasers. In a gaseous ion laser, the laser action takes place in a plasma tube where the gas is held at low pressures between two highly reflecting spherical mirrors. This assembly can act as an optical resonator when an electric discharge carries a current between two electrodes in the plasma tube. A side view of the laser plasma tube is shown in Figure 4. The discharge current excites gas atoms to ions in excited states by multiple-collision electron impact. The excited ions spontaneously emit photons between upper and lower energy levels of the excited-state energy manifold, and the emitted photons can cause induced or stimulated emission between the same energy levels in other excited gaseous ions. The laser action occurs because the electrical discharge produces a population inversion. Thus, more species are in the upper levels than in the lower energy levels, creating an energy distribution that is inverted compared to the thermal distribution as given by the Boltzmann equation. As long as the population inversion exists, the stimulated downward emission transitions will exceed stimulated upward transitions (absorptions) and electromagnetic radiation builds up in intensity. The light wave propagates down the plasma tube, with the light intensity growing exponentially with distance along the optic axis of the plasma tube, and is reflected back and forth between the mirrors. If this amplified light is much greater than losses due to scattering, diffraction, and nonperfect reflection at the mirrors, an equilibrium standing-wave oscillation is established, between the mirrors. Thus, when the laser action is established, the plasma tube is acting as an optical resonator cavity. One of the mirrors is made partially transmitting so that the amplified light beam can exit from the tube.

The desirable properties of laser radiation are derived from the lasing process. The light amplification produces the intense beam which is necessary for a Raman source since Raman scattering is inherently a weak process. The process of stimulated emission from single atoms produces photons in other atoms which are exact copies producing preferred modes of electromagnetic radiation of a given frequency, phase, direction of propagation, and electric polarization. Thus, a prism or other dispersing device allows one to select one highly monochromatic plasma lasing line among the many which are produced in a typical lasing plasma tube. Table I shows a collection of discrete lines, their wavelengths and wavenumbers, and relative output powers for a number

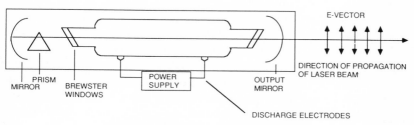

FIGURE 4. Side view of laser plasma tube.

TABLE I. Some Lasing Lines for Continuous-Wave Lasers

Gas laser	Wavelength in air (nm)	Wavenumber in air (cm^{-1})	Typical power[a]
Ar-ion	351.1	28,480.9	s[b]
	363.7	27,488.5	s[b]
	457.9	21,837.2	m
	465.8	21,468.9	w
	472.7	21,155.7	w
	476.5	20,987.0	m
	488.0	20,492.2	s
	496.5	20,140.7	m
	501.7	19,931.6	m
	514.5	19,435.1	s
	528.7	18,914.7	m
Kr-ion	350.7	28,511.0	vs[c]
	356.4	28,056.6	m[c]
	406.7	24,585.7	s[s]
	413.1	24,205.5	vs[c]
	468.0	21,365.7	s[c]
	482.5	20,724.7	w
	520.8	19,200.1	m
	530.9	18,837.2	m
	568.2	17,599.8	m
	647.1	15,453.9	s
	676.4	14,783.2	m
	752.5	13,288.1	m[b]
	799.3	12,510.6	w[b]
He–Ne	632.8	15,802.8	w
He–Cd	441.6	22,640.5	w

[a] vs: very strong (>1000 mW); s: strong (>500 mW); m: moderate (>100 mW); w: weak (< 100 mW).
[b] Requires special optics.
[c] Requires special optics and a high-power laser.

of gas lasers. Each laser line has, in fact, a band profile with a finite frequency bandwidth consisting of several modes of closely spaced frequencies. By using a mode-selecting etalon within the cavity, a linewidth of less than 0.001 cm^{-1} can be isolated; however, such highly resolved excitation lines have not been used in SERS experiments. The typical bandwidth is around 0.1 cm^{-1}. In addition to monochromaticity, selected laser lines will be highly coherent (in phase), self-collimated (unidirectional), and plane polarized. The latter property arises because gas laser tubes are fitted with Brewster angle end windows (refer to Figure 4) which minimize reflective losses for plane-polarized light in the transverse magnetic (TM) direction, i.e., the so-called p-polarization direction. Here, the electric vector is parallel to the vertical plane of the light emerging in a normal direction from the face of a Brewster window (Figure 4). An important type of spectral analysis of SERS vibrational lines,

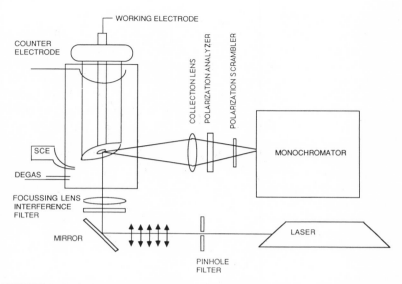

FIGURE 5. Optical setup about the electrochemical cell in a 90° collection geometry.

depolarization ratio studies, makes use of the plane-polarized nature of the exciting laser light.

2.4. Optical Setup and Depolarization Ratio Measurements

Two types of optical configurations have been used for SERS studies, the 90° geometry, as shown in Figure 5, and the backscattering geometry, as shown in Figure 6. In the 90° geometry, a focusing lens is used to focus the laser beam at a spot on the SERS working electrode, the incident light making an

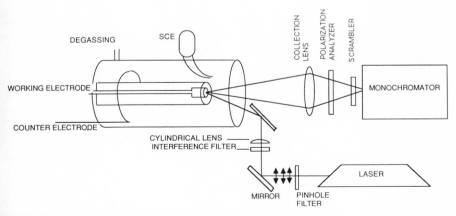

FIGURE 6. Optical setup about the electrochemical cell in a backscattering collection geometry.

angle of 90° with the optic axis of the monochromator. In this configuration, the electrochemical cell must have an optically flat cell bottom through which the excitation light enters and an optically flat side through which the SERS scattered light exits (Figure 5). To facilitate alignment, the cell can be mounted on an x-y-z-Φ micropositioner which has provisions for translations as well as rotation in the vertical plane that the optic axis of the monochromator makes with the entrance slit. An interference filter can be used to eliminate unwanted lasing lines, if the laser is not fitted with a prism device. Also a pinhole filter can be used to separate out plasma emission lines which are not propagating along the optic axis of the plasma tube. Both filters are placed in the optical path before the focusing lens. Since some spectrally reflected light might enter the entrance slit, it is good practice to use both types of filters with the 90° optical configuration. The pinhole filter may not be necessary with a SERS pretreated electrode since a highly roughened surface is not a good reflector. For experiments on mildly roughened or smooth electrodes, it is, however, quite necessary. This would be the case if one wanted to observe resonance Raman scattering from the interface of a smooth electrode using a SERS instrument.

In the backscattering geometry, only one side of the electrochemical cell must have an optically flat window (Figure 6). In this configuration, the focusing lens may be a cylindrical lens which produces a line image. An angle of incidence of around 45° can be used between the incident laser light and the optic axis of the monochromator. This geometry has some advantages over the 90° geometry when the reflected light does not come near the entrance slit of the monochromator. In some backscattering geometries, the spectrally reflected light does enter the entrance slit, and in this case a pinhole filter should be used to remove the plasma emission lines. For situations where the solution contains Raman interferences such as an organic solvent, a resonance Raman-active species, or a fluorescing species, the SERS working electrode which is parallel to the cell exit window in the backscattering configuration can be placed within a few millimeters of the window to greatly reduce the solution pathlength. In either optical configuration, the angle of incidence can be varied by rotating the electrode orientation with the micropositioner.

The main consideration for the collection lens is that it should have as low an f number as feasible in order to collect as much scattered light as possible. Also, the ratio of the collection lens radius to the distance between the lens and the entrance slit should be equal to the ratio of the radius of the first optical element in the monochromator to its distance to the entrance slit. This condition will ensure that the scattered light will fill the spectrometer optics, giving optimum sensitivity and resolving power. A scrambler, such as a calcite wedge, should be placed in front of the entrance slit to scramble the polarization, since monochromators have different sensitivities to light of different polarizations. For depolarization ratio studies, a polarization analyzer is placed between the collection lens and the scrambler.

With the polarization analyzer in place, polarization studies can be made by measuring the scattered light with the analyzer oriented so that only light is passed which has its electric vector parallel to the scattering plane. Then the measurement is made with the analyzer rotated exactly 90° so that only plane-polarized light with the electric vector perpendicular to the scattering plane is passed. The scattering plane can be defined as the plane made by the direction of propagation of the incident laser light and the direction of propagation of the observed scattered light. A polaroid sheet which can be rotated 90° serves as a suitable analyzer. The depolarization ratio, ρ, is defined as a ratio of the measured intensities with the analyzer in one orientation and then the other, and there are various types of depolarization ratios depending on the polarization of the incident light.

Consider a scattering geometry diagram, Figure 7, with electrode-fixed coordinates X, Y, Z with Z normal to the electrode surface and laboratory-fixed coordinates a, b, c. The coordinates Y and b are normal to the plane of the paper and are not shown. For 90° collection geometry, the laser light can be directed along the a-axis and the scattering light detected along the c-axis so that the ac-plane is the scattering plane. If the electric vector of the plane-polarized incident light is parallel ($\parallel$) to the scattering plane ($\parallel c$), it will be denoted p-polarized, as previously mentioned. If, on the other hand, the electric vector is perpendicular to the scattering plane ($\parallel b$), it will be denoted s-polarized or transverse electric light. For s-polarized incident laser light, the depolarization ratio is

$$\rho_s(\theta) = I_{ps}/I_{ss} \tag{5}$$

where the first subscript on the intensity indicates the polarization of the analyzed light with respect to the scattering plane and the second that of the incident light with respect to the scattering plane. The angle θ is the angle between the direction of propagation of the incident light and the direction of propagation of the observed scattered light. In order to ensure that the incident light has the correct polarization, a polaroid sheet can also be placed

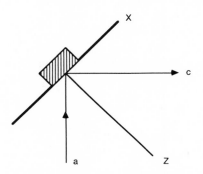

FIGURE 7. Scattering geometry diagram.

in the path of the incident light. For p-polarized incident laser light, the depolarization ratio is

$$\rho_p(\theta) = I_{sp}/I_{pp} \tag{6}$$

For the 90° collection geometry, it is usual to measure ρ for SERS as defined by Eq. (5), whereas for a backscattering collection geometry, Eq. (6) is usually used. For a backscattering geometry, the incident light is directed along the c-axis (Figure 7) and the scattering plane is the bc-plane. In this case, the angle θ is usually 45° or less. Since there are several ways of making a depolarization measurement, the type of ρ measured should always be defined.

2.5. Electrochemical Cell, Instrumentation, and Pretreatment

The electrochemical setup should be a three-electrode cell consisting of a SERS-active working electrode, an inert counter electrode, and a reference electrode. In most cases for aqueous electrolyte solutions, the counter electrode is a platinum wire and the reference electrode is a saturated calomel reference (SCE). If a deleterious species is produced at the counter electrode, it could be separated from the working and reference electrodes by a separate compartment with a ground glass frit separator. In most of the commonly used electrochemical SERS cells, such a separation has not been found to be necessary: however, if SERS is to be used for the study of electrode processes, the use of a two-compartment cell might be a good precaution.

Many different three-electrode cell designs have been used.[1,85,86] A cell which we have found convenient for use with a 90° scattering geometry is illustrated in Figure 8. This cell has an optically flat bottom and side with

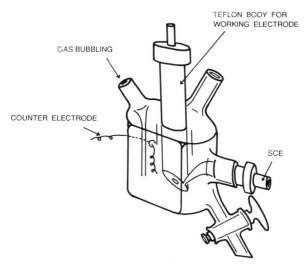

FIGURE 8. Electrochemical cell for SERS with 90° collection geometry.

provisions for inert gas (N_2, Ar, or He) bubbling for deoxygenation, a port for adding chemical species of flushing through blank electrolyte, and a port with a stopcock for removing solution while the cell is still positioned in the spectrometer. The latter features are handy for flushing out the cell with blank electrolyte after adsorption has occurred from the test solution. This can be accomplished under potential control of the working electrode without changing the optical alignment. Such a technique may be used to separate the Raman scattering of an irreversibly adsorbed species from that of an intense resonance Raman scatterer in solution. The entrance and exit ports also facilitate pretreatment in a different solution from the test solution. The working electrode can be constructed from a 99.999% pure SERS-active metal wire which is embedded in a Teflon cylinder. The electrode is cut at a 45° angle to facilitate the 90° scattering angle with the laser beam entering from the bottom of the cell. Another type of cell has been used with a backscattering geometry, as shown in Figure 9. This type of cell requires only an optically flat side window, and the working electrode surface is parallel to the optical window. When the solution contains an RR-active species, the electrode is positioned as close to the window as possible. The small thickness of the solution pathlength (a few millimeters) between the electrode and the window greatly reduces the amount of Raman scattering from solution and absorption of the scattered light in the solution.

The electrochemical instrumentation necessary for a SERS setup includes a potentiostat and a waveform generator. The latter should be capable of generating a potential pulse and a triangular waveform on top of an initial potential. For an Ag SERS electrode, two types of oxidation–reduction cycle (ORC) pretreatment techniques have been used, i.e., either a double potential

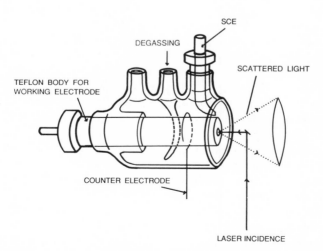

FIGURE 9. Electrochemical cell for SERS with backscattering geometry.

step pretreatment or a triangular sweep pretreatment. Usually the electrode is first polished with a finely divided alumina slurry, cleaned with highly pure water, and ultrasonicated to remove any adhering alumina. The initial potential is set negative to the onset of Ag oxidation, and the electrode potential is either stepped or swept to a potential where Ag oxidation occurs. In a halide-containing electrolyte, insoluble silver halide is formed on the electrode, which is then reduced back to Ag upon either reversing the potential sweep or stepping back to a potential negative to the region of stable silver halide. Figure 10 shows the nature of the cyclic voltammogram for pretreatment in electrolytes containing Cl^-, Br^-, and I^-. The potential range for observing SERS of around 0.9 V is shifted in the negative direction as the silver halide formed during the pretreatment becomes more insoluble.

In a 0.1 M KCl electrolyte, a double potential step sequence of -0.1 V to $+0.22$ V and back to -0.1 V versus SCE, with a dwell time of 1 to 5 s at $+0.22$ V, gives strong SERS spectra between 0.0 V and -0.9 V on Ag. The charge recovery is normally better than 99%. One ORC produces an Ag surface which usually yields intense SERS spectra. The amount of charge passed seems to affect the intensity of the SERS scattering. If more than 50 mC cm^{-2} is applied, the intensity of the SERS signal has been found to decrease in some systems.[28] A coulometer can be used to monitor the charge, and if desired, a device (integrator/comparator system) can be constructed which will control the amount of anodic charge passed.[28]

When the anion of the supporting electrolyte does not form an insoluble salt with Ag^+ ion, a potential limit of around $+0.5$ V versus SCE on the oxidizing part of the ORC cycle is sufficient to yield SERS. Here the deposition step involves plating out, on back-diffusion, of a soluble Ag^+-ion species. A similar procedure can be used with other SERS-active metal substrates. For example, with an Au electrode in 0.1 M KCl, a succession of triangular potential sweeps (20–25) from -0.3 V to 1.1 V versus SCE with a sweep rate of 0.5 V s^{-1} was found necessary to yield intense SERS spectra. It is also necessary to hold the potential for 1 to 1.5 s at the positive limit during each cycle.[87,88] The SERS intensity progressively increased after each cycle when 5 mC cm^{-2} of cathodic charge was passed during a cycle. Another method of SERS activation is to electrodeposit the metal from a solution containing its cation. For example, an Au electrode can be made SERS active by plating Au from $AuCl_4^-$ solutions.[89]

The pretreatment process roughens the electrode, producing sites of large-scale roughness (hemispherical nodules 25 to 500 nm in diameter) and molecular-scale roughness (possibly adatoms or adatom clusters). The large-scale roughness features have been associated with the electromagnetic enhancement mechanism and the adatom sites with the chemical or electronic resonance mechanism. The pretreatment procedure can be accomplished in the presence or absence of the laser light. An additional enhancement of as much as 10-fold occurs with laser illumination during pretreatment on Ag.

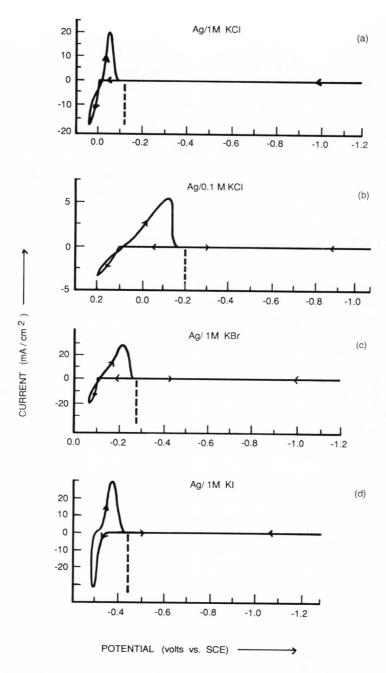

FIGURE 10. Voltammograms of Ag electrodes in different electrolytes: (a) 1 M KCl, (b) 0.1 M KCl, (c) 1 M KBr, and (d) 1 M KI. The corresponding voltammograms in the various electrolytes in the presence of 0.05 M pyridine are identical. The dotted vertical line indicates the starting potential for SERS scans. Adapted from Ref. 8 with permission.

The effect of the laser illumination seems to be the photoreduction of the AgX (X = halide) layer, producing Ag nuclei growth sites for the formation of Ag microstructures during the electrochemical reduction stage of the ORC.[90–92] Both large-scale and molecular-scale sites could be increased in such a process. The reproducibility of SERS intensities is affected by the manner in which the pretreatment is carried out. With pretreatment in the dark, about ±10% intensity reproducibility can be obtained for SERS bands, if each replication is made with freshly polished electrodes and the optical alignment is fixed, or if an internal standard is used to adjust the optical alignment. The use of laser illumination during pretreatment leads to larger irreproducibility. For a given test solution and SERS substrate, the optimum pretreatment conditions will vary, and some experimentation with voltage conditions, anodic charge passed, and laser illumination conditions during pretreatment will obviously be necessary.

If one uses an electrochemical instrument with provisions for current-potential (i versus V) measurement in the SERS setup, the i versus V curves can be followed during various phases of the SERS measurement. Thus, when a triangular potential waveform is used for pretreatment, simultaneous SERS intensity versus V curves and i versus V curves can be measured. Also, if the SERS intensity versus triangular potential sweep measurements are made after potential step pretreatment, both types of curves can again be followed (Figure 11). Figure 12 is a diagram of such a SERS instrument. With a scanning monochromator, the Raman-shifted wavenumber is fixed and both the potential sweep and Raman intensity recording are simultaneously triggered. With an optical multichannel analyzer (see below), Raman intensity–wavenumber-potential plots can be obtained.

2.6. The Monochromator and Detection System

The SER scattered light consists of the intense Rayleigh band at the excitation frequency of the laser, normally in the visible region, and the much weaker (10^{-2} to 10^{-4}) bands of the Raman process slightly shifted in frequency from the Rayleigh band. The spectrometer must separate the SERS bands while minimizing the stray light from the intense Rayleigh band. Two types of spectrometer designs have been used to obtain SERS spectra. One type is the conventional temporal scanning monochromator, with which the spectrum is obtained by sequentially scanning a dispersive device, e.g., a grating, so that the spatially dispersed bands move across the exit slit of the monochromator linearly with time. A second type of dispersing system, which has recently become popular, is a spectrograph with optical multichannel detection, where a portion of the spatially dispersed bands of the spectrum is focused on the exit port and detected in a multichannel mode. We will discuss each type of spectral recording in turn.

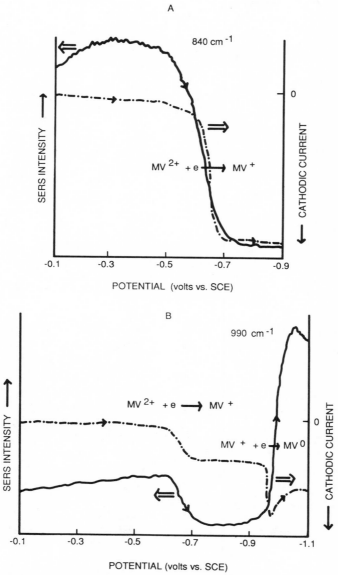

A

840 cm^{-1}

$MV^{2+} + e \longrightarrow MV^+$

POTENTIAL (volts vs. SCE)

B

990 cm^{-1}

$MV^{2+} + e \longrightarrow MV^+$

$MV^+ + e \longrightarrow MV^0$

POTENTIAL (volts vs. SCE)

FIGURE 11. Plot of simultaneously measured SERS intensity and cathodic current versus electrode potential: —— SERS intensity, — · — cathodic current. (A) Reduction of methyl viologen dication, MV^{2+}, to the radical cation MV^+. The 840-cm^{-1} band is characteristic of MV^{2+} only. It is seen that the drop in the SERS intensity closely follows the rise of the cathodic current. (B) The 990-cm^{-1} SERS band is characteristic of the MV^0 species, and it is seen that the rise in its intensity follows the second cathodic reduction wave at ~1.0 V versus SCE. The drop in the SERS intensity in the −0.6-V to −0.9-V region is due to the formation of the blue-colored MV^+ which is absorbing. MV^{2+} concentration is 0.01 M in 0.1 M KCl. Scan rate: 5 mV s^{-1}.

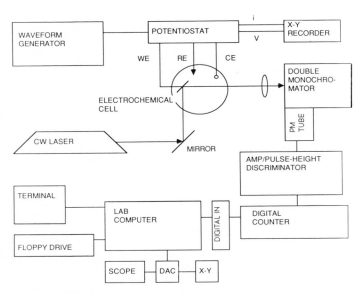

FIGURE 12. Block diagram of a computer-interfaced scanning monochromator SERS instrument with photon counting detection.

When a scanning monochromator is used to disperse the SER scattered light, a cooled photomultiplier tube serves as the detector. In this case, a double monochromator, usually with the Czerny–Turner mounting, is used in order to reduce stray light (Figure 13). With a double monochromator, only about 10^{-10} of the Rayleigh scattering will appear as grating scatter at a Raman shift of $-100 \, \text{cm}^{-1}$ from the Rayleigh band. For any larger Stokes-shifted (lower-energy) band, a still lower amount of stray scatter will occur with a double monochromator. For special experiments which examine Raman scattering closer than $100 \, \text{cm}^{-1}$ to the Rayleigh band, a triple monochromator would be necessary. The resolution of the monochromator is an important parameter of the spectral measurement and always should be reported when giving experimental details. If the grating monochromator has a focal length of 50 cm or greater, the slit width alone limits the resolving power of the instrument. The spectral slit width or bandpass (in cm^{-1}) of the instrument is the product of the physical width of the slits (mm) and the linear reciprocal dispersion $(\text{cm}^{-1} \, \text{mm}^{-1})$. The physical slit widths usually are controlled manually by a micrometer setting, and for a 100-μm slit width with a linear dispersion of $20 \, \text{cm}^{-1} \, \text{mm}^{-1}$, the resolution would be $2 \, \text{cm}^{-1}$. The resolution of the monochromator will be a function of the absolute wavelength at which the measurement is made and thus will vary with this setting. A practical definition of resolution is the observed bandpass, which can be measured as the width at half-maximum height at the exit slit of an infinitely narrow source line of a Hg or Xe discharge lamp which is incident on the monochromator

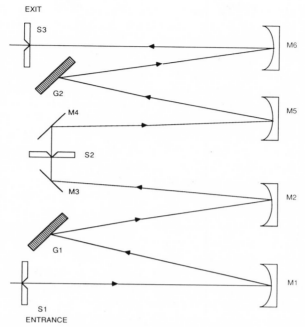

FIGURE 13. Schematic of the optics of a double monochromator. There are three slits marked S1, S2, S3; two plane gratings marked G1 and G2; four concave mirrors, M1, M2, M5, M6; and two 45° mirrors, M3 and M4. The middle slit S2 connects the two Czerny–Turner grating monochromators and should be set slightly wider than S1 and S2 to allow for different alignments in the two monochromators.

entrance slit. Such a lamp is handy for measuring the spectral resolution and calibrating the absolute wavelength of the Raman bands.

With a scanning monochromator using a photomultiplier (PM) tube, the most sensitive detection system which has normally been used in SERS instruments is photon counting. The PM tube used should have a high quantum efficiency and a very low dark current signal. In the photon counting method, the output anode pulses of the PM tube are fed to a low-noise amplifier whose output then goes to a pulse-height discriminator (Figure 12). The pulse-height discriminator level is set to pick up the largest number of signal pulses while rejecting background (dark-current) pulses. Amplifier–discriminator circuitry is available which also shapes the pulses so that they can be counted by logic-level digital counters which can be interfaced to a laboratory digital computer. The spectrum can then be sorted on magnetic medium and displayed through an oscilloscope monitor or in hardcopy form on a digital plotter or $X-Y$ recorder. The time constant in this type of measurement is the dwell time of the digital counter, which can be set under software control. For accurate measurement of band positions, a relationship exists between the time constant, the bandpass, and an upper limit on the spectral scan rate. This limit is given by the following expression

$$[\text{time constant(s)}][\text{scan rate}(cm^{-1}\,min^{-1})]/\text{bandpass}(cm^{-1}) \leq 15 \qquad (7)$$

Thus, for a time constant of 1 s and a bandpass of 3 cm^{-1}, the scan rate should not be above 45 $cm^{-1}\,min^{-1}$ for good resolution, i.e., accurate band positions. A typical SERS spectrum recorded with a scanning monochromator is shown in Figure 14.

The scanning spectrometer with a PM tube uses single-wavelength detection and thus operates in the simplex mode of optical spectrocopy. While this type of instrument is very sensitive, the simplex mode of recording a spectrum consumes a rather large amount of time. With a scan rate of 50 $cm^{-1}\,min^{-1}$, it takes 20 min to record a 1000-cm^{-1} region of a Raman spectrum. Optical multichannel detection spectrographs operate in the multiplex mode and allow a 1000-cm^{-1} region to be recorded in the seconds time domain. With this type of instrument, the dispersed image of the spectral lines is focused on a multichannel detector which is placed at the exit port of the spectrograph. With multiplex detection, the photon intensity at multiple wavelengths is recorded simultaneously. Two types of multichannel detectors have been used in SERS instruments—a *S*ilicon *I*ntensified *T*arget (SIT) vidicon (i.e., imager) and the silicon photodiode array (SiPDA). The SIT vidicon is a mosaic of discrete diodes allowing two-dimensional recording, while the SiPDA is a linear array of discrete diodes. These detectors can be cooled, and if purchased with an intensifier, sensitivities in the range of a good PM tube can be attained. Such a linear diode array device with 1024 discrete diodes spaced over 25.4 mm

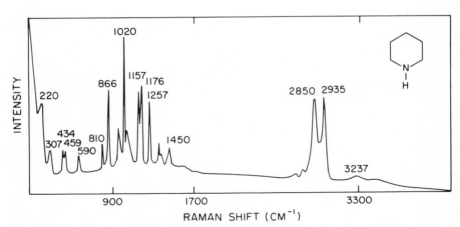

FIGURE 14. SERS spectrum of 0.05 *M* piperidine in 0.1 *M* KCl observed from an Ag electrode after double potential step pretreatment.[117] Full-scale sensitivity: 5×10^4 counts s^{-1}; 488.0-nm laser excitation; resolution: 2 cm^{-1}; scan rate: 50 $cm^{-1}\,min^{-1}$. Time constant: 0.5 s; photon counting detection.

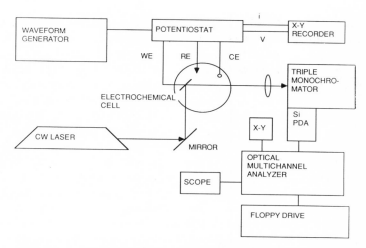

FIGURE 15. Optical multichannel analyzer (OMA) SERS instrument.

allows a SERS spectrum covering about 1000 cm^{-1} (around 500 nm absolute wavelength) to be recorded in a matter of seconds. The limit of recording speed is a few milliseconds in a nongated mode. Diode array detectors with gateable intensifiers allow spectral recording in the nanosecond time domain with nanosecond pulsed lasers.

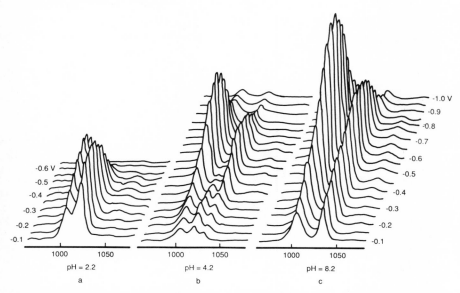

FIGURE 16. SERS spectrum of pyridine in the 950- to 1050-cm^{-1} region as a function of potential in three pH regions.[149] Recorded with a Spex Triplemate spectrometer, an intensified photodiode linear array detector, and a Tracor Northern optical multichannel analyzer (OMA).

Normally, a special type of spectrograph is used with a multichannel detector in Raman studies. In order to suppress the grating scatter from the Rayleigh light, a triple monochromator is used where the front end is a double monochromator in a subtractive dispersion mode. This stage produces a nondispersed spectrum while rejecting nearly all of the Rayleigh scattered light. The final stage is a single grating monochromator which disperses the spectrum on a flat focal plane which is coupled to the detector. The dispersed photons which strike a multichannel detector produce a charge pattern in the diodes which can be periodically read out into an interfaced computer. The complete instrument (Figure 15) is usually referred to as an optical multichannel analyzer (OMA). The use of an OMA device for recording SERS spectra not only has the advantage of fast recording of a single spectrum but also allows successive spectra to be recorded as a function of time for kinetic studies or as a function of electrode potential for three-dimensional recordings of SERS intensity–wavenumber–potential plots (Figure 16). Thus, this instrument expands the use of SERS for a wide variety of time-resolved studies.

3. Theory of the Electromagnetic Enhancement in SERS

3.1. The Electromagnetic Enhancement for Spherical Particles

In Eq. (1) for the detected integrated SERS intensity, the electromagnetic enhancement effect was incorporated in two factors, $L^2(\omega_L)L^2(\omega_S)$. We now consider the principal origin of these enhancement factors. The intensity of the scattered radiation from a single isolated stationary molecule can be expressed using Eqs. (1) and (2) as

$$I = (d\sigma/d\Omega)L^2(\omega_L)L^2(\omega_S)I_L K_E \qquad (8)$$

where $K_E = \Omega Q T_M T_O$ gathers the instrumental factors in one parameter. The intensity of the incident plane-polarized laser radiation is

$$I_L = cE_0^2/8\pi \qquad (9)$$

where E_0 is the time-averaged electric field of the incident laser radiation and c is the speed of light. Thus, Eq. (8) can be expressed in terms of the electric field E_0 using Eq. (9), giving

$$I = cK_E(d\sigma/d\Omega)L^2(\omega_L)L^2(\omega_S)E_0^2/8\pi \qquad (10)$$

3.1.1. Electrostatic Boundary Value Problem for a Metal Sphere

One source of the electromagnetic (EM) enhancement comes from the interaction of the uniform exciting field E_0 with the metallic particle. The metallic particle has a high conductivity which is expressed by its internal dielectric function ε_i, and it is assumed to be surrounded by a homogeneous medium of dielectric function ε_0 (Figure 17). Both dielectric functions depend on the frequency of the oscillating radiation, i.e., $\varepsilon = \varepsilon(\omega)$. If the metal particle is modeled as a sphere of radius a which is small with respect to the wavelength of the laser light, i.e., $\lambda \gg 20a$, the theoretical treatment of the electromagnetic enhancement factors is reduced to an electrostatic problem. This treatment for $\lambda \gg 20a$, with a single metal particle interacting with an initially uniform electric field of incident light at large distances from the dielectric particle, is called the long-wavelength or Rayleigh approximation in light scattering theory. The laser radiation is directed along the z-axis and the molecule is located at a position r' (Figure 17).

The problem of finding the electric field at r' is treated by solving the Laplace equation, $\nabla \Phi = 0$, for the electrostatic potential, Φ, with the appropriate boundary conditions. The electric field is then found from the negative gradient of the electrostatic potential. This is a correct procedure for the problem if the time-varying field of the incident light is suitably averaged as it will be in our case. The boundary conditions are:

(a) the potential at infinity

$$(\Phi = -E_0 r \cos \theta),$$

(b) the equality of the tangential electric fields at $r = a$

$$-1/a|\partial \Phi_{in}/\partial \theta|_{r=a} = -1/a|\partial \Phi_{out}/\partial \theta|_{r=a}$$

(c) the equality of the normal electric displacements at $r = a$

$$-\varepsilon_i|\partial \Phi_{in}/\partial r|_{r=a} = -\varepsilon_0|\partial \Phi_{out}/\partial r|_{r=a}$$

where Φ_{in} and Φ_{out} are the potentials inside and outside the dielectric sphere, respectively.

The solutions are given in a standard electrodynamics textbook, and so the potential outside the sphere is found to be[93]

$$\Phi = -E_0 r \cos \theta + g(a^3/r^2)E_0 \cos \theta \qquad (11)$$

with

$$g = [\varepsilon_i(\omega_L) - \varepsilon_0]/[\varepsilon_i(\omega_L) + 2\varepsilon_0] \qquad (12)$$

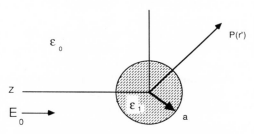

FIGURE 17. Spherical metal particle geometry.

The potential generated outside the sphere contains two contributions, i.e., one from the incident field (first term) and another from the field of an electric dipole located at the center of the sphere with polarizability ga^3 oriented in the direction of the incident field (second term). Thus the local electric potential at position r', in the small-sphere approximation, is equivalent to that generated by the applied incident field and a reflected field produced by the metal sphere.

The incident field polarizes the metal sphere so that there develop surface charges of opposite sign on either side of the sphere which alternate with frequency ω_L as the electric field of the incident light changes sign. This mode is called a localized dipolar surface plasmon (DSP). In the long-wavelength approximation, modes from higher multipoles need not be considered. Such a localized plasmon mode is distinct from a surface plasmon which propagates along a dielectric surface when excited by an electromagnetic wave. It should be noted also that a similar treatment can be given for a spherical cavity of radius a surrounded by a dielectric medium ε. The results for such a cavity can be obtained from the above dielectric sphere problem simply by replacing $\varepsilon_i \rightarrow 1/\varepsilon_i$ in the equations generated in the metal sphere treatment.[93] This fact is of some interest since a cavity enhancement effect also has been proposed.[94]

Returning to the problem at hand, the average field in the radial direction, $\bar{E}_n$, at position r' is found from the potential, as given by Eq. (11), by taking the derivative with respect to r, i.e., $E_n = -\partial\Phi/\partial r$. The root mean square value comes from integration over the elements of solid angle $d\Omega = \sin\theta \, d\theta \, d\varphi$ with limits of $\theta = 0$ to π and $\varphi = 0$ to 2π

$$\bar{E}_n = \langle E_n^2 \rangle^{1/2} = (1/3)^{1/2} E_0 [1 + 2g(a/r)^3] \tag{13}$$

Therefore, an induced molecular dipole, μ, oriented perpendicular to the sphere surface will be enhanced because the field normal component has an additional term. This EM enhancement is $\bar{E}_n/E_0 = L_n^2(\omega_L)$. In order to calculate the tangential electric field components (there are two), Eq. (11) is differentiated according to $E_t = -(1/r)(\partial\Phi/\partial\theta)$, and integration over the elements of solid angle gives

$$\bar{E}_t = \langle E_t^2 \rangle^{1/2} = (2/3)^{1/2} E_0 [1 - g(a/r)^3] \tag{14}$$

For vibrational modes oriented parallel to the tangential field, the EM enhancement is $\bar{E}_t^2/E_0^2 = L_t^2(\omega_L)$.

There is also a further EM enhancement effect to consider which is caused by the oscillating molecular dipole at position r' inducing a dipole (actually higher multipoles are also induced) in the metal sphere. Thus, the metal sphere acts as an antenna for the near field of the oscillating molecular dipole, and the emitted Raman radiation from the molecule at frequency ω_S (Raman-shifted frequency) is then enhanced by the presence of the metal particle. The dipole moment of the entire system, the so-called emission dipole of the molecule plus metal particle system, includes the effect of both the enhanced laser electric field and the antenna effect. The antenna effect, to a good approximation, has the same form as $L^2(\omega_L)$; however, the frequency of light is now the Stokes-shifted (Raman) frequency, ω_S, giving a factor $L^2(\omega_S)$. This frequency is manifested in the dielectric function which becomes

$$g_0 = [\varepsilon(\omega_S) - \varepsilon_0]/[\varepsilon(\omega_S) + 2\varepsilon_0] \quad (15)$$

3.1.2. Enhancement Factors for a Spherical Geometry

For a molecule adsorbed on the surface of a metal sphere ($r = a$) with its vibrational mode oriented normal to the surface, the average DSP enhancement is

$$\bar{G}_n = L^2(\omega_L)L^2(\omega_S) = (1/9)(1 + 2g)^2(1 + 2g_0)^2 \quad (16)$$

Note that in this treatment of the EM enhancement, we have not considered the effect which the induced dipole in the metal particle has on the local field felt by the molecule. This is the so-called image enhancement effect and contributes an additional factor so that the total EM enhancement, $\bar{G}_n$, becomes

$$\bar{G}_n = L^2(\omega_L)L^2(\omega_S)/(1 - \Delta) = (1/9)(1 + 2g)^2(1 + 2g_0)^2/(1 - \Delta) \quad (17)$$

For our purposes, we will consider that $\Delta \ll 1$ as in Ref. 95.

From Eq. (12) it is apparent that when $\varepsilon_i(\omega_L) = -2\varepsilon_0$, the value of g becomes infinite, which would maximize G. This situation will occur at a given value of ω_L; thus the system may be tuned into resonance by changing the excitation frequency, ω_L. However, the dielectric functions are complex variables so that a zero in the denominator of g giving an infinite enhancement is, in fact, not possible since ε_i should be expressed as $\varepsilon_i = \varepsilon_1 + i\varepsilon_2$. Assuming that the surrounding medium dielectric constant $\varepsilon_0 = 1$, the resonance condition for the spherical particle becomes $\varepsilon_1(\omega_L) = -2$. For this condition, the localized DSP of the metal sphere is said to be excited. The value of g then is $(-3 + \varepsilon_2 i)/i\varepsilon_2$, and for small ε_2, $|g^2| \sim 9/\varepsilon_2$. Thus, the value of the imaginary coefficient of the dielectric function will determine the size of the enhancement at resonance, and the smaller ε_2 is, the larger will be the DSP enhancement. When the Raman-shifted frequency ω_S is not too far from the exciting ω_L, an approximation for the average DSP enhancement from a spherical metal particle is obtained when $g = g_0$ so that Eq. (16) becomes

$$\langle G_M \rangle \approx (16/9)g^2 g_0^2$$

or

$$\langle G_M \rangle \approx \left| \frac{-3}{(\varepsilon_2/\varepsilon_0)} \right|^4$$

where $\varepsilon_1/\varepsilon_0 = -2$ is the resonance condition. It should be noted that for the case where the Raman band is shifted far from ω_L, the enhancements will be smaller because both g and g_0 cannot give the resonance condition.

Experimental values for ε_1 and ε_2 can be found for Ag, Au, and other metals at optical frequencies.[96,97] Table II gives dielectric components for Ag and Au relative to water. For water at optical frequencies, $\varepsilon_0 = 1.77$, and thus the resonance condition is $\varepsilon_1 = -2\varepsilon_0 = -3.54$, which occurs for Ag at 382 nm. The imaginary component ε_2 is fairly constant and for Ag is approximately 0.3 in the 300- to 500-nm wavelength region, according to the data of Johnson and Christy.[96] This value gives a $\langle G_M \rangle$ of 1.7×10^5 for Ag at 382 nm with a molecule oriented perpendicular to the surface. Kerker[98] has pointed out that there is some disagreement in the experimental values of ε_2, with larger values reported in the later study of Hagemann et al.,[97] leading to smaller enhancement factors[98] (Table II). The unaveraged $G_M = 9\langle G_M \rangle = 1.6 \times 10^6$, which shows that surface averaging reduces the expected enhancement. At 500 nm, $\varepsilon_1 = -10$, and letting $\langle G_M \rangle \approx (1/9)(1 + 2g)^4$ gives only 50 for the enhancement. Thus, the DSP enhancement with a spherical metal particle cannot explain the 10^5 to 10^6 enhancements found with 488- and 514-nm laser radiation on Ag. However, other metal particle geometries will give large EM enhancements at 500 nm as shown below.

It should be recognized for the small-sphere model that if the molecule is located at a distance from the surface of the particle, the enhancement would be diminished by the factor (a^3/r^3), and near the dipolar surface plasmon resonance

$$\langle G_M \rangle \approx (16/9)g^4(a/r)^{12}$$

Thus, if $a = 500$ nm and $r = 550$ nm so that the molecule is 50 nm from the surface, the enhancement only decreases by 0.32. It can be seen that the EM enhancement decays fairly slowly with distance from the metal sphere.

TABLE II. Some Dielectric Coefficients for Ag and Au Relative to Water[a]

Energy (eV)	Wavelength (nm)	Ag $\varepsilon_1{}^b$	Ag $\varepsilon_2{}^c$	Au $\varepsilon_1{}^b$	Au $\varepsilon_2{}^c$
1.50	828	−18.5	0.26 (1.77)	−14.6	0.92 (2.43)
2.00	621	−9.72	0.28 (1.27)	−6.04	0.77 (1.82)
2.40	518			−2.09	1.59 (1.99)
2.50	497	−5.40	0.18 (0.83)		
2.60	489			−1.04	2.60 (2.98)
3.00	414	−2.93	0.14 (0.58)	−0.96	3.23 (4.11)
3.25	382	−1.96	0.11 (0.48)		
3.30	376			−0.85	3.1 (4.16)
3.50	355	−1.13	0.16 (0.34)		
3.60	345			−0.68	3.18 (4.14)

[a] Adapted from Kerker.[98]
[b] ε_1 values from Johnson and Christy;[96] these do not differ significantly from those from Hagemann et al.[97]
[c] First column from Johnson and Christy[96] and second column in parentheses from Hagemann et al.[97]

3.2. The Electromagnetic Enhancement for a Prolate Metal Spheroid

A more general EM model (Figure 18), which may be a better representation of electrochemically pretreated SERS surfaces, is to allow the metal particle to be a hemispheroid protruding from a conducting plane with the molecule along the z-axis at a distance h from the spheroidal particle surface. The incident electric field is parallel to the z-axis. The spheroidal cross section has a semimajor axis of length a and a semiminor axis of length b with a focal point given by

$$f = (a^2 - b^2)^{1/2} \tag{18}$$

The aspect ratio a/b defines the nature of the spheroid such that the particle becomes a sphere for $a/b = 1$, an oblate spheroid for $a/b < 1$, and a prolate spheroid for $a/b > 1$.

3.2.1. Electrostatic Boundary Problem for a Prolate Metal Spheroid

The electrostatic problem is again to solve the Laplace equation but now we must consider the spheroidal geometry. If we let the particle be a prolate spheroid, it is convenient to solve the Laplace equation in prolate spheroidal coordinates ξ, η and φ. This is an orthogonal coordinate system where ξ and η are orthogonal spheroidal coordinates and φ is an azimuthal coordinate. At the surface of the spheroidal particle, $\xi = \xi_0 = a/f = 1/[1 - (b/a)^2]^{1/2}$, and

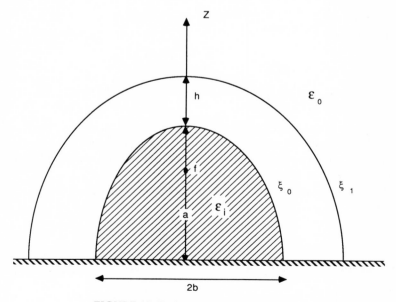

FIGURE 18. Prolate metal particle geometry.

at a position of a molecule located above the metal particle, $\xi_1 = (a + h)/f$ (Figure 18). In this coordinate system, considering azimuthal symmetry and only the dipolar term, it can be shown that the electrostatic potential outside the spheroid is given by[54,95]

$$\Phi_{out} = AQ_1(\xi)\eta - E_0 f \xi \eta \tag{19}$$

where $Q_1(\xi)$ is the Legendre function of the second kind. Inside the spheroid, the potential is given by[54,95]

$$\Phi_{in} = B\xi\eta \tag{20}$$

In these expressions, A and B are coefficients to be determined. This determination can be accomplished by requiring that Φ be continuous across the surface, i.e.,

$$\Phi_{in} = \Phi_{out}$$

at $\xi = \xi_0$, and equating the normal electric displacement vectors along the surface, i.e.,

$$-(\varepsilon_i/h_\xi)(\partial\Phi_{in}/\partial\xi) = -(\varepsilon_0/h_\xi)(\partial\Phi_{out}/\partial\xi)$$

at $\xi = \xi_0$, where h_ξ is the metric coefficient[99] for the ξ prolate spheroidal coordinate which cancels out of the expression. Substituting the resulting coefficients A and B into Eq. (19) gives the desired potential

$$\Phi_{out} = E_0 f\left[\frac{(\varepsilon_0/\varepsilon_i - 1)Q_1(\xi_0)\eta\xi_0}{(\varepsilon_0/\varepsilon_i)\xi_0 Q_1'(\xi_0) - Q_1(\xi_0)} - \xi\eta\right] \tag{21}$$

where $Q_1'(\xi_0)$ is the derivative of Q_1 with respect to ξ. The electric field in the normal direction is defined as

$$E_n = -(1/h_\xi)(\partial\Phi_{out}/\partial\xi)$$

where $h_\xi = f[(\xi^2 - \eta^2)/(\xi^2 - 1)]^{1/2}$, giving

$$E_n^2 = E_0^2\eta^2\left(\frac{\xi^2 - 1}{\xi^2 - \eta^2}\right)\left[1 + \frac{(1 - \varepsilon_i/\varepsilon_0)\xi_0 Q_1'(\xi_1)}{(\varepsilon_i/\varepsilon_0)Q_1(\xi_0) - \xi_0 Q_1'(\xi_0)}\right]^2 \tag{22}$$

for the square of the normal field. Thus, for a molecule oriented in the normal direction, the DSP enhancement for a prolate spheroid is $L_n^2(\omega_L) = E_n^2/E_0^2$ as given by Eq. (22). The enhancement at the tip of the spheroid along the major axis is given by Eq. (22) with $\eta = \pm 1$.

It is of interest to calculate the surface-averaged squared field, which is defined by

$$\langle|E|^2\rangle = \int_{-1}^{1} d\eta \int_{0}^{2\pi} d\varphi\, h_\eta h_\varphi |E|^2 \tag{23}$$

where the spheroidal metric coefficients are defined as[99]

$$h_\eta = [(\xi^2 - \eta^2)/(1 - \eta^2)]^{1/2}$$
$$h_\varphi = f[(\xi^2 - 1)(1 - \eta^2)]^{1/2}$$

Performing the necessary integrations and normalizing with respect to the area of the spheroid gives

$$\langle |E_n|^2 \rangle = E_0^2 \frac{[\xi_0^2(\xi_0^2 - 1)^{1/2}\sin^{-1}(1/\xi_0) - (\xi_0^2 - 1)]Q_n^2}{(\xi_0^2 - 1)^{1/2} + \xi_0^2\sin^{-1}(1/\xi_0)} \qquad (24)$$

where Q_n is the term in square brackets in Eq. (22). The surface-averaged normal field enhancement $\langle L_n^2(\omega_L) \rangle = \langle |E_n|^2 \rangle / E_0^2$ will be significantly less than the unaveraged DSP enhancement $L_n^2(\omega_L)$.

A similar treatment gives the tangential electric field from the definition

$$E_t = -(1/h_\eta)(\partial\Phi_{\text{out}}/\partial\eta)$$

Then, using this relationship with Eq. (21) at $\xi = \xi_0$ results in

$$E_t^2 = E_0^2\xi_0^2 \frac{(1 - \eta^2)}{(\xi_0^2 - \eta^2)}\left[1 + \frac{(1 - \varepsilon_i/\varepsilon_0)\xi_0 Q_1(\xi_0)}{(\varepsilon_i/\varepsilon_0)Q_1(\xi_0) - \xi_0 Q_1'(\xi_0)}\right]^2 \qquad (25)$$

The surface-averaged tangential field is found from Eqs. (23) and (25):

$$\langle |E_t|^2 \rangle = \frac{E_0^2\xi_0^2[2\sin^{-1}(1/\xi_0) + (\xi_0^2 - 1) - \xi_0^2\sin^{-1}(1/\xi_0)]Q_t^2}{[(\xi_0^2 - 1)^{1/2} - \xi_0^2\sin^{-1}(1/\xi_0)]} \qquad (26)$$

where Q_t is the term in square brackets in Eq. (25). It is seen from Eq. (25) that at the tips of the spheroid ($\eta = \pm1$), the tangential field is zero.

3.2.2. Enhancement Factors for Prolate Spheroidal Geometry

For the purpose of calculating the DSP electromagnetic enhancement, we first consider the unaveraged normal enhancement at the tips of the prolate spheroid ($\eta = \pm1$). Equation (22) expressed in terms of $L_n^2(\omega_L)$ becomes

$$L_n^2(\omega_L) = \left[1 + \frac{(1 - \varepsilon_i/\varepsilon_0)\xi_0 Q_1'(\xi_1)}{(\varepsilon_i/\varepsilon_0)Q_1(\xi_0) - \xi_0 Q_1'(\xi_0)}\right]^2 \qquad (27)$$

where the Legendre functions are given by

$$Q_1(\xi) = \{(\xi/2)\ln[(\xi + 1)/(\xi - 1)]\} - 1 \qquad (28)$$

and

$$Q_1'(\xi) = \{(1/2)\ln[(\xi + 1)/(\xi - 1)]\} - \xi/(\xi^2 - 1) \qquad (29)$$

For the total enhancement, G_n, the effect of the metal particle on the emitting Raman radiation must also be included; thus

$$G_n = L_n^2(\omega_L)L_n^2(\omega_S)$$

where for the $L_n^2(\omega_S)$ term, the metal dielectric function is taken at the Raman-shifted frequency ω_S. If we let $\omega_L \approx \omega_S$ and take the molecule as being oriented parallel to symmetry axis z and at the surface of the prolate metal particle so that $\xi_1 = \xi_0$, the total electromagnetic enhancement can be approximated by

$$G_n = \left[\frac{(\varepsilon_i/\varepsilon_0)[Q_1(\xi_0) - \xi_0 Q_1'(\xi_0)]}{(\varepsilon_i/\varepsilon_0)Q_1(\xi_0) - \xi_0 Q_1'(\xi_0)} \right]^4 \qquad (30)$$

It should be remembered that the dielectric functions are dependent on the exciting frequency, ω_L. It is clear that the DSP resonance occurs when there is a zero in the denominator of Eq. (30), i.e., when

$$\text{Re } \varepsilon_i(\omega_L) = \varepsilon_0 \xi_0 Q_1'(\xi_0)/Q_1(\xi_0) \equiv \varepsilon_1^R(\omega_L) \qquad (31)$$

We can calculate the value ε_1^R which gives the dipolar surface plasmon resonance for the aspect ratios $a/b = 2$ and 3. With $a/b = 2$, $\xi_0 = 1/[1 - (b/a)^2]^{1/2} = 1.15$, and from Eqs. (28), (29) and (31) using $\varepsilon_0 = 1.77$ we find that $\varepsilon_1^R = -8.42$. With $a/b = 3$, $\xi_0 = 1.06$, and $\varepsilon_1^R = -14.6$. Earlier we found the value of $\varepsilon_1^R = -3.54$ for the Ag sphere ($a/b = 1$) in water. Thus, for $a/b = 1, 2, 3$, we find $\varepsilon_1^R = -3.54, -8.42$, and $-14, 6$, respectively. If we compare these values with the experimental dielectric data for Ag,[96] we can find the wavelength at which the resonance condition occurs. It is seen to move from 382 nm ($a/b = 1$) to 480 nm ($a/b = 2$) to 580 nm ($a/b = 3$), which shows that as the eccentricity increases, the dipolar surface resonance shifts to longer wavelength.

At the surface plasmon resonance, Eq. (30) can be written

$$G_M \approx \left[\frac{\varepsilon_1^R[Q_1(\xi_0) - \xi_0 Q_1'(\xi_0)]}{\varepsilon_2^R Q_1(\xi_0)} \right]^4 \qquad (32)$$

and this relationship represents the maximum enhancement at the tip of the prolate spheroid when the dipolar surface plasmon resonance condition holds. For such a condition, the metal dielectric components are given by ε_1^R and ε_2^R evaluated at the frequency of the dipolar surface plasmon resonance. Calculations for $a/b = 2$, $\xi_0 = 1.155$ give $Q_1 = 0.522$, $Q_1' = 2.16$, and $G_M = 6.90 \times 10^8$ from Eq. (32).

As the aspect ratio a/b increases, the total enhancement at the tip increases, because the electric field at the tip is magnified. This is called the lightning rod effect and is given by the $(Q_1 - \xi_0 Q_1')/Q_1$ ratio in Eq. (32). The spheroidal coordinate ξ_0 can be approximated by $\xi_0 \approx 1 + (1/2)(b/a)^2$ for $a/b > 1$, and as a/b increases, $\xi_0 \to 1$. This is the condition for a tip resonance. Equation (32) can then be explicitly expressed in terms of the aspect ratio a/b by using the approximations

$$\xi_0 = 1 + (1/2)(b/a)^2, \qquad Q = \xi_0 \ln(2a/b) - 1,$$

$$Q' = \ln(2a/b) - (a/b)^2 - \tfrac{1}{2}$$

with

$$\varepsilon_1^R = \frac{\varepsilon_0[\ln(2a/b) - (a/b)^2 - (\tfrac{1}{2})]}{\ln(2a/b) - \{1/[1 + (\tfrac{1}{2})(b/a)^2]\}} \qquad (33)$$

and thus the maximum enhancement, G_M^{Tip}, for $\omega_L \approx \omega_S$ with the molecule located at the tip of the prolate spheroid becomes

$$G_M^{\text{Tip}} \approx \left[\frac{\varepsilon_0[(a/b)^2 - \tfrac{1}{2}][\ln(2a/b) - (a/b)^2 - (\tfrac{1}{2})]}{\varepsilon_2^R[\ln(2a/b) - 1]^2} \right]^4 \qquad (34)$$

The real part of the dielectric function at resonance, Eq. (33), is negative and increases as a/b gets large, i.e., as the spheroid becomes more eccentric. Equation (34) is a good approximation for $a/b \geq 3/1$. For a sphere $(a/b = 1)$, we have already calculated $G_M = 1.6 \times 10^6$; however, this occurred at $\omega_L = 382$ nm. On the other hand, for an aspect ratio $a/b = 2$, $G_M = 6.90 \times 10^8$, which occurs at 480 nm. So, as a/b increases, the dipolar resonance frequency moves to longer wavelengths [see Eq. (31)] and the maximum enhancement increases. For $a/b \geq 3$, we can use Eq. (34) to investigate the trend with increasing aspect ratio. For $a/b = 3$, $\lambda_M = 590$ nm, $\varepsilon_2^R \approx 0.4$, and $G_M = 4.5 \times 10^{10}$; whereas for $a/b = 4$, $\lambda_M \approx 710$ nm, $\varepsilon_2^R \approx 0.5$, and $G_M = 2.1 \times 10^{11}$. It results that the dipolar surface plasmon resonance and lightning rod effect can lead to extremely large enhancements for excitation throughout the visible region. However, these enhancements are within an order of magnitude of each other because even though $|\varepsilon_1^R|$ and the tip resonance become larger as a/b increases, the value of the imaginary component of the dielectric constant, ε_2^R, which is in the denominator of Eq. (34), also becomes larger as the excitation shifts towards the red.[96,97]

It should be pointed out at this juncture that a given laser excitation frequency fixes ε_1^R and thus fixes the value of a/b at which the surface plasmon resonance occurs. Molecules located at spheroids with smaller or larger values of a/b will not be enhanced as much because the dipolar surface plasmon is off resonance. Thus, at a given laser excitation, only spheroids of one aspect ratio will show the maximum enhancement. Since in the actual experimental situation, there are various possible molecular orientations and a distribution of bump sizes on an electrochemically pretreated SERS metal surface and in a colloidal system, only a fraction of the sites will show the maximum enhancement. Furthermore, all molecules will not be located at the tip of the metal particles but will be distributed over the metal surface. It follows that the actual EM enhancement will be an average over the spheroid surface of isolated particles as well as an average over the aspect ratio of the particle distribution. These two effects greatly lower the net EM enhancement.

As an example of the effect of averaging over the surface of the spheroid, we can compare the maximum enhancement at $a/b = 2$ as given by Eq. (32)

with the normalized surface-averaged enhancement as calculated for $\langle |E_n|^2 \rangle / E_0^2$ using Eq. (24). The resonance condition with $a/b = 2$ gives $\xi_0 = 1.154$ and $\varepsilon_1^R = -8.42$ with $\varepsilon_2^R \approx 0.3$. Then Q_n^2 (Eq. 24) is calculated to be 2.63×10^4 and $\langle |L_n|^2 \rangle = (0.24) \times (2.63 \times 10^4)$. Assuming that $\omega_L \approx \omega_S$ gives $\langle |L_n|^2 \rangle^2 = 4.0 \times 10^7$, which can be compared to $G_M = 6.9 \times 10^8$ from Eq. (32), i.e., a factor of ~ 17 smaller. As the aspect ratio gets larger, the effect of averaging over the surface of the spheroid becomes more important. At $a/b = 3$, the average enhancement is ~ 25 times less than the maximum tip enhancement. The above calculations are only for the case where the exciting laser light is at the resonance wavelength and there is a small difference between ω_L and ω_S. It should be emphasized again that there will be a distribution of bump sizes, leading to much lower DSP enhancements than given by these calculations.

3.3. Electrodynamic Effects

Another important factor, the electrodynamic effect, tends to lower greatly the actual overall EM enhancement. We have only considered the small-particle limit, but this is usually not the experimental situation. Scanning electron microscopy (SEM) shows that many particles are substantially larger than the small-particle limit of around 10 nm for the 400- to 700-nm wavelength region of visible exciting light. One would have to use transmission electron microscopy (TEM) to characterize a surface with bump sizes of 10 nm or less. Studies using SEM techniques show that electrochemically pretreated SERS surface have particle sizes in the 20 to 500-nm range. For these larger particles, a treatment based on the Rayleigh approximation ($\lambda/a \gg 20$) is no longer valid. In this case, an electrodynamic treatment based on Maxwell's equations rather than the electrostatic treatment using Laplace's equation is required. Such calculations have, indeed, been made using numerical solutions for the scattering problem.[100,101] In such a treatment, multipole terms higher than the dipolar term become important. The result is that $L^2(\omega)$ decreases rapidly for radii above 10 nm.[100] Thus, for an aspect ratio of $a/b = 2$ with $a = 100$ nm, the numerical calculations show a 980-fold decrease in $L^2(\omega)$ over that calculated from the electrostatic limit. This means an 8.1×10^3 decrease in the overall EM enhancement. Considering that surface averaging also decreases the tip enhancement by about a factor of around 17, we see that a maximum tip enhancement calculated within the Rayleigh approximation will be decreased by a factor of around 10^5 for the more exact treatments for large spheroids. Thus, if all metal particles involved in the SERS scattering were spheroids with an aspect ratio of $a/b = 2$ and a semimajor axis of 100 nm, the total DSP enhancement[8] would be around 1×10^3. Although we have seen that exact calculations can be made using EM enhancement models, it is also clear from the heterogeneity of bump sizes on a typical SERS surface that these are idealized calculations. Indeed, an additional factor which we have not considered is the electromagnetic interaction of spheroidal bumps which are separated by a distance comparable to their size. Interactions among metal particles will introduce additional effects on the resonance shapes. It can be concluded that it is impossible to calculate a highly accurate EM enhancement for a real surface. The best estimate is that the EM effect gives a 10^2 to 10^4 overall SERS enhancement.[102] Since the observed SERS enhancement is 10^5 to 10^6, there is good reason to invoke an additional mechanism such as a charge transfer resonance Raman effect to explain the total SERS enhancement.

4. The Chemical Enhancement in SERS

When electromagnetic radiation interacts with matter, the oscillating electric field causes the matter to oscillate at the same frequency as the radiation. This response is expressed in terms of the induced dipole moment, which may be written

$$\mu = \alpha \cdot E + E \cdot \chi \cdot E \tag{35}$$

where $E = E_0 \cos \omega t$ is the electric field due to light of frequency ω, α is the polarizability, and χ is the third-order susceptibility. This expression clearly may be expanded to include higher-order terms. Note that since the dipole moment is a vector quantity, as is E, α is then a second-rank tensor (i.e., has nine components due to all possible pairs of products of three spatial coordinates) and χ is a third-rank tensor (with 27 components). We may understand the polarizability physically if we note that, for example, the term $\alpha_{xy}E_y$ in Eq. (35) represents the fact that an electric field with a component along the y-axis can induce a component of the dipole moment along the x-axis. For our purposes, the most important term is the polarizability since most normal and surface-enhanced processes of interest depend only on this term.

The oscillations induced through the polarizability cause the matter to radiate light in all directions at the same frequency as the incident light. This process is known as Rayleigh scattering. We may understand this process more easily if we consider classically an electron of charge $-e$ bound elastically to an equilibrium position at which there is a charge of $+e$. In the presence of an oscillating electric field, the classical equation of motion is

$$md^2r/dt^2 + kr = -eE_0 \cos \omega t$$

Here r is the displacement of the electron from the origin, m its mass, and k is the force constant binding the electron. Letting $\omega_0 = \sqrt{(k/m)}$, the steady-state solution to this equation is

$$r = -eE_0 \cos \omega t/[m(\omega_0^2 - \omega^2)] = -(\alpha/e)E_0 \cos \omega t$$

where we obtain an expression for the classical polarizability α by recognizing that the classical expression for the dipole moment is $\mu = er$. An electron with an acceleration a will radiate an amount of energy equal to $2e^2a^2/3c^3$ in one second, where c is the speed of light. Since $a = (d^2r/dt^2)$ averaged over time, we obtain as a classical expression for the radiated energy $E_0^2\alpha^2\omega^4/3c^3$. We must then average over all directions in space, obtaining for the scattered intensity $I = 8\pi\omega^4 I_L\alpha^2/9c^4$, in terms of the intensity of the exciting light $I_L = cE_0^2/8\pi$. In addition, if the matter itself has natural oscillation frequencies ω_{IF} (such as the normal modes of a molecule or phonon modes of a crystal), some light also will be radiated at the sum and difference frequencies, $\omega \pm \omega_{IF}$, where $\omega = \omega_L$ and $\omega_{IF} = \omega_S$ in the notation of Section 3. This process is known as Raman scattering. In order to conserve total energy in this process, the matter is left either in a state of lower energy (for the sum, called the anti-Stokes spectrum), or higher energy (for the difference, called the Stokes spectrum). The total intensity of scattered radiation in photons per molecule per second, for a transition from the initial state I to the final state F is given by the expression:

$$I = 8\pi[(\omega \pm \omega_{IF})^4/9c^4]I_L \sum_{\sigma,\rho} |\alpha_{\sigma\rho}|^2 \qquad (36)$$

We may use second-order perturbation theory to obtain the expression for the $\rho\sigma$ component of the polarizability (ρ and σ refer to x, y, or z) for transitions between states I and F, in terms of the quantum mechanical wave functions associated with the molecular states as

$$\alpha_{\sigma\rho} = \sum_{K \neq I,F} \left\{ \frac{\langle I|\mu_\sigma|K\rangle\langle K|\mu_\rho|F\rangle}{E_K - E_I - \hbar\omega} + \frac{\langle I|\mu_\rho|K\rangle\langle K|\mu_\sigma|F\rangle}{E_K - E_F + \hbar\omega} \right\} \qquad (37)$$

In this equation, μ_σ and μ_ρ are the dipole moment operators for the polarization directions σ and ρ, E_I and E_F are the energies of the initial and final states, while the E_K are the energies of all the other states with the exception of I and F. The notation $\langle I|\mu_\rho|K\rangle$ represents the dipole moment matrix element. It is quantum mechanical shorthand for the integral over the dipole moment operator times the product of the wave function ψ for the two states involved (in this case, I and K) and is

$$\langle I|\mu_\rho|K\rangle = \int \psi_K\mu_\rho\psi_I \, d\tau \qquad (38)$$

where the integration is carried out over the entire range of the variables x, y, z for both the nuclei and the electrons and we assume the wave functions are real. In the case where the matter involved in the scattering process is a molecule, the wave function ψ is to be considered vibronic (vibrational and electronic) and the sum extends over all the vibronic molecular states. Note that the first term in Eq. (37) represents the process which leaves the molecule in a higher state (Stokes) while the second term refers to the process which leaves it in a lower state (anti-Stokes).

For most molecules of interest to us, we may make the zero-order Born–Oppenheimer approximation, which takes advantage of the fact that nuclei are much more massive than electrons and thus move more slowly. This enables us to separate the nuclear from the electronic motions by writing the vibronic function as a product of an electronic function and a vibrational one. Using the shorthand notation $|I\rangle = |I_e\rangle|i\rangle$, in which we use the subscript e to indicate a function of the electronic coordinates (which depends only weakly on the nuclear coordinates), while the lower case i indicates functions of the nuclear coordinates. Note also that since the final sate is usually an excited vibrational level of the ground electronic state, we may choose $|F\rangle = |I_e\rangle|f\rangle$. We may then integrate over the electronic coordinates

$$M_{KI}(Q) = \langle I_e|\mu|K_e\rangle \tag{39}$$

where Q is the internuclear separation for a diatomic molecule and M the purely electronic transition moment between states K and I. For polyatomic molecular spectra, it is necessary to remember that nuclear motions may be combined into normal modes, each of which oscillates with a characteristic frequency ω_{IF}, and thus we choose Q to represent motion along one of these normal modes. If we consider only slight displacements from equilibrium for these modes, then $Q = Q_0 + (Q - Q_0)$, Q_0 being the equilibrium internuclear separation, and we may expand the electronic matrix element as

$$M_{KI}(Q) = M_{KI}(Q_0) + (\partial M_{KI}/\partial Q)_0(Q - Q_0) + \cdots \tag{40}$$

Substituting this expansion into Eq. (37) and recognizing that $E_K - E_I = \hbar\omega_{KI}$, we obtain

$$
\begin{aligned}
\alpha_{\sigma p} = {} & \sum_{K \neq I} \sum_k \left[\frac{M_{KI}^\sigma(Q_0) M_{KI}^\rho(Q_0)}{\hbar(\omega_{KI} - \omega)} + \frac{M_{KI}^\rho(Q_0) M_{KI}^\sigma(Q_0)}{\hbar(\omega_{KI} + \omega)} \right] \langle i|k\rangle\langle k|f\rangle \\
& + \sum_{K \neq I} \sum_k M_{KI}^\rho(Q_0) \left[\frac{\partial M_{KI}^\sigma}{\partial Q} \right]_0 \left[\frac{\langle f|Q|k\rangle\langle k|i\rangle + \langle f|k\rangle\langle k|Q|i\rangle}{\hbar(\omega_{KI} - \omega)} \right. \\
& \left. + \frac{\langle f|Q|k\rangle\langle k|i\rangle + \langle f|k\rangle\langle k|Q|i\rangle}{\hbar(\omega_{KI} + \omega)} \right] + \cdots
\end{aligned}
\tag{41}
$$

where, for notational convenience, we shall now shift the origin of the normal coordinates to Q_0, so that $Q - Q_0$ has been replaced by Q. Note that i and f represent vibrational levels of the ground electronic state I_e, while k represents vibrational levels of excited electronic states K_e. Matrix elements of the type $\langle k|i\rangle$ correspond to Franck–Condon overlap integrals. The two terms in Eq. (41) each contribute to the overall scattering. However, in certain limits, one or the other is dominant.

4.1. Normal Raman Scattering

We first consider the most common case, where the frequency of the incident light is considerably less than that needed to induce an optically allowed transition. This means $\omega_{KI} \gg \omega$, and thus the denominator in both terms of Eq. (41) essentially may be considered constant. Letting $\Omega^{-1} = \omega_{KI}/(\omega_{KI}^2 - \omega^2)$, we may remove this term from under the summations. This enables us to utilize the completeness relationship, which is usually expressed symbolically as $\sum |k\rangle\langle k| = 1$, where the sum extends over all states. Equation (41) then becomes

$$\alpha_{\rho\sigma} = 2M^{\rho}(Q_0)M^{\sigma}(Q_0)(\hbar\Omega)^{-1}\langle i|f\rangle + 2M^{\rho}(Q_0)[\partial M^{\sigma}/\partial Q]_0(\hbar\Omega)^{-1}\langle i|Q|f\rangle \tag{42}$$

Due to the fact that vibrational wave functions in the same electronic state are orthogonal, $\langle i|f\rangle = 0$ unless $i = f$. The latter case is simple Rayleigh scattering and there is no frequency shift. Furthermore, in the harmonic oscillator approximation, it may be shown that

$$\langle i|Q|f\rangle = \begin{cases} 0 & \text{if } i \neq f \pm 1 \\ \sqrt{(i+1)/2} & \text{if } i = f + 1 \\ \sqrt{(i/2)} & \text{if } i = f - 1 \end{cases} \tag{43}$$

and we therefore expect scattered radiation at frequencies shifted from the incident frequency by plus or minus one vibrational quantum, or Raman scattering at the fundamental vibrational frequency. Observe also that the first term in Eq. (42), which gives rise to Rayleigh scattering, will be considerably more intense than the second term since the latter is proportional to the derivative of an electronic dipole moment matrix element. Indeed, experimentally it is found that Rayleigh intensities are often three orders of magnitude more intense than Raman intensities. This proves that the approximation in Equation (40) is valid.

4.2. Resonance Raman Scattering

We now examine the case in which the frequency of the exciting light is near that of one of the molecular optical transitions, namely, $\omega \approx \omega_{KI}$ in Eq. (41). Note that in order to prevent obtaining an infinite intensity, we have to remember that all oscillations are in reality damped. Classically, we account for this by adding a frictional term to the harmonic oscillator equations. Quantum mechanically, this appears through the uncertainty principle as applied to energy levels. Every excited state has only a finite lifetime τ due to the fact that there are numerous processes available to remove the molecule from that state, including radiation. This causes an uncertainty in the exact energy of the state in question, $\Delta E_{KI} = \hbar\Delta\omega_{KI}$. The uncertainty relation is then $\Delta E_{KI}\tau = \hbar/2$, from which we obtain $\Gamma \equiv \Delta\omega_{KI} = (2\tau)^{-1}$. This leads us to infer that the spectral line associated with the $I \to K$ transition will be broadened by an amount inversely proportional to the lifetime of the excited state. This effect is incorporated into our scattering equations by replacing the resonance denominator by $\omega_{KI} - \omega + i\Gamma$. This added damping factor is usually treated as phenomenological, with Γ determined spectroscopically as an experimental measure of the lifetime of the state in question. Note that the fact that this term is imaginary is not a problem since the intensity is proportional to the absolute magnitude of the polarizability squared. In addition, there is no fear that our theory will give us an infinite intensity at resonance.

However, near resonance, the approximation made in the previous section, namely that the denominators were almost constant, is no longer valid. In this case, we recognize that most of the scattering will now be obtained from the first term alone since we have a rapidly converging Taylor series in which, when the factors $\langle i|k\rangle\langle k|f\rangle$ are not small, the second term is several orders of magnitude less than the first. In the usual notation we write $\alpha_{\sigma\rho} = A + B + C$, where

$$A = \hbar^{-1} \sum_{K_e} M^{\rho}_{KI}(Q_0) M^{\sigma}_{KI}(Q_0) \left[\sum_k \frac{\langle i|k\rangle\langle k|f\rangle}{(\omega_{KI} - \omega + i\Gamma)} + \frac{\langle i|k\rangle\langle k|f\rangle}{(\omega_{KI} + \omega + i\Gamma)} \right] \quad (44)$$

This comes from the first term in Eq. (41). The terms B and C from the second term will be considered in Section 4.3. Since we are presumed near a resonance with a particular vibronic state, one term in these sums will dominate. The scattering intensity is then:

$$I = \frac{8\pi(\omega \pm \omega_{IF})^4 I_L}{9c^4\hbar^2} \sum_{\rho,\sigma} [M^{\rho}_{KI}(Q_0) M^{\sigma}_{KI}(Q_0)]^2 \frac{|\langle i|k\rangle\langle k|f\rangle|}{(\omega_{KI} \pm \omega)^2 + \Gamma^2} \quad (45)$$

where $\pm$ refer to Stokes or anti-Stokes radiation. We can see that the scattered intensity will depend on which particular excited state is in resonance with the exciting light. Normal Raman scattering intensity varies only slightly with excitation frequency, and consequently it is not too important to choose any particular one. Thus, any fixed frequency laser such as an argon-ion laser or a He–Ne laser may be used. However, for resonance Raman work, it would be preferable to use a tunable laser such as a dye laser which can be tuned to particular frequencies. Note also that the intensity of scattered radiation will not be a smoothly varying function of incident frequency, since it will depend on the magnitude of the Franck–Condon overlap factors connecting the excited vibrational level with that of the ground state. These overlap factors are not zero as in Eq. (42) because the vibrational levels involved belong to *different* electronic states and are therefore not orthogonal. Furthermore, the selection rules obtained in Eq. (43) for normal Raman scattering no longer hold due to the intrusion of only one chosen excited state. Consequently, resonance Raman spectra tend to be rich in overtones ($f = i + 2, i + 3, \ldots$) in contrast with the normal spectra, which are largely confined to fundamental frequencies. It is useful also to note that since resonance Raman spectra arise from the first term in the expansion of $M(Q)$, we expect a considerable increase in intensity in comparison with normal Raman spectra. This is borne out by observation, three to four orders of magnitude difference often being obtained. However, since the resonance Raman spectrum often coincides with the fluorescence spectrum, it may be obscured by the latter, so that the advantage afforded by the increased intensity must often be forgone by moving the exciting light off resonance to obtain a weaker, but observable normal Raman spectrum. Other techniques to avoid interference from fluorescence are available. One is to take advantage of the different time scales, by discriminating against the slower fluorescence electronically.

Note that when the Franck-Condon factors in Eq. (44) are small, it is also possible to obtain resonance Raman intensity from the second term in Eq. (41), when similar considerations apply. However, in this case we must consider contributions from the Herzberg-Teller effect, which is described in the next section.

4.3. Herzberg-Teller Corrections

It has been found that the above theory is quite useful in predicting the correct overall behavior of matter in the presence of electromagnetic radiation. However, in order to predict correctly Raman intensities, it has been found necessary to refine the theory by accounting for small deviations in the electronic wave functions with nuclear motion. In the framework of the Herzberg-Teller theory,[81,82] it is assumed that the corrected electronic wave functions may be obtained by the use of a first-order perturbation expansion as a linear combination of the complete set of zero-order Born-Oppenheimer functions, discussed above. Since here we are mainly interested in the normal Raman effect, we shall consider only corrections to the second term in Eq. (41). If we first examine corrections to the state K, the resulting expression for the derivative of the transition moment with motion along a normal mode is:

$$(\partial M_{KI}/\partial Q)_0 = \sum_{S \neq K} h_{KS} M_{SI}(Q_0)/(E_S^0 - E_K^0) \qquad (46)$$

where h_{KS} is the KS matrix element of the derivative of the Hamiltonian operator with respect to the normal coordinate:

$$h_{KS} = \langle K|\partial H/\partial Q|S \rangle \qquad (47)$$

This matrix element represents the coupling of states K and S through terms in the Hamiltonian, which has previously been neglected in the Born-Oppenheimer approximation. Since we have expanded the dipole moment matrix element, which is responsible for the intensities of optical transitions, we can see that the effect of this expansion is to allow states to "borrow" intensity from one another. A transition which may be forbidden in the zero-order theory may obtain intensity from a nearby state to which a transition is allowed. Note that in Eq. (46), the sum runs over all the excited states S, except the state K. The E^0 are the zero-order energies, and we may write the denominator as $\hbar(\omega_S - \omega_K) = \hbar\omega_{SK}$. Clearly, efficient borrowing depends on a nonzero coupling term h_{KS} and a small denominator (thereby favoring nearby states).

On substituting Eq. (46) into Eq. (41), note that the result involves sums over pairs of excited states, and since the wave functions are assumed real, so that $M_{KS} = M_{SK}$ [see Eq. (38)], we may see that in this summation there will be pairs of terms which differ only by interchange of K and S.[81] We may then take advantage of the identity:

$$\sum_{S \neq K} \frac{\omega_{KI}}{(\omega_{KI} - \omega_{SI})(\omega_{KI}^2 - \omega^2)} + \frac{\omega_{SI}}{(\omega_{SI} - \omega_{KI})(\omega_{SI}^2 - \omega^2)}$$

$$= 2 \sum_{S < K} \frac{\omega_{KI}\omega_{SI} + \omega^2}{(\omega_{KI}^2 - \omega^2)(\omega_{SI}^2 - \omega^2)} \tag{48}$$

to obtain the expression:

$$B = -\frac{2}{\hbar^2} \sum_{K \neq I} \sum_{S < K} \frac{(\omega_{KI}\omega_{SI} + \omega^2)}{(\omega_{KI}^2 - \omega^2)(\omega_{SI}^2 - \omega^2)}$$

$$\times h_{KS}[M_{IK}^{\rho}(Q_0)M_{SI}^{\sigma}(Q_0) + M_{IK}^{\sigma}(Q_0)M_{SI}^{\rho}(Q_0)]\langle i|Q|f \rangle \tag{49}$$

It was this equation that was derived by Albrecht[81] in his definitive study of Raman intensities. Note that by considering

$$(\partial M_{KI}/\partial Q)_0 = \sum_{S \neq I} h_{IS}M_{SI}(Q_0)/(E_S^0 - E_I^0) \tag{50}$$

thereby including Herzberg–Teller mixing of the ground state with intermediate states, we obtain another term in the expression for α:

$$C = -\frac{2}{\hbar^2} \sum_{K \neq I} \sum_{S > I} \frac{(\omega_{KI}^2 \omega_{KS}^2 + \omega^2)}{(\omega_{KI}^2 - \omega^2)(\omega_{KS}^2 - \omega^2)}$$

$$\times h_{IS}[M_{SK}^{\rho}(Q_0)M_{KI}^{\sigma}(Q_0) + M_{SK}^{\sigma}(Q_0)M_{KI}^{\rho}(Q_0)]\langle i|Q|f \rangle \tag{51}$$

This term was not considered important by Albrecht since he was considering only molecular states, and thus only transitions from the ground state, in which case the latter expression would have been inconsequential. This is due to the rather large energy gap between ground and first excited states for molecules. In our case, however, where we will allow the possibility that the intermediate state is a metal state and these states may lie close to the molecular ground state, then this term may not be ignored.

4.4. Surface-Enhanced Raman Spectroscopy: A Charge Transfer Theory

We are now in a position to understand how the presence of a metal surface near a molecule can enhance the intensity of Raman scattering from that molecule through a charge transfer process.[84] We imagine that the molecule is weakly bound to the surface at some equilibrium distance along the axis perpendicular to the surface plane. When the molecule and the surface are far apart, it is assumed that no interaction exists between them. As the molecule approaches the surface, there are both physical and chemical forces through which they affect each other. Firstly, the surface may act as a mirror where the molecular dipole moment induces an image dipole of opposite polarity on the opposite side of the surface. These two dipoles tend to attract each other by means of a dipole–dipole interaction. We call the molecule physisorbed. A second interaction is related to the possible creation of a weak chemical bond between molecule and metal. This is usually more likely if the molecule has a lone pair of electrons such as in a nitrogen heterocycle. It is also likely that this interaction is enhanced either by the interaction with metal adatoms or adclusters on the surface or by the existence of special surface sites which can more readily form chemical bonds with the adsorbed molecule. In this case, we call the molecule chemisorbed. Generally speaking, chemisorbed molecules are more strongly bound to the surface than those that are physisorbed, but in either case the interaction tends to be weak compared to normal chemical bonding. Note that both physisorption and chemisorption can coexist. Although all molecules to some extent may be physisorbed, only molecules with certain chemical attributes can be chemisorbed. For the charge transfer effect to be sizable, we shall consider that the molecules involved are chemisorbed.

We should say a few words concerning the nature of molecular and metallic energy levels. Generally, polyatomic molecules have sets of rather discrete levels, grouped according to electronic state. Within each electronic state is a series of more closely spaced vibrational levels. The combined electronic and vibrational state is called a vibronic state, and these states were discussed in the previous section under the Born–Oppenheimer approximation. Metals, on the other hand, tend to have large numbers of states bunched together in bands. Since there are so many of these states within a small range of energies and since the properties of adjacent states do not differ significantly, we usually measure them by means of a density of states, which is assumed to be a function of energy, $\rho(E)$. Metals often have several bands of energy levels. The inner or core electrons usually are tightly bound to the individual nucleus and still retain some of their atomic character. Valence electrons in a metal, by contrast, form bands which spread over the entire metal. The electrons in this band are quite mobile and are responsible for the ability of the metal to conduct electricity. For this reason, this band is called the conduction band. Often this band is not completely filled with electrons, and the highest filled level of this band is called the Fermi level, E_F. In Ag and certain other metals, the conduction bands are derived from the $4s$ and $4p$ atomic orbitals. In an electrode immersed in an electrolyte solution, variation of applied potential V may add or remove electrons from the conduction band, thus changing the Fermi level. If $E_F(0)$ is the Fermi level at an arbitrary zero voltage, then the Fermi level is given by $E_F = E_F(0) + eV$.

In any complete quantum mechanical treatment, we could not consider the energy levels of the molecule to be distinct from those of the metal. We would have to speak of the levels of the molecule–metal system. The correct Hamiltonian for such a system would be written:

$$H = H_{mol} + H_{met} + H' \tag{52}$$

where H_{mol} and H_{met} are the Hamiltonians of the molecule and the metal when they are far apart and presumably not interacting (zero order). H' is the interaction energy, which clearly is a function of the molecule–metal distance. Generally, this term is considered to be small (first order), and both the molecule and the metal still retain the overall character they have in the separated system. Their wave functions and energy levels are only slightly perturbed by the interaction energy. For example, the correct wave functions to first order are a linear combination of the molecule and metal (zero-order) functions ψ_{met}^0 and ψ_{mol}^0, which may be expressed:

$$\psi_{mol} = a\psi_{mol}^0 + b\psi_{met}^0 \tag{53a}$$

$$\psi_{met} = a'\psi_{mol}^0 + b'\psi_{met}^0 \tag{53b}$$

in which the mixing coefficients are functions of the molecule–metal distance through the term H'. Since the interaction is weak, we have $b \ll a$ and $a' \ll b'$ so that we may speak of the energy levels of the molecule–metal system as either molecule or metal energies even though they may not be considered separately, each sharing some of the character of the other.

A molecule–metal system can be involved in a charge transfer type resonance Raman process if the exciting radiation is in partial resonance with the energy difference between metallic and molecular states involved in a photon-stimulated charge transfer. If we consider the above case of very weak chemisorption (virtually no mixing of metal and molecule wave functions), an adatom or other point defect site (steps, kinks, vacancies) is necessary to relax the momentum selection rule which forbids photon-stimulated charge transitions from bulk metal states, i.e., the charge transfer matrix elements would be allowed for the case where the metal states are localized adatom or adcluster states. In a metal- (Ag) to-molecule charge transfer, the process can be described[8] as: (1) a metallic electron in an sp-conduction band absorbs a photon, the electron being promoted to the vacant sp-band above the Fermi level, leaving a hole below the Fermi level in the filled sp-band; (2) the electron either tunnels to a temporary negative molecular complex or crosses over to an excited state of the molecule–metal system, which is the charge transfer acceptor level of the chemisorbed molecule; (3) the electron returns to the metal, recombines with the hole (assumed to be stationary), and in the process reradiates a Raman-shifted photon; and (4) the molecule is left in the excited vibrational state. If, in fact, there is some mixing of the zero-order metal and molecular wave functions, then the momentum selection rule has no meaning. In such a case, the charge transfer process takes place between the surface adatom and the chemisorbed molecule, i.e., within the adatom–molecule complex. The charge transfer resonance energy is the difference between the adatom surface state level and the charge transfer acceptor level, both of which depend on the interaction energy H'.

A similar set of steps can be used to describe the molecule-to-metal charge transfer process[59]: (1) a ground-state electron (in the charge transfer donor level of the chemisorbed molecule) absorbs a photon and is vertically excited; (2) it either tunnels to a vacant level above the Fermi level of the metal, forming a temporary positive molecule complex, or crosses over to the lowest unoccupied level of the adatom surface state of the coupled molecule–adatom system; (3) the electron returns to the molecular ground electronic state and in the process reradiates a Raman-shifted photon; and (4) the chemisorbed molecule is left in an excited vibrational state. The charge transfer energy is the difference between the molecular ground state and the Fermi level in a weakly coupled molecule–metal system, or between the HOMO and the LUMO of a more strongly coupled molecule–adatom complex.

The broad background continuum which extends out to 4000 cm^{-1} can be explained by charge transfer processes in the adatom–molecule complex. If the hole produced in the metal conduction band by metal-to-molecule charge transfer loses a continuum of energy, subsequent electron–hole pair radiative recombination leads to an inelastic continuum. On the other hand, loss of a continuum of energy in the electron excited into the conduction band upon molecule-to-metal charge transfer also can lead to the inelastic continuum on subsequent radiative recombination with the unoccupied orbital on the molecule. It is also possible to envisage an enhanced inelastic continuum resulting from electron–hole pair excitations from adatoms by loss of a continuum of energy in either the hole or the excited electron. In this case, the electron–hole excitation process does not have to involve the molecule.

We now consider a derivation of a SERS charge transfer enhancement in terms of the polarizability components. If we view the molecule–metal distance to be a normal mode of the combined system and assume that the zero-order wave functions of the molecule or metal have been chosen so as to satisfy the Born–Oppenheimer conditions, then we may follow the Herzberg–Teller theory, adopting Eq. (40) and Eqs. (46)–(51).

We consider the situation in which the frequency of the exciting light is far from any molecular resonance ($\omega \ll \omega_K$). Then in Eq. (44) we may remove the frequency-dependent terms from the summation over all vibrational states k. Utilizing the closure relation $\sum |k\rangle\langle k| = 1$, we see that in Eq. (44), $\sum \langle i|k\rangle \times \langle k|f\rangle = \langle i|f\rangle$, which vanishes in the harmonic oscillator approximation unless $i = f$. Thus, far from a molecular resonance, the term A contributes only to the Rayleigh line. If, alternately, we consider in term A those states in which either K or I is actually a metal state (Figure 19a), we might expect resonant contributions to the intensity, and we cannot carry out the sum over vibrational states if the laser width is less than vibrational spacings. In the latter case, a

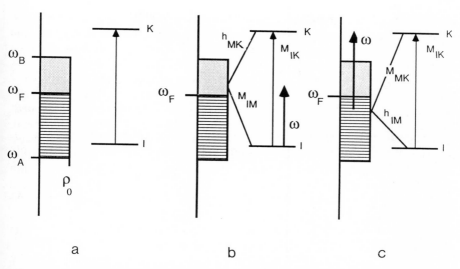

FIGURE 19. (a) Energy level scheme for molecule–metal system.[84] The (discrete) molecular levels are $|I\rangle$ and $|K\rangle$, between which a transition is assumed to be allowed. The (continuous) metal levels of the conduction band of the metal are shown on the left. The conduction band ranges between ω_A and ω_B and is assumed to have a constant density of states ρ_0. The filled levels range up to ω_F, the Fermi level, and are depicted by lines, while the unfilled levels are depicted by gray areas. (b) Scheme for molecule-to-metal charge transfer transitions between the ground molecular state and unfilled levels of the metal. The transition borrows intensity from the allowed transition by means of vibronic coupling between metal levels and the excited molecular level through the matrix element h_{MK}. (c) Scheme for metal-to-molecule charge transfer transitions between filled levels of the metal and the excited molecular state. The transition borrows intensity from the allowed transition by means of vibronic coupling between metal levels and the ground molecular state through the matrix element h_{IM}.

single vibration will dominate the sum. We have for $|k\rangle = |M\rangle$

$$A_f = (2/\hbar) \sum_M M^\sigma_{MI} M^\rho_{MI} \langle i|k\rangle\langle k|f\rangle \frac{\omega_{MI} + \omega_f}{(\omega_{MI} + \omega_f)^2 - \omega^2} \tag{54}$$

while for $|I\rangle = |M\rangle$ we obtain

$$A_k = (2/\hbar) \sum_M M^\sigma_{KM} M^\rho_{KM} \langle i|k\rangle\langle k|f\rangle \frac{\omega_{KM} + \omega_k}{(\omega_{KM} + \omega_k)^2 - \omega^2} \tag{55}$$

where ω_k and ω_f are the frequencies of a particular excited- or ground-state vibration, respectively. Term A_f represents resonant molecule-to-metal charge transfer from the molecular ground state to one of the unfilled metal levels M, while A_k represents resonant metal-to-molecule charge transfer from a filled metal state M to an excited state K. The sum over vibrational states k has been removed.

Considering the terms B and C, we now let the intruding states be those of the metal and utilize the letter M to replace S when referring to metal states. As discussed above, we must really consider two situations. One is when the metal states are connected by the interaction matrix element h_{KM} to various excited molecular states (Figure 19b)

$$B = -(2/\hbar^2) \sum_{K \neq I} \sum_{M < K} [M^\sigma_{KI} M^\rho_{MI} + M^\rho_{KI} M^\sigma_{MI}] \frac{(\omega_{KI}\omega_{MI} + \omega^2)h_{KM}\langle 1|Q|f\rangle}{(\omega^2_{KI} - \omega^2)(\omega^2_{MI} - \omega^2)}$$

$$(56)$$

If it is presumed that we are far from a resonance for a free molecule ($\omega \ll \omega_{KI}$), we see that the polarizability becomes quite large for $\omega \approx \omega_{MI}$. This represents a charge transfer transition from the molecular ground state to one of the unfilled ($> E_F$) levels of the metal conduction band. In the limit where the molecule is far from the metal, this transition is quite weak, but when the molecule–metal distance is small, this transition borrows intensity from the allowed molecular transition between the states I and K by the dipole transition matrix element M_{IK}. Note that this is no longer a normal Raman transition, because it is between real states of the system, and we might expect the intensities to be more on the order of the resonance Raman effect. In fact, this is exactly what is observed. Note, however, that in spite of this, Eq. (56) involves only vibrational levels of the ground state $\{i$ and $f\}$ in the factor $\langle i|Q|f\rangle$, and thus in contrast to the resonance Raman effect, we expect to see no overtones in the spectrum, and we should not expect erratic behavior as a function of excitation frequency. These expectations are borne out also by observation. What we do expect is that as a function of applied potential, we should see a resonance shape. We expect the resonance maximum to occur at $[E_F(0) + eV] - E_I = \hbar\omega$, or that the voltage maximum will increase with increasing excitation frequency. This is exactly what has been observed in molecules without low-lying π^* states. We take this effect to be diagnostic for molecule-to-metal charge transfer transitions.

However, surface enhancement has been observed in molecules with a low-lying π^* state. In these cases, interaction between the metal and the excited molecular levels explains the observed enhancement:

$$C = -(2/\hbar^2) \sum_{K \neq I} \sum_{M > I} [M^\sigma_{MK} M^\rho_{KI} + M^\rho_{MK} M^\sigma_{KI}] \frac{(\omega_{KI}\omega_{KM} + \omega^2)h_{IM}\langle i|Q|f\rangle}{(\omega^2_{KI} - \omega^2)(\omega^2_{KM} - \omega^2)}$$

$$(57)$$

Now we see that a resonance is predicted when $\omega \approx \omega_{MK}$. The transition is a charge transfer between the filled levels of the metal conduction band ($< E_F$) to low-lying excited levels K of the molecule (Figure 19c). We once again expect to observe a resonance shape in the plot of intensity versus applied potential, but now $E_K - [E_F(0) + eV] = \hbar\omega$, and the voltage maximum will decrease with increasing excitation frequency. This is exactly what has been observed for a series of molecules possessing a low-lying excited state and may be considered a test of metal-to-molecule charge transfer. Note that in this picture, it is the interaction between the metal and the ground electronic state through h_{IM} (Figure 19c) which causes the vibronic coupling.

We may now carry out the sum over metal states in both B and C by recognizing that the metal states are so closely spaced that the sum may be replaced by an integral over energy (or frequency). Furthermore, we see that since adjacent metal states will have essentially the same wave function, we may write such terms as $M_{IM}h_{MK} = M_{IM}^0 h_{MK}^0 \rho(\omega_{MI})$, where the terms with superscript 0 are independent of energy and $\rho(\omega_{MI})$ is just the metal density of states. We then have

$$A_f = (2/\hbar) M_{MI}^{\sigma 0} M_{MI}^{\rho 0} \langle i|k\rangle\langle k|f\rangle \int_{\omega_{FI}}^{\omega_{BI}} \frac{(\omega_{MI} + \omega_f)\rho_u(\omega_{MI}) \, d\omega_{MI}}{(\omega_{MI} + \omega_f)^2 - \omega^2} \qquad (58)$$

$$A_k = (2/\hbar) M_{MK}^{\sigma 0} M_{MK}^{\rho 0} \langle i|k\rangle\langle k|f\rangle \int_{\omega_{KA}}^{\omega_{KF}} \frac{(\omega_{KM} + \omega_k)\rho_f(\omega_{KM}) \, d\omega_{KM}}{(\omega_{KM} + \omega_k)^2 - \omega^2} \qquad (59)$$

$$B = -(2/\hbar^2) \sum_{K \neq I} [M_{KI}^{\sigma} M_{MI}^{\rho 0} + M_{KI}^{\rho} M_{MI}^{\sigma 0}] \frac{h_{KM}^0 \langle i|Q|f\rangle}{(\omega_{KI}^2 - \omega^2)}$$

$$\times \int_{\omega_{FI}}^{\omega_{BI}} \frac{(\omega_{KI}\omega_{MI} + \omega^2)}{(\omega_{MI}^2 - \omega^2)} \rho_u(\omega_{MI}) \, d\omega_{MI} \qquad (60)$$

$$C = -(2/\hbar^2) \sum_{K \neq I} [M_{MK}^{\sigma 0} M_{KI}^{\rho} + M_{MK}^{\rho 0} M_{KI}^{\sigma}] \frac{h_{IM}^0 \langle i|Q|f\rangle}{(\omega_{KI}^2 - \omega^2)}$$

$$\times \int_{\omega_{KA}}^{\omega_{KF}} \frac{(\omega_{KI}\omega_{KM} + \omega^2)}{(\omega_{KM}^2 - \omega^2)} \rho_f(\omega_{KM}) \, d\omega_{KM} \qquad (61)$$

Transitions in B are molecule-to-metal transitions, and therefore ρ_u represents the density of unfilled metal states above the Fermi level, while C represents metal-to-molecule transitions and ρ_f refers to the density of filled states. We shall take the simplest possible function for the density of states. We choose it to be a constant within the conduction band and zero outside:

$$\rho_u(\omega_{MI}) = \begin{cases} \rho_0 & \text{for } \omega_{FI} < \omega_{MI} < \omega_{BI} \\ 0 & \text{elsewhere} \end{cases}$$

$$\rho_f(\omega_{KM}) = \begin{cases} \rho_0 & \text{for } \omega_{KA} < \omega_{KM} < \omega_{KF} \\ 0 & \text{elsewhere} \end{cases}$$

where ω_A and ω_B are the lower and upper limits of the conduction band, ω_F is the Fermi level, and $\omega_{FI} = \omega_F - \omega_I$, $\omega_{BI} = \omega_B - \omega_I$, etc. Note that this assumption represents the 0 K limit. For finite temperatures, we may obtain a better fit by inserting the Fermi function. Additional improvements could come from a more realistic density-of-states function for a particular metal (see below). It is not obvious, for example, that a sharp cutoff at the band edges is correct. However, in order to simplify the mathematics and since we are mainly interested in the behavior near ω_{FI}, we feel this is a good starting point. We may further simplify the mathematics by considering only borrowing from a single state K, eliminating the sum over K. It is a simple matter to rectify this if needed. Defining κ_A as the coefficients of the integrals in A_k, A_f and $\kappa_{B,C}/(\omega_{KI}^2 - \omega^2)$ as the coefficient of the integrals in B and C, respectively, and carrying out the integrations we obtain

$$A_f = \kappa_A \rho_0 \left[\ln \left| \frac{\omega_{FI} + \omega_f - \omega}{\omega_{BI} + \omega_f - \omega} \right| + \ln \left| \frac{\omega_{FI} + \omega_f + \omega}{\omega_{BI} + \omega_f + \omega} \right| \right] \qquad (62)$$

$$A_k = \kappa_A \rho_0 \left[\ln \left| \frac{\omega_{KF} + \omega_k - \omega}{\omega_{KA} + \omega_k - \omega} \right| + \ln \left| \frac{\omega_{KF} + \omega_k + \omega}{\omega_{KA} + \omega_k + \omega} \right| \right] \qquad (63)$$

$$B = \frac{\kappa_B \rho_0}{2} \left[\frac{1}{\omega_{KI} - \omega} \ln \frac{|\omega_{FI} - \omega|}{|\omega_{BI} - \omega|} + \frac{1}{\omega_{KI} + \omega} \ln \frac{|\omega_{FI} + \omega|}{|\omega_{BI} + \omega|} \right] \qquad (64)$$

$$C = \frac{\kappa_C \rho_0}{2} \left[\frac{1}{\omega_{KI} - \omega} \ln \frac{|\omega_{KA} - \omega|}{|\omega_{KF} - \omega|} + \frac{1}{\omega_{KI} + \omega} \ln \frac{|\omega_{KF} + \omega|}{|\omega_{KA} + \omega|} \right] \qquad (65)$$

Notice that all these terms have logarithmic singularities. In this region of frequency then, we should include a phenomenological damping factor $i\Gamma$ where Γ is the inverse of some characteristic damping time in the conduction band. Considering only the dominant term, we then expect the surface-enhanced Raman spectroscopy lineshape in each case to be

$$I \propto \{\ln |\omega_{FI} + \omega_f - \omega + i\Gamma|\}^2 \qquad (66a)$$

$$I \propto \{\ln |\omega_{KF} + \omega_k - \omega + i\Gamma|\}^2 \qquad (66b)$$

$$I \propto \{\ln |\omega_{FI} - \omega + i\Gamma|\}^2 \qquad (66c)$$

$$I \propto \{\ln |\omega_{KF} - \omega + i\Gamma|\}^2 \qquad (66d)$$

These terms give the correct dependence on excitation frequency and V_{MAX}. Since $h\omega_{FI} = E_F(0) + eV$, where $E_F(0)$ is the Fermi level at an arbitrary zero applied potential with respect to a reference electrode, we can see that for metal-to-molecule transfer [Eqs. (66b) and (66d)] we predict a positive slope for V_{MAX} against ω, while for molecule-to-metal transfer [Eqs. (66a) and (66c)] we predict negative slopes. This is exactly what is observed.[34,69-76] Note that resonances are also predicted [in Eqs. (62)–(65)] for $\omega = \omega_{KA}$ or ω_{BI}. These "band edge" resonances should be relatively independent of voltage. To our knowledge, such resonances have never been observed and, in fact, may be weak or very broad if the band edge is not sharp (see below).

As a test of the lineshapes, we have fitted Eqs. (62–65) to a variety of experimental data, using Γ as an adjustable parameter and choosing ω_{KI} be the lowest-lying allowed molecular transition frequency.[84] The resonance frequency is selected so that the peak of the observed profile matches the calculated curve, and the surface concentration of adsorbed species is assumed to be relatively independent of potential in this region. In Figure 20, we show

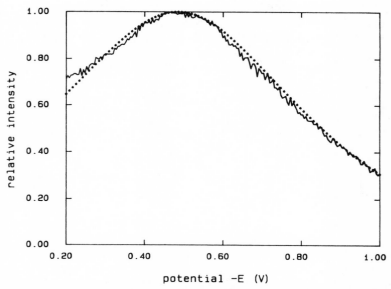

FIGURE 20. Intensity-voltage profile for the 1020-cm^{-1} line of piperidine on a silver electrode.[84] Dots show the best fit of Eq. (64) using a linewidth parameter $\Gamma = 0.3$ eV.

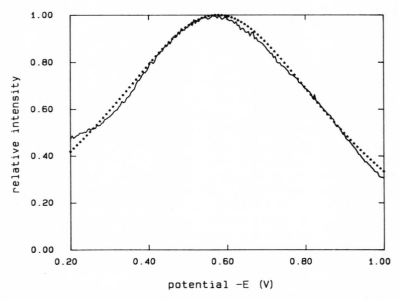

FIGURE 21. Intensity-voltage profile for the 1008-cm^{-1} line of pyridine on a silver electrode.[84] Dots show the best fit of Eq. (65) using a linewidth parameter $\Gamma = 0.3$ eV.

a fit of the predicted intensity–voltage profile for the case of molecule-to-metal transfer. The example chosen is the 1020-cm^{-1} line of piperidine on a silver electrode. For the optimum fit, we use a linewidth Γ of 0.3 eV and $\omega_{KI} = 9$ eV. Similarly, in Figure 21, we show a fit for the case of metal-to-molecule transfer using the 1008-cm^{-1} line of pyridine. We choose $\Gamma = 0.3$ eV and $\omega_{KI} = 4$ eV. In both cases the fit may be termed excellent. The value for Γ is consistent with typical metal damping times and is not too sensitive to the choice of ω_{KI}.

A word should be added concerning the effect of variations in the density-of-states function. If we consider, for example, the following integration by parts:

$$\int_{-\infty}^{+\infty} \frac{\rho(\omega_{MI})\, d\omega_{MI}}{\omega_{MI} - \omega + i\Gamma} = \rho(\omega_{MI}) \ln\,(\omega_{MI} - \omega + i\Gamma)|_{-\infty}^{+\infty}$$

$$+ \int_{-\infty}^{+\infty} \frac{\partial\rho(\omega_{MI})}{\partial\omega_{MI}} \ln\,(\omega_{MI} - \omega + i\Gamma)\, d\omega_{MI} \quad (67)$$

we see that [assuming $\rho(\omega)$ vanishes at $\pm\infty$] there is an increase in intensity predicted whenever $(\partial\rho/\partial\omega)$ is large. For the step function in the above examples, we then obtained logarithmic resonances. However, even when the density function changes less rapidly, we expect it to make some contribution to the enhancement. This may explain the fact that the observed slopes of plots of V_{MAX} against ω are not unity as predicted by the charge transfer theories. Another possible explanation for nonunity slopes is that the metal states involved are adatom surface states which see less of the electrode potential change than the Fermi level.

Let us now examine the circumstances under which the various terms given by Eqs. (62)–(65) contribute to the SERS intensity. Most generally, we expect either molecule-to-metal transfer, in which case A_f and B must be considered, or metal-to-molecule transfer, in which case A_k and C must be considered. It is unlikely that charge transfer in both directions would occur simultaneously. In either case, notice that due to the term $\langle i|k\rangle\langle k|f\rangle$ in A_k or A_f, these terms should only contribute to Raman transitions $(i \rightarrow f)$ which are totally symmetric (assuming $\langle i|$ is totally symmetric). However, intense overtones are possible. The terms B and C have a factor $\langle i|Q|f\rangle$, which enables both totally symmetric and non–totally symmetric vibrations. In the harmonic oscillator approximation, we expect no overtones to be allowed, although they would be weakly allowed if slight anharmonicities are included.

Very few observations of overtones in SERS have been reported, and even those that have are quite weak.[103] Though this calls into question the magnitude of the contribution of the A terms, it does not rule them out. Since non–totally symmetric vibrations are often as intense as totally symmetric ones in SERS, it is likely that the B and C terms are at least as important, if not more so, than the A terms. It is probably safe to regard totally symmetric vibrations as having possible contributions from both A_f (or A_k) and B (or

C) terms, while non–totally symmetric vibrations can have contributions only from either B or C.

5. Overall Enhancement Equations for Surface Raman Scattering

5.1. Effect of Concentration in a Pure EM Surface Effect

If the general Raman cross section factor, $d\sigma/d\Omega$, in Eq. (8) is replaced with the polarizability tensor as given in Eq. (36), the equation for the enhanced scattering intensity of an individual molecule can be written as

$$I = \frac{8\pi(\omega \pm \omega_{IF})^4}{9c^4} K_E I_L L^2(\omega) L^2(\omega_{IF}) \Sigma_{\sigma,\rho} |\alpha_{\sigma\rho}|^2 \tag{68}$$

where all of the parameters have been previously defined in Eqs. (8) and (36). The source of enhanced scattering is contained in the factors

$$L^2(\omega) L^2(\omega_{IF}) \Sigma_{\sigma,\rho} |\alpha_{\sigma\rho}|^2 \tag{69}$$

Furthermore, in order to calculate the total intensity of the collected scattered light, the number of molecules which undergo enhanced scattering must be included in the scattering equation. If the only enhancement mechanism is through the electromagnetic factors, $L^2(\omega) L^2(\omega_{IF})$, a calculation of the scattering intensity would have to take into account the distribution of molecules as a function of distance, r, from the surface of the metal particle, since $L(r)$ is a function of distance. This effect would introduce the factor

$$\Lambda = A \int_{r_0}^{r} L^2(\omega, r) L^2(\omega_{IF}, r) N(r) \, dr \tag{70}$$

in Eq. (68), where A is the cross-sectional area of the laser beam, assumed for simplicity to be uniform with r, r_0 is at the surface of the metal particle, and $N(r)$ is the distribution of molecules with respect to r. In the vicinity of an electrode, a charged molecule would have a distance distribution related to the static double-layer potential dependence on r. Also, if an electrode reaction occurs, there would be a concentration gradient depending on the electrode process.

Thus, for a pure EM enhancement of nonoptically absorbing molecules, the scattering intensity equation becomes

$$I_{\text{SEMERS}} = (8\pi/9c^4)(\omega \pm \omega_{IF})^4 K_E I_L \Lambda \Sigma_{\sigma,\rho} |\alpha_{\sigma\rho}^{\text{NR}}|^2 \tag{71}$$

where the sum is over the polarizability tensor components, $\alpha_{\sigma\rho}^{\text{NR}}$, for a normal Raman process, and SEMERS denotes surface *EM* enhanced *R*aman *s*cattering. It is doubtful in the electrochemical environment that I_{SEMERS} is measurable above the background scattering, which is mainly due to solvent scattering. However, under UHV conditions or with an adsorbate/air interface, a pure EM enhancement can be observed.

5.2. Overall Enhancement Equation for SERS

We will define SERS as the case where the major enhancement contributions come from EM effects and an adsorption-induced charge transfer (CT) resonances. In this case, the contribution to enhanced scattering of molecules not adsorbed on the metal surface should be small with respect to those in the first layer since, firstly, the EM enhancement factors are larger on the surface and, secondly, the charge transfer resonance can only occur at the surface, i.e., for a chemisorbed species. The CT polarizability can be 10 to 10^3 times larger than the NR polarizability. Accordingly, we write the SERS scattering intensity equation as

$$I_{SERS} = (8\pi/9c^4)(\omega \pm \omega_{IF})^4 K_E I_L L^2(\omega, r_0) L^2(\omega_{IF}, r_0) \Sigma_{\sigma,\rho} |\alpha_{\sigma\rho}^{CT}|^2 \Gamma \qquad (72)$$

where the α^{CT} polarizability components are given by the A_f and A_k terms of Eqs. (54) and (55) and the B and C terms of Eqs. (56) and (57), and Γ is the surface concentration of adsorbed species in molecules per cm^2. Note that the total measured surface-enhanced intensity

$$I_T = I_{SERS} + I_{SEMERS}$$

is in fact the sum of two contributions, one from molecules chemisorbed on the surface and one from those molecules either physisorbed or further away but still close enough to be enhanced by the EM effect. We have postulated that for chemisorbed species which can undergo a charge transfer type resonance, $I_{SERS} \gg I_{SEMERS}$, so that $I_T \approx I_{SERS}$. On the other hand, in the case of a RRS process occurring near a surface, molecules which are both adsorbed and in solution may significantly contribute to enhanced scattering.

In order to determine experimentally the enhancement produced by the *surface* effect, the enhanced scattering intensity is divided by the scattering intensity of molecules totally outside the enhancement range of the metal surface and the ratio normalized for the number of scattering molecules involved. For an electrode/solution interface with a nonoptically absorbing species, the SERS enhancement ratio, G_{SERS}, is

$$G_{SERS} = (I_{SERS}/I_{sol})(NV/\Gamma A) \qquad (73a)$$

where N is the solution concentration in molecules per unit volume, V the solution volume element intercepted by the exciting laser beam, and ΓA the number of surface molecules involved in enhanced scattering. Note that we have assumed that the same laser power is used to measure both I_{SERS} and I_{sol}; otherwise, a correction for the power ratio must be made. Also, for more accurate evaluations, the integrated Raman intensities over the band profiles should be used. However, measurements of G_{SERS} values tend to be very approximate because it is difficult to estimate the factor $(NV/\Gamma A)$ due to the difficulty in determining Γ.

Electrochemical techniques such as chronocoulometry, integrated cyclic voltammetry curves, or differentiating double-layer capacity measurements as a function of bulk concentrations can be used to estimate ΓA; however, these types of determinations may still be in error since they measure the total number of surface species, while the number of molecules at active sites which exhibit both the EM and CT effects may be much smaller. Thus, the estimates of $G_{SERS} \approx 10^5$ to 10^6 found at pretreated Ag, for example, may err on the low side if special active sites are involved in SERS. It has been estimated that only 3% of the surface sites are SERS active.[34,35]

The experimental enhancement ratio, G_{SERS}, for the nonoptically absorbing molecules also can be expressed in terms of theoretically defined parameters:

$$G_{SERS} = \frac{L^2(\omega, r_0) L^2(\omega_{IF}, r_0) \Sigma_{\sigma,\rho} |\alpha_{\sigma\rho}^{CT}|^2}{\Sigma_{\sigma,\rho} |\alpha_{\sigma\rho}^{NR}|^2} \tag{73b}$$

5.3. Enhanced Scattering in a Surface-Enhanced Resonance Raman Process

When molecules can undergo a RRS process near a rough SERS-active metal surface, the process can be called *surface-enhanced resonance Raman scattering*, SERRS. In this case, molecules can be enhanced by an EM effect as well as by a resonance Raman effect if the exciting light is within the absorption band of the molecule and the plasmon resonance of the metal microstructure. Then Eq. (71) becomes

$$I_{SERRS} = (8\pi/9c^4)(\omega \pm \omega_{IF})^4 K_E I_L \Lambda \Sigma_{\sigma,\rho} |\alpha_{\sigma\rho}^{RR}|^2 \tag{74}$$

where Λ is defined by Eq. (70) and α^{RR} components are those appropriate for the RR process (Eq. 44). The ratio I_{SERRS}/I_{sol} can be approximated by

$$I_{SERRS}/I_{sol} = \Lambda/NV \tag{75}$$

where it has been assumed that the polarizability components are the same in solution and hear the surface of the metal protrusion. In SERRS, the only *surface* enhancement for Raman scattering near the metal bump is electromagnetic in nature. Therefore, lower enhancements ratios are expected than in SERS, where the additional enhancement ratio $|\alpha^{CT}/\alpha^{NR}|$ can be on the order of 10 to 10^3. In fact, the SERRS case may even show lower enhancements than predicted by Eq. (75) if the values for the polarizability components α^{RR} near the metal surface are lower than in the bulk, i.e., they do not cancel. We will consider this effect later.

Normally, it is difficult to make a reasonable calculation using the right-hand side of Eq. (75) for a comparison with the experimentally obtained ratio. This is because (a) a uniform cross-sectional area as a function of r is an oversimplification, (b) an accurate expression for $N(r)$ is not readily available, and (c) Raman-scattered light can be absorbed by the colored species in the bulk. A simpler experimental situation is where a monolayer of a colored species, e.g., a dye, is irreversibly adsorbed on a rough Ag or Au surface in the absence of any dye in solution. Then we can express the SERRS enhancement as

$$G_{SERRS} = L^2(\omega, r_0) L^2(\omega_{IF}, r_0) (\Sigma |\sigma^{RR}|^2_{surf} / \Sigma |\alpha^{RR}|^2_{sol})$$ (76)

since the optical absorption process is the same in solution as one the surface, and one might expect the polarizability factors to cancel. However, the damping processes from nondadiative decay mechanics may be more efficient for a molecule on a metal surface than in the free state.[103]

As previously noted (Eq. 44), the RR polarizability tensor components contain a resonance denominator ($\omega_{KI} \pm \omega - i\Gamma$), where ω_{KI} is related to the electronic energy difference between the ground state and excited state and Γ is a molecular damping constant. This Γ is not to be confused with the surface coverage, for which, by convention, the same symbol is used. On a rough Ag or Au surface, Γ will contain an additional damping component, Γ_{surf}, due to the damping caused by coupling to the electronic plasmon resonance. Thus, for SERRS, Γ is equal to $\Gamma_1 + \Gamma_{surf}$, where Γ_1 is the damping constant for decay processes which are not related to the presence of the metal surface. It has been suggested[104] that $\Gamma_{surf} = 10\Gamma_1$, and thus $\Gamma_1/\Gamma \approx 1/11$.

For a Raman resonance when $\omega \approx \omega_{KI}$, the denominator of the polarizability tensor component is of the order $11\Gamma_1$. It follows that the surface Raman polarizability component may be reduced by approximately a factor of eleven over that in solution. This would lead to a G_{SERRS} factor (Eq. 76) which would be *circa* ten times smaller than would be predicted if the polarizability components on the surface and in solution are equal. It should be pointed out, too, that a similar metal damping process would also take place for the charge transfer enhancement process.

An additional effect to consider for SERRS is that of the fluorescence which can occur simultaneously with a RRS process. This is the so-called delayed fluorescence,[104] which is a radiative decay process for molecules which have relaxed from their initial excited vibrational state of the excited electronic state to the lowest vibrational level of the excited state. Delayed fluorescence can be additionally damped on a rough Ag surface for the same reasons that the RRS process is additionally damped, i.e., by coupling to the surface plasmon resonances. On the other hand, the electromagnetic enhancement factors $L^2(\omega) L^2(\omega_{IF})$, will also cause an enhancement of the fluorescence in a similar manner to the Raman process. However, the decay of excited molecules by the surface channel will tend to mitigate the EM fluorescence enhancement effect. Two cases have been discussed in the literature[104]: (a) the case of a molecule with a fluorescence quantum efficiency near unity in the free state, i.e., QE ≈ 1, and (b) the case with QE $\ll 1$. If QE is low (<0.01), the fluorescent emission is estimated to be enhanced by ca. 10; whereas, for QE ~ 1, a surface quenching of fluorescent emission by ca. 10^{-1} has been estimated.[104] Note that whereas a SERRS process can be enhanced by a factor larger than 10^3, the fluorescence process is only slightly enhanced or may, in fact, be quenched by the surface. Thus, fluorescence background may be greatly lowered on roughened SERS surfaces, producing an additional increase in the signal-to-noise ratio for the surface RR process compared to the solution RR process. In some cases, it is impossible to observe a RR spectrum in solution because of the fluorescence background but a well-defined spectrum can be obtained on SERS surfaces. Lippitsch was the first to illustrate this phenomenon with bile pigments using SERRS on Ag colloids.[105] For molecules like laser dyes, with high fluorescence quantum yields, huge SERRS enhancements can be obtained, e.g., rhodamine 6G has an apparent enhancement of 10^{10} on Ag colloids.[106]

6. Symmetry Considerations for SERS

6.1. Vibrational Selection Rules for SERS

According to the classical model of the Raman process, the electromagnetic field of the exciting light induces a dipole moment, μ, in the molecule which is proportional to the electric field $\mathbf{E}$. The proportionality factor, α, which is called the electric polarizability is

$$\mu = \alpha \mathbf{E} \tag{77}$$

where μ and $\mathbf{E}$ are vectors and α is a tensor. Equation (77) represents three equations:

$$\begin{aligned}
\mu_x &= \alpha_{xx}E_x + \alpha_{xy}E_y + \alpha_{xz}E_z \\
\mu_y &= \alpha_{yx}E_x + \alpha_{yy}E_y + \alpha_{yz}E_z \\
\mu_z &= \alpha_{zx}E_x + \alpha_{zy}E_y + \alpha_{zz}E_z
\end{aligned} \tag{78}$$

for the three components of the dipole moment, μ_x, μ_y, and μ_z. The proportionality factors depend on the field direction e.g., α_{xy} is the contribution of the polarizability to the dipole moment in the x-direction, μ_x, due to a component of the electric field, E_y, in the y-direction. In matrix form, Eq. (77) becomes

$$\begin{bmatrix} \mu_x \\ \mu_y \\ \mu_z \end{bmatrix} = \begin{bmatrix} \alpha_{xx} & \alpha_{xy} & \alpha_{xz} \\ \alpha_{yx} & \alpha_{yy} & \alpha_{yz} \\ \alpha_{zx} & \alpha_{zy} & \alpha_{zz} \end{bmatrix} \begin{bmatrix} E_x \\ E_y \\ E_z \end{bmatrix} \tag{79}$$

where μ and $\mathbf{E}$ are column vectors and α, the polarizability tensor, is a real symmetric square matrix, such that $\alpha_{xy} = \alpha_{yx}$, $\alpha_{xz} = \alpha_{zx}$, and $\alpha_{yz} = \alpha_{zy}$. In consequence, one need only consider six polarizability components out of the nine.

Raman vibrational selection rules are related to the symmetry of the polarizability components. Considering only the fundamental vibrations (ground vibrational state to first excited vibrational state transitions), a vibration of a given symmetry will be Raman active if it belongs to the same symmetry species as one of the six components of the polarizability tensor.[107] These components transform as the product of two rectangular coordinate axes, e.g. x^2, y^2, z^2, xy, xz, and yz. On the other hand, if the vibration has the same symmetry as a dipole moment component, which is related to the translational axis x, y, or z, the vibration will be infrared (IR) active.[107] The selection rules are obtained by using group theory to determine if transition moments are nonzero.[108] For the Raman effect, the transition moment to consider is

$$\langle i | \alpha_{\sigma\rho} | f \rangle = \int \psi_i \alpha_{\sigma,\rho} \psi_f^* \, d\tau \tag{80}$$

In this expression, $\sigma, \rho = x, y$, or z, where $\alpha_{\sigma\rho}$ is one of the polarizability components and $\langle i |$ and $| f \rangle$ are the initial and final vibrational wave functions. Each of the factors in Eq. (80) belongs to an irreducible representation Γ, and according to group theory, in order for the integral in Eq. (80) to be nonvanishing, the direct product $\Gamma_i \times \Gamma_\alpha \times \Gamma_f$ must contain the totally symmetric representation. Thus, for a fundamental vibration to be Raman active, $\Gamma_i \Gamma_f$ must transform under symmetry operations[108] in the same manner as one of the $\alpha_{\sigma\rho}$ components. This is the basis of the normal Raman vibrational selection rule.

It is easy to determine which symmetry representations of the vibrations of a molecule will be Raman active by looking them up in group theory character tables for a given point group. The symmetry representations which appear in the character tables form the irreducible representation of the point group. Of course, it is more difficult to assign symmetry representations to observed vibrational bands, and a detailed normal mode analysis is usually needed for these assignments. For pyridine, which is represented by the C_{2v} point group, the irreducible representations are denoted a_1, a_2, b_1, and b_2. The character table shows that a_1 transforms as x^2, y^2, z^2, a_2 as xy, b_1 as xz, and b_2 as yz. Therefore, all of the fundamental modes in pyridine are Raman allowed. When consideriing SERS vibrational selection rules in terms of the vibronic theory presented for the charge transfer mechanism in Section 4.4, it can be shown that the same vibrational selection rules apply as for normal Raman scattering.[84] Vibrations must belong to the same symmetry species as one of the polarizability components represented in the character tables of the point group in order to be SERS active by this mechanism. It should be emphasized that for a chemisorbed molecule, the symmetry of the point group is that of the metal–molecule system. Because of the surface bond, it is possible that the scattering system in SERS will have a lower symmetry than the molecule in the bulk of the solution.

This lowering of symmetry for SERS appears to be the case for the pyrazine system, which, because it has two nitrogens in the 1- and 4-positions of the heterocyclic ring (Figure 22), has D_{2h} symmetry. The character table (Figure 22) shows that the representations a_g, b_{1g}, b_{2g}, and b_{3g} transform as the binary product of two translational coordinates and are Raman active, whereas the b_{1u}, b_{2u}, and b_{3u} representations transform as one of the translational coordinates and are IR active; i.e., the Raman bands are mutually exclusive of the IR bands. This mutual exclusion will always be true for a point group with a center of symmetry. The *gerade* representations (denoted by the subscript g), which are symmetric with respect to the inversion center, will be Raman active and the *ungerade* representations (denoted by the subscript u), which are antisymmetric with respect to the center, will be IR active.[107] However, if the pyrazine is attached to a surface Ag atom, the symmetry point group becomes C_{2v} and IR modes become Raman active.[109]

Several research groups have seen normally forbidden Raman lines in SERS for pyrazine.[17,73,109-111] The SERS spectrum of pyrazine in Figure 23, obtained from an electrochemical cell, exhibits both g and u modes. We have interpreted this effect as a lowering of symmetry caused by a surface bond[109]; however, an alternate explanation which does not require a surface complex also has been considered.[17,112] This explanation involves a higher-order term in the induced dipole expression, which can be written[17]

$$\mu_i = \sum_j \alpha_{ij} E_j + (\tfrac{1}{3}) \sum_j A_{ijk}(\partial E_i / \partial k) \tag{81}$$

If the second term in this equation can become comparable to the first term

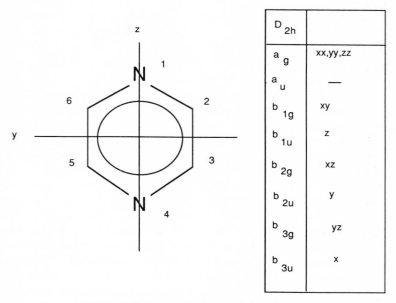

D_{2h}	
a_g	xx,yy,zz
a_u	—
b_{1g}	xy
b_{1u}	z
b_{2g}	xz
b_{2u}	y
b_{3g}	yz
b_{3u}	x

FIGURE 22. Symmetry characteristics of pyrazine.

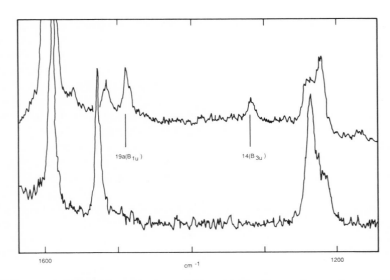

FIGURE 23. Normal Raman and SERS spectra of pyrazine between 1200 cm^{-1} and 1600 cm^{-1}. The lower trace is the NR spectrum with intense features at 1237, 1519, and 1597 cm^{-1}. The upper trace is the SERS spectrum at $E = -0.4$ V versus SCE. Note the appearance of the bands at 1317 and 1488 cm^{-1}, both of which are Raman forbidden transitions in D_{2h} symmetry.

because a large field gradient, $(\partial E_i/\partial k)$, exists at the metal, then new Raman selection rules are obeyed. Under such conditions, the molecular vibrations which belong to the same irreducible representations as the tensor components A_{ijk} will be allowed in Raman scattering. Furthermore, it is true that all of the representations which transform as x, y, and z will also transform as some components of A_{ijk} so that all IR-active modes will be allowed by electric field gradient-induced transitions. This is the alternate explanation for IR-active modes in SERS spectra. On the other hand, there are modes which are both Raman and IR inactive but which transform as a component of A_{ijk}, and these also can be active if the field gradient is important. A clear decision between the two explanations for the appearance of forbidden modes in SERS can be made in favor of field gradient-induced transitions if modes which are silent in both IR and Raman and which transform as a component of A_{ijk} appear in the SERS spectra.

6.2. Surface Selection Rules in SERS

The directional properties of the local field electromagnetic enhancement have been used to deduce the orientation of small molecules adsorbed on Ag colloids and electrodes. Moskovits and Suh[113] and Creighton[114] used these properties to deduce "surface selection rules," which essentially are the modifications of band intensities due to the normal and tangential enhancement components of the local field of a molecule near a metal surface. Here the surface is modeled as a sphere within the long-wavelength (small-sphere) approximation, $\lambda > 20a$. These arguments were used to deduce the orientation of phthalazine at the surface of a Ag colloid[113] and the orientation of pyridine at a SERS-activated electrode.[114] In the first case, it was deduced that the phthalazine molecules were standing up on the metal surface, as opposed to lying down with the π-orbitals interacting with the Ag surface. In the second case, the pyridine molecules were found to be lying down on the Ag electrode surface (Figure 24) at -0.5 V versus SCE. We present later additional orientation studies for pyridine at a SERS-active Ag electrode, which show that both orientations (Figure 24) are possible as a function of electrode potential.

Before we illustrate these surface selection rules according to the formulation of Moskovits and Suh,[113] it should be pointed out that attempts to apply the treatment to SERS spectra from cold-deposited silver films were not as successful.[113] Also, in this treatment, the surface is modeled as an individual metal sphere; however, SERS-active colloids are large aggregates with randomly adhering spheres,[113] and the surface of a SERS electrode is more like a set of prolate hemispheroids connected by a conducting plane. However, a small-sphere model is a good first-order approximation.

To illustrate these surface selection rules, recall from expression (69) that the SERS intensity is proportional to

$$|\alpha_{\sigma\rho}|^2 L^2(\omega) L^2(\omega_{IF})$$

where $\omega \equiv \omega_L$ and $\omega_{IF} \equiv \omega_S$ in the notation of Section 3. For simplicity, let $L^2(\omega) \equiv L^2$ and $L^2(\omega_{IF}) \equiv L'^2$. From Eqs. (13) and (14), with the molecule at the surface of the sphere ($r = a$),

$$L_n^2 = \langle E_n^2 \rangle / E_0^2 = (1/3)(1 + 2g)^2 \tag{82}$$

and

$$L_t^2 = \langle E_t^2 \rangle / E_0^2 = (1/3)(1 - g)^2 \tag{83}$$

where Eq. (83) is for the EM field polarized in only one tangential direction. These equations give the electromagnetic enhancement effect at a metal sphere

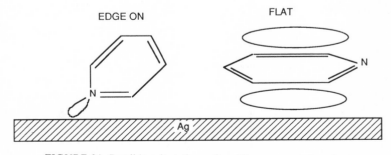

FIGURE 24. Possible orientations of pyridine on a metal surface.

for the normal and tangential components at the frequency of the incident beam, ω. Since there will be also an effect on SERS intensity from the scattered beam, ω_{IF}, the normal and tangential components of these enhancement factors are needed and can be expressed as

$$L_n'^2 = (1/3)(1 + 2g')^2 \tag{84}$$

$$L_t'^2 = (1/3)(1 - g')^2 \tag{85}$$

One might then expect that three classes of vibrational modes would coupled into different products of the enhancements factors as

$$
\begin{aligned}
&\alpha_{zz}: L_n^2 L_n'^2 \\
&\alpha_{xz}, \alpha_{yz}: \tfrac{1}{2}[L_n^2 L_t'^2 + L_t^2 L_n'^2] \\
&\alpha_{xx}, \alpha_{yy}, \alpha_{xy}: L_t^2 L_t'^2
\end{aligned}
\tag{86}
$$

where x, y, z are surface-fixed coordinates with z normal to the surface.

For a molecule with C_{2v} symmetry, such as phthalazine or pyridine, the irreducible representations are

$$
\begin{aligned}
&a_1: X^2, Y^2, Z^2 \\
&a_2: XY \\
&b_1: XZ \\
&b_2: YZ
\end{aligned}
\tag{87}
$$

for the molecule-fixed coordinates. If the molecule is standing up, then in terms of surface-fixed coordinates, $X \rightarrow x$, $Y \rightarrow y$, and $Z \rightarrow z$; however, if the molecule is lying down, then $X \rightarrow z$, $Y \rightarrow y$, and $Z \rightarrow x$. Thus, in the latter case, it can be seen that b_1 transforms as α_{xz} and b_2 as α_{xy}. According to the relationship in Eq. (86) and Table III, the ratio of band intensities for modes of b_1 symmetry to those for modes b_2 symmetry for the molecule lying down should depend on the enhancement factors roughly as L_n^2 / L_t^2. On the other

TABLE III. EM Enhancement Factors as a Function of Surface Orientation for C_{2v} Symmetry

C_{2v}	$\alpha_{\sigma\rho}$	Standing-up (edge-on)	Lying-down (face-on)
a_1	xx	$L_t^2 L_t'^2$	$L_n^2 L_n'^2$
	yy	$L_t^2 L_t'^2$	$L_t^2 L_t'^2$
	zz	$L_n^2 L_n'^2$	$L_t^2 L_t'^2$
a_2	xy	$L_t^2 L_t'^2$	$\sim L_n^2 L_t^2$
b_1	xz	$\sim L_n^2 L_t^2$	$L_n^2 L_t'^2$
b_2	yz	$\sim L_n^2 L_t^2$	$L_t^2 L_t'^2$
$\dfrac{(a_2/b_1) \text{ blue}}{(a_2/b_1) \text{ red}}$		>1	$=1$
$\dfrac{(b_2/b_1) \text{ blue}}{(b_2/b_1) \text{ red}}$		$=1$	>1

hand, if the molecule is standing up, then b_1 transforms as α_{xz} and b_2 as α_{yz}, and since according to Eq. (86) and Table III, these modes couple into the same enhancement factor products, the band ratios should be independent of the enhacement factors.

Now g (and g' as well) in Eqs. (82)–(85) are functions of frequency, and ratios such as L_n^2/L_t^2 are also functions of exciting frequency. To the blue of the localized surface plasmon peak, the ratio L_n^2/L_t^2 progressively decreases, but to the red of the peak, L_n^2/L_t^2 gradually increases. In fact, for metals, since Re $(\varepsilon) \to -\infty$ as the excitation frequency approaches the infrared Eq. (83) shows that $L_t^2 \to 0$ in the infrared, while Eq. (82) shows that L_n^2 is finite in the infrared. These results are the basis of the surface selection rules for IR surface spectroscopy, where the electric field of the light incident on metals must be concentrated in the normal direction. They also emphasize that the tangential component of the field will decrease with respect to the normal component as the excitation source is moved towards the red region of the visible spectrum.

If a molecule of C_{2v} symmetry is lying down, the ratio of the band intensities for modes with b_1 symmetry to those for modes with b_2 symmetry should increase with decreasing excitation frequency to the red of the resonance peak. For a molecule of C_{2v} symmetry standing up, this ratio would be independent of frequency. In the SERS spectrum of phthalazine, the 1594-cm^{-1} band is assigned to b_1 symmetry and the 754-cm^{-1} band to b_2, and the ratio of these band intensities with 488-nm and 602-nm exciting light is found to be independent of exciting frequency. On this basis, it was concluded that the molecule is standing up.[113] Also, the ratio of the 805 (a_1) to the 754 (b_2) cm^{-1} bands increases from blue to red as predicted for a molecule standing up, i.e., it should scale roughly as L_n^2/L_t^2. Thus, by examining various intensity ratios such as $(a_2/b_1)_{\text{blue}}/(a_2/b_1)_{\text{red}}$, etc., the orientation of the molecule can be deduced. The expected trends for these relations for C_{2v} symmetry are given in Table III. It should be noted that dividing the ratio with blue exciting light by that with red exciting light removes effects from the chemical SERS enhancement mechanism.

Similar trends in the ratio of bands of different symmetry with frequency can easily be calculated from the enhancement factors given in Section 3 for other point groups and also for a prolate spheroid as well as for a sphere. In principle, it is possible to test the "surface selection rules" for a variety of molecules at both colloids and electrodes.

We have made intensity ratio measurements for 0.05 M pyridine in 0.1 M KCl at a pretreated Ag electrode as a function of potential.[115] The bands compared and assignments used were 388 cm^{-1} (a_2): 414 cm^{-1} (b_1) and 1155 cm^{-1} (b_2): 944 cm^{-1} (b_1). It was found that

$$(a_2/b_1)_{488}/(a_2/b_1)_{647} = 0.98 \text{ and } (b_2/b_1)_{488}/(b_2/b_1)_{647} = 1.60 \text{ at } -0.6 \text{ V}$$

versus SCE, which indicates (Table III) that, in agreement with the results of Creighton,[114] (at -0.5 V) the molecule is lying down on the rough surface at

−0.6 V, versus SCE. On the other hand, at −0.9 V versus SCE, the former ratio
is 2.8 and the later ratio is 1.3, indicating, within the experimental error of the
measurements, that the pyridine is standing up at this potential. At these more
negative potentials, a change in orientation has occurred. It should be pointed
that at −0.4 V versus SCE, both ratios are close to unity, showing that here
the analysis is inadequate. Clearly, it is possible that the preferred orientation
is inclined to the surface and possibly rotated about the C_2 axis. An attempt
at a more refined orientation analysis using a rotation matrix and calculated
enhancement factors was not in good agreement with experimental results.[114]
EM field effects, as used above, also could be calculated with a rotation matrix
for a more refined orientation analysis.

7. Effects of Electrode Potential in SERS

7.1. Effect of Electrode Potential on SERS Intensities

For SERS in an electrochemical environment, the measured scattering
intensity has been shown to be a function of electrode potential. The electrode
potential, V, is applied between the SERS-active working electrode, e.g., Ag
or Au, and a reference electrode, e.g., an SCE; however, it is the potential
drop between the electrode surface and the adjacent solution which affects
molecules within the enhancement range. Observed effects of electrode potential on intensities are seen in plots of intensity versus potential (Figures 20
and 21). Here, normalized intensities have been fit to the I-V lineshape derived
from the Herzberg-Teller terms in the charge transfer theory. It was assumed
in this fit that the surface concentration did not change with potential.

Other factors, of, course, can affect the lineshape. Loss of active sites
(Figure 1) occurs at potentials negative to the point of zero charge (ca. −1.0 V
versus SCE on Ag) and can lead to loss of SERS intensity. However, this loss
occurs at the most negative end of the I-V lineshape and only affects the
negative potential extremity of the lineshape. When the I-V profile is rescanned
after potential quenching, if there is any intensity left, the potential profile,
although lower, does not drastically change in shape (Figure 1b).

In considering how SERS intensities can be affected by electrode potential
in the absence of a redox process, it is to be expected that for many molecules
the surface concentration, Γ, can change with potential, i.e., $\Gamma = \Gamma(V)$. It
should be mentioned that SERS-active surfaces stabilize adsorption so $\Gamma(V)$
may be less affected by potential on a rough surface than on a smooth surface.
Figure 25 shows that the position of the I versus V profile depends on the
vibrational mode. Obviously, if different bands show very different I versus
V curves, there are other factors in addition to the surface concentration which
depend on potential and which affect intensities. Changes in orientation of
surface molecules as a function of electrode potential have been considered

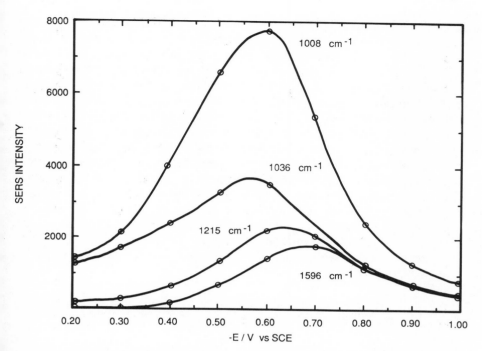

FIGURE 25. SERS intensity versus potential for pyridine bands of different symmetry at an Ag electrode. The 1008-cm^{-1} and 1036-cm^{-1} bands are totally symmetric a_1 modes and peak near -0.6 V. The 1215-cm^{-1} and 1596-cm^{-1} bands can be assigned to b_1 modes and peak at more negative potentials. Each point is obtained from a spectrum of the band at the given potential recorded with an OMA instrument. The plotted points have the background removed.

as another source of the potential dependence of I versus V, as previously discussed in Section 6.2. The net result of these potential-dependent effects on SERS intensities is that the intensity is proportional to three potential-dependent factors: $T(V)\alpha^{CT}(V)\Gamma(V)$, where $T(V)$ is a matrix which depends on the orientation of the scatterer. This result makes it difficult to extract $\Gamma(V)$ from SERS I versus V plots. However, if $T(V)$ and α^{CT} are independent of surface concentration, it is still possible to get the surface coverage, Γ/Γ_{max}, by measuring band intensities with respect to the variation of bulk concentrations.

A possible explanation for the shift in the potential of maximum intensity, V_{MAX}, with vibrational mode (shown in Figure 25) comes out of the I-V lineshape analysis of the SERS charge transfer theory of Section 4.4. It is observed (Figure 25) that V_{MAX} is more positive for totally symmetric vibrations than for non–totally symmetric vibrations. The dependence of the SERS intensity on V is given in Eqs. (66) by expressions of the form

$$I \propto |\alpha^{CT}|^2 \propto \{\ln |\omega^{CT} + \omega_i - \omega + i\Gamma|\}^2 \tag{88}$$

where ω^{CT} is an energy difference and ω_i is a vibrational frequency which only appears in polarizability components from the A terms and not from the B and C terms. Note that in Eqs. (66a) and (66b) the resonances occur at $\omega = \omega_{FI} + \omega_f$ or $\omega = \omega_{KF} + \omega_k$, while in Eqs. (66c) and (66d) the resonances occur at $\omega = \omega_{FI}$ or ω_{KF}, respectively. Since ω_f and ω_k are vibrational frequencies (ca. 0.2 eV), we should expect shifts of this order of magnitude for different vibrations. The observed shifts are about this size. Furthermore, the equations show that for metal-to-molecule charge transfer, the totally symmetric modes should have a V_{MAX} at a more positive potential than the non–totally symmetric modes. The reverse should be observed for molecule-to-molecule charge transfer. Figure 25 shows that for pyridine (metal-to-molecule transfer) this prediction is observed experimentally, since the totally symmetric a_1 modes peak near -0.6 V while non–totally symmetric a_2, b_1, and b_2 modes are shifted in the negative direction. On the other hand, in piperidine[116] (molecule-to-metal transfer), the non–totally symmetric a″ modes peak at ca -0.5 V and the totally symmetric a′ modes peak at ca. -0.8 V, again consistent with what would be predicted from Eqs. (66a) and (66c).

7.1.1. Charge Transfer Resonance Dependence on Potential and Excitation Frequency

Note that for molecule-to-metal charge transfer, $\hbar\omega_{FI} = E_F - E_I$ represents the energy difference between the Fermi level and the molecular ground state I (HOMO) of the molecule involved in the photon-stimulated charge transfer excitation (Figure 19). This energy state difference can be written as $\hbar\omega_{FI} = E_F(0) + eV - E_I$ where $E_F(0)$ is the Fermi energy level at an arbitrary zero potential with respect to a reference electrode. Of course, it is necessary to express all energy levels with respect to a common reference level. For metal-to-molecule charge transfer, $\hbar\omega_{KF} = E_K - [E_F(0) + eV]$. Considering only the B and C terms of the polarizability, the charge transfer resonance occurs when $\hbar\omega_{FI} = \hbar\omega$ for a molecule-to-metal excitation and when $\hbar\omega_{KF} = \hbar\omega$ for a metal-to-molecule excitation, where $\hbar\omega$ is the laser excitation energy. For a given ω, the intensity of the scattered light will depend on V and vice versa. Thus, the charge transfer resonance can be tuned by either the excitation frequency, ω, or the electrode potential, V. The overall excitation profile, I versus ω, at constant V will of course depend on both the charge transfer and electromagnetic resonances.

The shift in the I versus V profile with ω was shown in Figure 3 for 2-methylpyridine and imidazole. For 2-methylpyridine, V_{MAX} increases (becomes more positive) with increase in excitation energy, $\hbar\omega$, from 600 nm to 488 nm; whereas, for imidazole, V_{MAX} deceases with increasing $\hbar\omega$. We found[73] that for a series of substituted pyridines (case I), the slope $\delta V_{MAX}/\delta(\hbar\omega)$ was always positive, while it was always negative for a series of nitrogen heterocycles containing saturated nitrogens, such as imadazole (case II). Recognizing that the π-system of the pyridines could provide a charge transfer acceptor level, i.e., a LUMO level or the K excited state in Figure 19c, it is reasonable to conclude that case I molecules are involved in metal-to-molecule charge transfer. On the other hand, the nonbonded nitrogen electrons could provide the donor level or ground state I (shown in Figure 19b), indicating that case II molecules are involved in molecule-to-metal charge transfer. In general, the sign of the V_{MAX} versus $\hbar\omega$ slope for any given molecule can be used to determine the direction of a charge transfer process.

Furthermore, it was determined experimentally that V_{MAX} can be linearly correlated with the Hammett sigma function, σ_x, for case I molecules and with the pK_a for case II molecules.[73] For case I, a linear regression analysis gives $V_{MAX} = -0.58 + 0.45\sigma_x$ at 488 nm and $V_{MAX} = -0.78 + 0.51\sigma_x$ at 601 nm with a linear correlation coefficient of 0.96 in both cases. Thus, for the substituted pyridines, V_{MAX} shifts to more negative potential with increasing electron-donating power of the substituent (negative σ_x). For case II, the linear regression analysis gives $V_{MAX} = -0.97 + 0.037\,pK_a$ at 488 nm and $V_{MAX} = -1.01 + 0.074\,pK_a$ at 601 nm with linear regression coefficients of 0.81 and 0.92, respectively. In other words, for the saturated nitrogen heterocycles, V_{MAX} shifts to more positive potential with increasing pK_a.

These trends can be understood by considering the energy level diagrams in Figure 26. For case I, Figure 26a is the situation where there is a charge transfer resonance condition for metal-to-molecule charge transfer. In Figure 26b, for case I, the excitation energy $\hbar\omega_2$ has been lowered from the $\hbar\omega_1$ value of the original resonance condition, Figure 26a. In order to reestablish the resonance condition, the potential, V_2, must be moved in a negative direction (upwards in the diagram towards the vacuum level). This gives a positive slope for case I molecules. Figure 26c represents the situation for a different molecule, where a substituent of greater electron-donating power has resulted in an increase in the excited-state acceptor level. Now, for the same $\hbar\omega_1$ as in Figure 26a, the potential V_3 has to be moved in a negative direction to reestablish the charge transfer resonance. Thus, substituents with decreasing Hammett σ_x values (increasing electron-donating power) shift the charge transfer acceptor level towards the vacuum level.

For case II molecules, representing molecule-to-metal charge transfer, an increase in excitation energy, $\hbar\omega_2 > \hbar\omega_1$, on going from (a) to (b) in the lower half of Figure 26, leads to a negative $\delta V_{MAX}/\delta(\hbar\omega)$ slope. For a base containing a saturated nitrogen, an increase in pK_a means that the lone pair of the nonbonding electrons on the nitrogen would have better electron donor strength for interaction with a surface adatom. Assuming that the molecule binds weakly to an adatom or adcluster, the resulting splitting of the molecular orbitals of the adcluster–molecule ($E_H - E_N$) would depend on the pK_a of the molecule. As seen for case II molecules in Figure 26c, a lowering of E_N leads to a positive shift in V_3 at resonance compared to V_1 in Figure 26a,

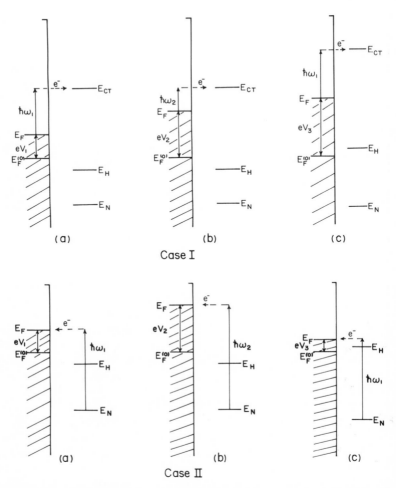

FIGURE 26. Energy level diagrams. Case I: Metal-to-adcluster–complex charge transfer. (a) Original resonance condition, (b) change in resonance voltage on a shift in $\hbar\omega_1$ to $\hbar\omega_2$, (c) change in resonance voltage on shift in E_{CT}. Case II: Adcluster–complex-to-metal charge transfer. (a) Original resonance condition, (b) change in resonance voltage on shift in $\hbar\omega_1$ to $\hbar\omega_2$, (c) change in resonance voltage on shift in E_N.[73]

which is consistent with the trend in V_{MAX} with pK_a. These observations are good evidence for the charge transfer mechanism in SERS.

7.1.2. Electric Fields Effects

Another way in which the electrode potential can alter SERS intensities is through an electric field effect which can mix closely lying excited vibrational levels. Assume that two excited vibrational levels with eigenfunctions $|f_1\rangle$ and

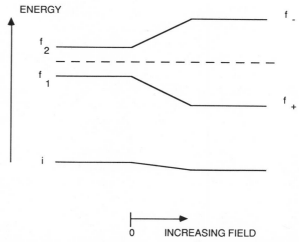

FIGURE 27. Effect of electric field on close-lying excited vibrational states.

$|f_2\rangle$ that give closely lying SERS bands are mixed by the static electric field in the double layer. Two new energy levels $|f_-\rangle$ and $|f_+\rangle$ are obtained (Figure 27). A calculation by the variational method[117] shows that these eigenfunctions are given by

$$|f_-\rangle = c_1|f_1\rangle + c_2|f_2\rangle$$
$$|f_+\rangle = c_2|f_1\rangle - c_1|f_2\rangle$$

The coefficients c_1 and c_2 which are determined in the calculation[117] depend on the electric field, which in turn can be represented by the potential drop across the compact double layer, and thus are a function of electrode potential, V. For the C and B polarizability components, SERS intensities are proportional to $\langle i|Q|f_+\rangle^2$ and $\langle i|Q|f_-\rangle^2$, and the intensities of the two bands would be potential dependent and would vary with potential in quite different ways.

The splitting between the levels $|f_-\rangle$ and $|f_+\rangle$ may nor may nor be close to the splitting between the original levels $|f_1\rangle$ and $|f_2\rangle$. In the former case, the frequency of the bands would not change very much from the field-free case, i.e., the NR spectrum of the molecule in bulk solution. However, in either situation, the SERS intensities of the bands would be dependent on electric field strength, which would be manifested in the matrix elements in the numerator of the polarizability components. This effect can be referred to as a SERS Stark effect.

7.2. SERS Intensities as a Function of Potential in the Presence of an Electrode Reaction

Previous considerations in Section 7.1 of the effect of electrode potential on SERS intensities have dealt with the situation where an electrode reaction is not taking place. It was pointed out that in this case it should be possible to measure relative intensities which should be equal to relative surface coverage. In the presence of an electrode reaction, relative SERS intensity measurements should also be proportional to relative surface concentrations and indicate redox behavior as a function of potential.

If the SERS scattering equation is written as $I = \text{const.} \times \alpha\Gamma$ where α includes all enhancement effects and potential-dependent effects besides those related to Γ, then for saturated surface coverage: $\Gamma = \Gamma'$, $I' = \text{const.} \times \alpha\Gamma'$, and $I = I'(\Gamma/\Gamma')$, showing that I/I' is equal to the relative surface concentration (Γ/Γ'). In principle, α is a function of potential so that in the absence or presence of a redox process, it is likely to change with potential and I' must be measured as a function of potential. One may avoid making a series of concentration-dependent studies to determine I' by referring intensities to values where the surface species are totally in the oxidized or reduced form, i.e., at potentials very positive or negative to the standard potential of the electrode reaction. There are, however, consequences of using this type of a relative intensity, which we will examine shortly.

Spectroelectrochemical studies refer to the measurement of variables which are related individually to the concentrations of the oxidized and reduced species and then relating these measured variables to electrochemical relationships like the Nernst equation. Because of the large number of bands in vibrational spectra, it should be relatively easy to find SERS bands which are proportional to oxidized and reduced species surface concentrations. If the surface concentrations are related to bulk concentrations by a Langmuir isotherm, it can be shown[118] for a reversible (Nernstian) redox process involving adsorbed species that

$$\Gamma_O/\Gamma_R = (\beta_O\Gamma_{O,S}/\beta_R\Gamma_{R,S}) \exp\left[(nF/RT)(E - E^{0'})\right] \tag{89}$$

where the subscripts O, R refer to the oxidized and reduced species, subscript S refers to saturated monolayer coverage, β is the adsorption coefficient, E is the electrode potential, and $E^{0'}$ is the formal redox potential of the couple O/R. At a potential $E = E_a$ where $\Gamma_O = \Gamma_R$,

$$E_a = E^{0'} - (RT/nF) \ln (\beta_O\Gamma_{O,S}/\beta_R\Gamma_{R,S}) \tag{90}$$

If one plots relative SERS intensities versus electrode potential for both $I_O/I_{O'}$ and $I_R/I_{R'}$, then the crossover potential E_x at which

$$I_O/I_{O'} = I_R/I_{R'}$$

is equal to

$$E_x = E^{0'} - (RT/nF) \ln [\beta_O(\Gamma_{O,S}/\Gamma_{O'})/\beta_R(\Gamma_{R,S}/\Gamma_{R'})] \tag{91}$$

where Γ' values are determined at the potential E_x. The correspondence between Eqs. (90) and (91) is quite close. In fact, if $\Gamma_{O,S} = \Gamma_{R,S}$ and $\Gamma_{O'} = \Gamma_{R'}$, then $E_x = E_a$ and the adsorption potential determined by SERS should be equal to that measured by conventional electrochemistry techniques, namely:

$$E_a = E_x = E^{0'} - (RT/nF) \ln (\beta_O/\beta_R) \tag{92}$$

This equivalence will only be true if SERS intensities are proportional to average surface concentrations and not surface concentrations at minority SERS-active specific sites. It should be pointed out that if the relative intensities

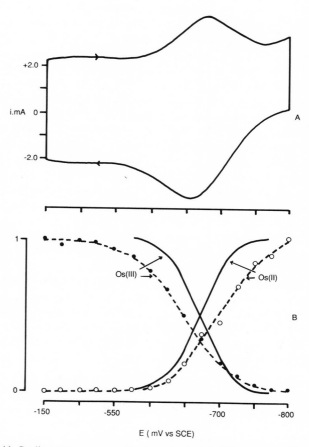

FIGURE 28. (A) Cyclic voltammogram for the adsorbed $Os^{III}(NH_3)_5py/Os^{II}(NH_3)_5py$ redox couple, and (B) plot of the relative intensities and surface concentrations for adsorbed $Os^{III}(NH_3)_5py$ and $Os^{II}(NH_3)_5py$ versus electrode potential. The electrode is roughened Ag and the medium is 0.1 M NaCl + 0.1 M HCl. The solid lines in (B) are relative surface concentrations obtained by integrating (A) and the dashed curves are relative peak intensities of the 1020- and 992-cm^{-1} SERS bands. Both surface concentrations and SERS intensities are those relative to the values at -0.15 and -0.80 V versus SCE. Reproduced from Ref. 120 with permission.

used are referred, not to intensities at saturated concentrations, but to intensities at potentials where the redox species are totally in one form or the other, then the logarithmic term in Eq. (91) will contain additional ratios of proportionality factors, α. The analysis will still be valid but deviations between E_x and E_a, merely related to the proportionality factors, will occur.

A SERS spectroelectrochemical plot of I/I' versus electrode potential E is shown in Figure 28B. Here, relative intensities are referred to intensities at potentials where the redox species are in one form or the other. It is seen that the crossover potential for relative SERS intensities (dashed lines) is very close to the crossover potential for surface charge (solid lines), where $\Gamma_O = \Gamma_R$ as determined by integrating the cyclic voltammogram (Figure 28A). These plots were made[119,120] for the redox coulle $Os(NH_3)_5py^{3+/2+}$, i.e., the pentaamminepyridineosmium cation couple.

The correspondence between the cyclic voltammetry and SERS determinations of the redox potential of the adsorbed species means that the surface sites responsible for SERS must not be significantly different in terms of redox behavior from those occupied by a majority of the adsorbate.[120] Other examples of agreement between SERS and electrochemically determined redox potentials on Ag are found for the $Ru(NH_3)_6^{3+/2+}$ couple[121] and for the methyl viologen MV^{2+}/MV^+ couple.[122]

Additional evidence that SERS spectra are representative of behavior at majority sites on polycrystalline Ag is the equivalence (shapes and relative band intensities) of the SERS spectrum of pyridine with the surface-unenhanced Raman spectrum of pyridine on smooth Ag.[123] The latter spectrum was observed at high laser power using OMA techniques. The main difference between the two spectra is the narrower bandwidths of the unenhanced spectrum. This result and the results of the above-reported spectroelectrochemical experiments indicate that SERS can be used to elucidate complex redox equilibrium behavior which would be representative of the entire surface. Of course, there are cases where a redox product may not be SERS active. Finally, it should be mentioned also that the Raman processes involved in SERS are extremely fast so that time-resolved SERS should be capable of following the fastest electrochemical dynamic processes of interest.

8. Application of SERS to Chemical Systems

The number of molecules and ions whose spectra have been observed with SERS is quite substantial, and we will present only a selective coverage of the literature in this section. A list of over 80 species which have been reported to give SERS at Ag electrodes has been compiled by Seki,[51] with literature citations through 1982. These spectra were obtained in aqueous solutions, usually with dilute KCl as the supporting electrolyte. This compilation also contains species whose SERS spectra have been observed at other metals and at a variety of different interfaces including colloidal suspensions, metal/vacuum or gas interfaces, and metal particles in an argon isolation matrix. The reader is referred to the compilation of Seki[51] for the older literature references.

8.1. Neutral Nitrogen-Containing Molecules on Ag and Cu Electrodes

The chemical species most commonly investigated with the SERS technique have been the neutral, polar nitrogen-containing molecules. The first SERS spectrum was obtained with pyridine, and this molecule has been repeatedly used as a test species for numerous SERS investigations.[51] In addition, a number of substituted pyridines have been studied (refer to Table IV). As seen from an analysis of the SERS of pyridine (Section 6.2), the surface binding in these systems could occur via the lone pair of electrons on nitrogen, in the σ-system, in edge-on fashion or via the π-electron system in a flat orientation (Figure 24). This orientation changes with potential for pyridine on an Ag electrode, and the same type of orientational change with potential has been inferred from an analysis of the SERS of phthalazine.[126]

Nitrogen-containing compounds which show SERS, other than substituted pyridines, are the atomatic and nonaromatic nitrogen heterocycles and compounds with amine substituents, as listed in Table V. Other categories of nitrogen-containing organic compounds are dyes (cyanines, azo dyes, etc.) and many molecules of biological significance. Many of these species absorb light in the 400- to 700-nm region and exhibit SERRS as well as SERS, depending on the laser excitation wavelength. Some of the biologically important nitrogen-containing molecules that exhibit SERS are derivatives of compounds tabulated in Tables IV and V. The SERS and SERRS of other biological compounds such as nitrogen-containing amino acids, nucleoside and nucleotide bases and nucleic acid bases, DNA, lipid molecules, macrocyclic ring compounds (e.g., porphyrins and phthalocyanines), and redox proteins have

TABLE IV. Pyridines Whose SERS Has Been Observed on Ag Electrodes[a]

Pyridine
2-Methyl-, 3-methyl-, and 4-methylpyridine
2,6-Dimethylpyridine (lutidine)
2,4,6-Trimethylpyridine
2-Cyano-, 3-cyano-, and 4-cyanopyridine
2-Acetyl-, 3-acetyl-, and 4-acetylpyridine
2-Pyridine-, 3-pyridine-, and 4-pyridinecarboxyaldehyde
N-Methyl-3-pyridine
N-Methyl-4-pyridine
3-Pyridyl- and 4-pyridylcarbinol

[a] References given by Seki.[51]

TABLE V. Other Nitrogen Compounds Whose SERS Has Been Observed on Ag [a]

Nitrogen heterocycles	Amines and related compounds
Pyrazine	Ammonia[45]
Pyrimidine	Aniline
Isoquinoline[51,134]	Dimethylaniline
Quinoline[135]	
Quinoxaline[133]	
2,2'-Bipyridine[51,144,146]	p-Nitrosimethylaniline
4,4'-Bipyridine[142,143]	p-Phenylenediamine
1,10-Phenanthroline	Benzylamine
Phthalazine (2,3-diazanapthalene)[113,126]	p-Aminobenzoic acid[125]
Mercaptobenzotriazole[134]	p-Nitrobenzoic acid
Benzotriazole[136]	
Imidazole[70]	Urea (Ag sols)[127]
1-Methylimidazole	Thiourea
Piperidine[117]	
Triethylenediamine (DABCO)[124]	
Quinuclidine[70]	
Piperazine[70]	

[a] Reference for all molecules without a citation can be found in Seki.[51]

been reviewed by Cotton.[128] Table VI lists some of these molecules. Since nitrogen-containing organic molecules comprise a large segment of organic chemistry and are widely distributed in living systems, these compounds represent an immense area for SERS investigations. When conditions are found

TABLE VI. Molecules of Biological Interest Whose SERS Has Been Observed [a]

Guanine, adenine, β-alanine
Histidine, tyrosine, tryptophan, phenylalanine
Leucine–isoleucine–valine(LIV)-binding protein
Leucine-specific (LS) protein
Adenine, adenosine, 5'-adenosine monophosphate
Polyadenylic acid and DNA
Polyriboadenylic acid
Nucleic acid bases
Hemoglobin
Myoglobin
Cytochrome c
Rhodopseudomonas sphaerodies—photosynthetic reaction center
meso-Tetra(4-carboxyphenyl)porphine
Tetrasodium meso-tetra(4-sulfantaphenyl)porphyin

[a] References given in review of Cotton.[128]

where biological activity is maintained on SERS-active substrates, these studies take on added significance.

The SERS of nitrogen-containing molecules can also be observed on Cu and Au electrodes.[51] SERS studies on Cu electrodes have been applied to investigate the nature of chemisorption for corrosion-inhibiting nitrogen heterocycles, e.g., benzotriazole,[129-131] benzimidazole,[131] and related compounds.[131] A SERS study[131] shows that with the exception of possible loss of protons, the molecules do not dissociate on adsorption and that the vibrational bands on the nitrogen heterocyclic part of the molecule are more affected by the surface than those of the benzene part of the molecule. Presumably, the former is bound directly to the surface. This may be a fairly general result since we found the same effect with isoalloxazines (flavins), which, like the corrosion inhibitors, have a benzene ring fused to a N-heterocyclic ring.[115]

Roughened Cu surfaces are also of interest as heterogeneous catalysts, such as in catalytic dehydroaminations. SERS was used to characterize the adsorption of m-toluidine, $CH_3C_6H_4NH_2$, which serves as a model compound for the interactions of organic amines with Cu catalysts.[132] This study is noteworthy since a normal coordinate analysis was used to show that the surface species was $R-NH \cdots Cu$ with surface binding from the nitrogen lone pair. The application of SERS as a probe for nitrogen compounds in corrosion science, heterogeneous catalysis, and biological processes, as well as for the study of redox mechanisms (Section 7.2), indicates the variety of areas of its application.

8.2. Anions and the Effect of Supporting Electrolyte at Ag Electrodes

The addition of a supporting electrolyte, such as a potassium halide salt, has the added effect in SERS studies of introducing a species which can adsorb on the metal surface and give SERS bands. For example, at Ag electrodes, in addition to molecular SERS bands, one finds bands for adsorbed halides.[137] The major low-frequency bands for the $Ag-X^-$ stretching vibration on Ag[41] are I^- ($115 \, cm^{-1}/90 \, cm^{-1}$), Br^- ($162 \, cm^{-1}/143 \, cm^{-1}$), and Cl^- ($240 \, cm^{-1}$). These surface bands are broader than the SERS bands for molecular vibrations which do not involve atoms directly bonded to the surface, and they also show vibrational frequency shifts with electrode potential ($21 \, cm^{-1}/V$ for $Ag-Cl^-$). Also, the band intensities decrease as the Ag electrode potential is moved in the negative direction. Since the point of zero charge (pzc) is around $-0.8 \, V$ versus SCE on polycrystalline Ag, there is a sizable potential range on Ag which favors specific anion adsorption. In the absence of a molecular species such as pyridine, the SERS surface band for Cl^- disappears at potentials more negative than ca. $-0.6 \, V$ versus SCE. In addition to the major surface band, other SERS bands are observed in pure aqueous potassium halide electrolyte. For example, in $1 \, M$ KCl at $-0.2 \, V$ versus SCE at Ag, bands are observed[41] at 8, 18, 110, 150, and 240 cm^{-1} and at 1610, 3498, and 3570 cm^{-1}. The latter group of bands are assigned to the vibrations of coadsorbed water. These water bands are sharper than those observed for bulk water, possibly indicating a more ordered structure at the interface. Also, the water band frequencies are affected by the nature of the halide and the supporting cation.[138] The very-low-frequency SERS bands have not been definitively assigned[139] and may be a libration of the adsorbed species, a surface complex of AgX_2^-, or, possibly, elastic vibrations of large-scale or atomic-scale metal particles.

The potential window for observation of SERS on Ag electrodes is dependent on the halide. On changing from Cl$^-$ to Br$^-$ to I$^-$, the potential for oxidation of Ag to AgX shifts in the negative direction; however, the stabilization of SERS-active sites also shifts in the negative direction on going to the heavier (more strongly adsorbed) halide. In the case of the I$^-$ ion, for example, SERS can be observed at potentials negative to -1.0 V versus SCE on Ag. Hydrogen ion reduction (ca. -1.4 V) eventually interferes with SERS at more negative potentials.

In addition to numerous studies of halides, which have been reviewed elsewhere,[4,8,139] other anions have been investigated with SERS on Ag electrodes, as indicated in Table VI. There is, in fact, an extensive literature on the SERS of CN^{-}[4,8,139] and SCN^{-}[4] on Ag. More complicated anions such as organic dicarboxylates[127] and EDTA[140] and inorganic anion metal complexes (Table VII) have also been investigated with SERS. In the dicarboxylate study, several types of surface bonding of the species on colloidal Ag have been indicated, e.g., through σ-bonding via carboxylate groups (phthalate and maleate) or π-bonding via either a single connecting C=C bond (fumarate) or a benzene ring (isophthalate and terephthalate).

8.3. Cationic Species at Ag Electrodes

Cationic species which have been observed with SERS are listed in Table VIII. Early references for SERS investigations of pyridinium, tetraalkylam-

TABLE VII. Anions Whose SERS Has Been Observed on Ag Electrodes ^{a}

Cl$^-$, Br$^-$, I$^-$
SO_4^{2-}, NO_3^-
$Ru(CN)_6^{4-}$, $Pt(CN)_4^{2-}$, MoO_4^-
N_3^-, CN$^-$, SCN$^-$
CO_3^{2-}, formate, acetate
EDTA
Dithizone
$Fe(CN)_6^{3-}/Fe(CN)_6^{4-}$ (Au electrode)[141]

^{a} References for all molecules without a citation can be found in Seki.[51]

TABLE VIII. Cations Whose SERS Has Been Observed on Ag Electrodes [a]

Pyridinium[51,147–149]
Tetramethylammonium, tetraethylammonium, and tetrabutylammonium
N-Methylpyridinium
Cetylpyridinium[150,151]
Cetyltrimethylammonium[150,151]
Methyl viologen, $MV^{2+/1+\,[153-155]}$
$Os(NH_3)_5pyridine^{3+/2+\,[119,120,143]}$
$Os(NH_3)_5pyrazine^{3+/2+}$, $Os(NH_3)_5bpy^{3+/2+}$, $Ru^{II}(NH_3)_5pyridine$, and $Ru^{II}(NH_3)_5pyrazine^{[120,143]}$
$Ru(NH_3)_6^{3+/2+\,[121]}$
$M^{III}(en)_2NCS$ and $M^{III}(NH_3)_{6-x}(NCS)_x$, where $M^{III} = Cr^{III}$, Ru^{III}, Co^{III}, Os^{III}, and $Rh^{III\,[145]}$
Ru-pyridyl complexes[146]
$Ru^{II}(b_{py})_3^{2+/1+}$ (nonaqueous medium)[152,163]

[a] References for all molecules without a citation can be found in Seki.[51]

monium ions, and N-methylpyridinium are found in the compilation of Seki.[51] SERS spectral observations for Cl^- and pyridinium indicate that these species are coadsorbed, and loss of the Cl^- band is accompanied by loss of pyridinium bands.[148,149] These bands disappear at ca. -0.6 V versus SCE. It is apparent that at potentials positive to the pzc, specific adsorption of halide is necessary to observe the SERS of pyridinium.[148,149] It is likely that there is some type of an ion-pair interaction between the pyridinium and the adsorbed halide. In the case of the tetraalkylammonium ions, which have no coordination sites for surface bonding, the SERS intensity also is abruptly attenuated when the halide desorbs. Loss of SERS activity may involve loss of the surface species and/or loss of SERS-active sistes.

In the pyridinium/pyridine system, we have found[149] that a ratio of band intensities for bands characteristic of the conjugate acid and base can be used to estimate changes in pH at the surface (assumed to be the outer Helmholtz plane). The SERS results show that, as expected, increased specific adsorption of Cl^- lowers the pH in the double layer. Surfactants with pyridinium and other cationic head groups have also been observed to give SERS. The SERS of cetylpyridinium ($C_{16}H_{33}$ alkyl chain) and cetylquinolinium ions on Ag and Au colloids and of cetylpyridinium and cetyltrimethylammonium on an Ag electrode have been observed.[151]

The SERS of methyl viologen cations, MV^{2+} and MV^+, and a wide variety of cationic metal complexes have also been observed (Table VIII). With the MV^+ radical cation, both a SERS and a RR process can be observed from the surface Raman spectrum.[154,155] The MV^{2+} and MV^+ cations seem to interact with the Ag surface via adsorbed halides. On adding an electron, the SERS carbon–carbon stretching frequency shifts from $1292\ cm^{-1}$ for MV^{2+} to $1352\ cm^{-1}$ for MV^+ for the carbons bridging the pyridine rings in the molecule. This upward frequency shift is indicative of increased electron density in the bridging carbon–carbon bond on electron transfer.

For the cationic metal ion complexes $M^{z+}(NH_3)_5L$, where z is 3 or 2 and L is a ligand, such as pyrazine or 4,4'-bipyridine, which has a free nitrogen coordinating site, SERS spectral data suggest that the complexes can bind directly to the metal surface via this free site.[120] Likewise, with substitution-inert metal isothiocyanate complexes $M^{III}L_xNCS$ (Table VIII), the SERS spectra indicate that the sulfur atom of NCS^- is bound to the Ag electrode and the nitrogen to the metal ion of the complex.[145] These examples of complexes with bridging ligands illustrate the type of molecular-level information that can be obtained with SERS concerning the interaction of redox species with electrodes.

8.4. Hydrocarbons at Ag Films and Au Electrodes

Since hydrocarbons are important reactants in many heterogeneous catalytic processes on metallic catalysts at both the metal/gas and the metal/electrolyte interfaces, the SERS of alkenes, alkynes, and aromatic hydrocarbons have been investigated at these interfaces. For the metal/gas interface, strong SERS for hydrocarbons has been observed on low-temperature Ag films under UHV conditions, while for the metal/electrolyte interface, pretreated Au electrodes give strong SERS for hydrocarbons adsorbed from aqueous solutions. The SERS of ethylene, propylene, butenes, acetylene, benzene, and benzene derivatives adsorbed on Ag films in the UHV has been discussed by Moskovits and Dilella[17] and Pockrand.[156]

Although the SERS of benzene has been reported at an Ag electrode,[157] it has been difficult to duplicate,[8,88] and absorbed hydrocarbons at Ag electrodes appear to give very weak SERS.[158] Nonpolar unsaturated hydrocarbon molecules can chemisorb to metal surfaces via π-bonding interactions, and these interactions seem to be stronger on Au than on Ag. In addition, the SERS-active sites seem to be more stable at lower adsorbate coverage on Au than on Ag. Alkenes and alkynes adsorb on pretreated Au electrodes from saturated aqueous solutions primarily by donation of electron density from π-orbitals to vacant Au orbitals.[158,159] This process leads to lower-frequency bands for the $C=C$ stretching modes.

The SERS spectra of 1,4-cyclohexadine,1,3-cyclohexadiene, phenylacetylene, and 2-butyne have been observed on Au electrodes.[158] Ethylene is found to electrooxidize,[159] while acetylene polymerizes to $trans$-polyacetylene.[158] The SERS of phenylacetylene has been observed on Ag and Cu electrodes.[160] SERS spectra have been interpreted to indicate that benzene is adsorbed in a flat orientation on an Au electrode.[88] For monosubstituted benzenes,[88] the SERS bands indicate attachment via the benzene ring for alkyl, benzoate, and aldehyde substituents, attachment via the benzene ring and halogen atom for the substituted halogen compounds, and surface binding via only the substituent groups for the nitrile ($-CN$) and nitro ($-NO_2$) group-substituted benzenes. These few initial SERS studies of unsaturated hydrocarbons adsorbed from aqueous solution indicate that SERS can be an effective surface probe for the study of nonpolar organics at an Au/electrolyte interface.

8.5. SERS under Nonstandard Conditions and in Nonaqueous Media

The usual conditions for the observation of SERS spectra in electrochemical cells have been for molecules adsorbed from aqueous solutions at close to standard temperature and pressure. Of course, there are numerous observations of SERS at the solid/UHV interface, where the pressure is quite low ($<2 \times 10^{-10}$ torr) and temperatures are maintained around 120 K. Some investigations have also been made of the applicability of using SERS to study interfaces at high pressures.[161,162]

SERS studies at 12 K for Li surfaces[17] and at liquid He temperatures using matrix-isolated Ag, Na, and K metal clusters[163,164] have been reported. Molecules such as CO and metal spheres formed by gas aggregation[165] can be trapped in a solid matrix at 4 K. The SERS bands of CO chemisorbed on Ag particles were observed in this way.[163] Ethylene and acetylene were co-condensed in argon and colloidal silver, and SERS spectra were obtained of the organic molecules adsorbed on the Ag particles.[50]

The use of nonaqueous media for SERS studies in electrochemical cells also has been a somewhat unusual matrix. The first papers on SERS from nonaqueous solution appeared in 1983.[166,167] One involved a study of 0.1 M pyridine in N',N'-dimethylformamide (DMF) with 0.05 M tetrabutylammonium perchlorate (TBAP) as the supporting electrolyte.[166] The anodization was performed with a single triangular potential sweep from -2.2 V to $+0.5$ V versus SCE. In the other paper, the occurrence of SERS from acetonitrile was definitively established with the tris(2,2'-bipyridine)ruthenium complex, $Ru(bpy)_3^{2+}$, in 0.1 M TBAP.[167] Pretreatment was obtained with one double-potential step from -0.30 V to $+0.33$ V versus SCE. In this investigation, the typical characteristic features of SERS were obtained. Chronocoulometry was used to show that monolayer coverage was obtained. The SERS intensities were found to scale with surface roughness; the spectrum was weakly potential dependent in the nonfaradaic region, and the enhancement factor was greater than 10^6. Other metal cations also showed SERS.[167] The SERS of ruthenium bipyridyl complexes also have been observed in propylene carbonate solvent.[152]

Other SERS studies at Ag electrodes have been made in acetonitrile solvent. In a study of tetraalkylammonium ions in aqueous solutions, it was mentioned that the tetraethylammonium cation shows SERS from acetonitrile solutions.[168] SERS has been observed for SCN$^-$,[169] adsorbed CH_3CN,[169] and Fe(III) protoporphyrin IX dimethyl ester,[170] all in acetonitrile. With nonaqueous solvents, surface Raman studies may give strong NR scattering from bulk solution and SERS from adsorbed solvent. The former interference can be reduced by using thin-layer cells, and both interferences can be mitigated by background subtraction or spectral difference recording. A general discussion of SERS from acetonitrile has been given by Irish et al.[171] with preliminary results for propylene carbonate. SERS bands for solvated Li^+, Na^+, trace H_2O (D_2O), and OH^- (OD^-) in acetonitrile were reported. SERS studies in alcohol[172] and pyridine[173] solvents also have been reported. From these results, it would appear that observations of SERS in acetonitrile, DMF, propylene carbonate, and various other nonaqueous solvents are possible. Such studies should be useful in nonaqueous battery research and nonaqueous organic and inorganic electrode mechanistic investigations.

Acknowledgments

We wish to acknowledge Joel I. Gersten of the Physics Department of City College of CUNY for discussions concerning the electromagnetic effect in SERS which were helpful in the preparation of this chapter. Thanks are also due to our students, I. Bernard, L. Sanchez, S. Sun, T. Lu, and J. Xu, whose unpublished results were used in this chapter. In addition, we are indebted to the National Science Foundation, the City University PSC-BHE Research Award Program, and the MBRS program of the National Institutes of Health for financial support for the portions of the work described in this chapter which come from our laboratory.

References

1. M. Fleischmann, P. J. Hendra, and A. J. McQuillian, *Chem. Phys. Lett.* **26**, 163 (1974).
2. D. L. Jeanmaire and R. P. Van Duyne, *J. Electroanal. Chem.* **84**, 1 (1977).
3. M. G. Albrecht and J. A. Creighton, *J. Am. Chem. Soc.* **99**, 5215 (1977).
4. R. L. Birke, J. R. Lombardi, and L. A. Sanchez, in: *ACS Advances in Chemistry Series* No. 201 (K. Kadish, ed.), pp. 69–107, American Chemical Society, Washington, D.C. (1982).
5. R. L. Birke and J. R. Lombardi, in: *Advances in Laser Spectroscopy* (B. Garetz and J. R. Lombardi, eds.), Vol. 1, pp. 143–153, Heyden and Sons, London (1982).
6. T. E. Furtak, in: *Advances in Laser Spectroscopy* (B. Garetz and J. R. Lombardi, eds.), Vol. 2, pp. 175–205, John Wiley and Sons, Chichester (1984).
7. A. Otto, in: *Light Scattering in Solids* (M. Cardona and G. Guntherodt, eds.), Vol. IV, Springer-Verlag, Berlin (1984).
8. R. K. Chang and B. L. Laube, in: *CRC Critical Reviews in Solid State and Materials Science*, Vol. 12, pp. 1–73, CRC Press, Inc., Boca Raton, Florida (1984).
9. R. L. Paul, A. J. McQuillan, P. J. Hendra, and M. Fleischmann, *J. Electroanal. Chem.* **66**, 248 (1975).
10. U. Wenning, B. Pettinger, and H. Wetzel, *Chem. Phys. Lett.* **70**, 49 (1980).
11. B. Pettinger, U. Wenning and H. Wetzel, *Surface Sci.* **101**, 409 (1980).
12. C. S. Allen, G. C. Shatz, and R. P. Van Duyne, *Chem. Phys. Lett.* **75**, 201 (1981).
13. M. L. A. Temperini, H. C. Chagas, and O. Sala, *Chem. Phys. Lett.* **79**, 75 (1981).
14. V. V. Marinyuk, R. M. Lazorenko-Manevich, and Ya. M. Kolotyrkin, *J. Electroanal. Chem.* **110**, 111 (1980).
15. G. Lauer, T. F. Schaaf, and J. T. Hunecke, *J. Chem. Phys.* **73**, 2973 (1980).
16. I. Pockrand, *Chem. Phys. Lett.* **85**, 37 (1982).

17. M. Moskovits and D. P. Dillella, in: *Surface Enhanced Raman Scattering* (R. K. Chang and T. E. Furtak, eds.), pp. 243–273, Plenum Press, New York (1982).
18. P. A. Lund, R. R. Smardzewki, and I. E. Tevault, *Chem. Phys. Lett.* **89**, 508 (1982).
19. W. Krasser, *International Conference on Raman Spectroscopy*, Ottawa, Canada, 1980, p. 420.
20. T. Lopez-Rios, C. Petternkoffer, I. Pockrand, and A. Otto, *Surface Sci.* **121**, L541 (1982).
21. M. Fleischmann, P. R. Graves, I. R. Hill, and J. Robinson, *Chem. Phys. Lett.* **95**, 322 (1983).
22. R. E. Naaman, S. J. Buelow, O. Cheshovsky, and D. R. Herschbach, *J. Phys. Chem.* **84**, 2692 (1980).
23. L. A. Sanchez, R. L. Birke, and J. R. Lombardi, *Chem. Phys. Lett.* **79**, 219 (1981).
24. T. H. Wood and M. V. Kelin, *Solid State Commun.* **35**, 263 (1980).
25. I. Pockrand and A. Otto, *Solid State Commun.* **38**, 1159 (1981).
26. I. Pockrand and A. Otto, *Solid State Commun.* **37**, 109 (1981).
27. T. H. Wood, *Phys. Rev. B* **24**, 2289 (1981).
28. R. P. Van Duyne, in: *Chemical and Biochemical Applications of Lasers* (C. B. Moore, ed.), Vol. 4, pp. 101–185 Academic Press, New York (1979).
29. R. L. Birke, J. R. Lombardi, and J. I. Gersten, *Phys. Rev. Lett.* **43**, 71 (1979).
30. H. A. Wetzel, H. Gerischer, and B. Pettinger, *Chem. Phys. Lett.* **80**, 392 (1981).
31. V. V. Marinyuk, R. M. Lazorenko-Manevich, and Ya. M. Kolotyrkin, in: *Advances in Physical Chemistry* (Ya M. Kolotyrkin, ed.), pp. 148–191, MIR Publishers, Moscow (1982).
32. S. H. Macomber and T. E. Furtak, *Chem. Phys. Lett.* **90**, 59 (1982).
33. T. Watanabe, N. Yanagihara, K. Honda, B. Pettinger, and L. Moerl, *Chem. Phys. Lett.* **96**, 649 (1983).
34. T. E. Furtak and D. Roy, *Phys. Rev. Lett.* **50**, 1301 (1983).
35. L. Moerl and B. Pettinger, *Solid State Commun.* **43**, 315 (1982).
36. B. H. Loo and T. E. Furtak, *Chem. Phys. Lett.* **71**, 68 (1980).
37. T. E. Furtak, G. Trott, and B. H. Loo, *Surface Sci.* **101**, 374 (1980).
38. V. V. Marinyuk, R. M. Lazorenko-Manevich, and Ya. M. Kolotyrkin, *Solid State Commun.* **43**, 721 (1982).
39. J. Billmann, G. Kovacs, and A. Otto, *Surface Sci.* **92**, 153 (1980).
40. K. Bunding, R. L. Birke, and J. R. Lombardi, *Chem. Phys.* **54**, 115 (1980).
41. M. Fleischmann, P. J. Hendra, I. R. Hill, and M. E. Pemble, *J. Electroanal. Chem.* **117**, 243 (1981).
42. B. Pettinger, M. Philpott, and J. Gordon, *Surface Sci.* **105**, 469 (1981).
43. S. H. Macomber, T. E. Furtak, and T. M. Devine, *Surface Sci.* **122**, 556 (1982).
44. T. T. Chen, J. F. Owen, and R. K. Chang, *Chem. Phys. Lett.* **89**, 356 (1982).
45. L. Sanchez, J. R. Lombardi, and R. L. Birke, *Chem. Phys. Lett.* **108**, 45 (1984).
46. C. Y. Chen and E. Burnstein, *Bull. Am. Phys. Soc.* **24**, 341 (1978).
47. E. Burstein, C. Y. Chen, and S. Lanquist, *Proc. US–USSR Symp. Inelastic Light Scattering in Solids* (J. Birman, H. Z. Cummins, and K. K. Rebane, eds.), p. 479, Plenum Press, New York (1979).
48. J. C. Tsang and J. R. Kirtley, *Solid State Commun.* **30**, 617 (1979).
49. I. Bernard, J. R. Lombardi, and R. L. Birke, unpublished results (1981).
50. K. Manzel, W. Schulze, and M. Moskovits, *Chem. Phys. Lett.* **85**, 183 (1982).
51. H. Seki, *J. Electron Spectrosc. Relat. Phenom.* **20**, 413 (1983).
52. M. Kerker, D.-S. Wang, and H. Chew, *Appl. Opt.* **19**, 4159 (1980).
53. J. I. Gersten and A. Nitzan, *J. Chem. Phys.* **73**, 3023 (1980).
54. J. I. Gersten and A. Nitzan, in: *Surface Enhanced Raman Scattering* (R. K. Chang and T. E. Furtak, eds.), pp. 89–107, Plenum Press, New York (1982).
55. F. J. Adrian, *Chem. Phys. Lett.* **78**, 45 (1981).
56. S. L. McCall, P. M. Platzman, and P. A. Wolff, *Phys. Lett.* **77A**, 381 (1980).
57. J. F. Evans, M. G. Albrecht, D. M. Ullevig, and R. M. Hexter, *J. Electroanal. Chem.* **106**, 209 (1980).
58. S. G. Schulz, M. Janik-Czachor, and R. P. Van Duyne, *Surface Sci.* **104**, 419 (1981).
59. J. I. Gersten, R. L. Birke, and J. R. Lombardi, *Phys. Rev. Lett.* **43**, 147 (1979).

60. E. Burstein, Y. J. Chen, S. Landquist, and E. Tosatti, *Solid State Commun.* **21**, 567 (1979).
61. F. W. King and G. C. Schatz, *Chem. Phys.* **32**, 245 (1979).
62. H. Ueba, *J. Chem. Phys.* **73**, 725 (1980).
63. C. Y. Chen, E. Burstein, and S. Landquist, *Solid State Commun.* **32**, 63 (1979).
64. A. Otto, *Surface Sci.* **75**, L392 (1978).
65. B. N. J. Persson, *Chem. Phys. Lett.* **82**, 561 (1981).
66. K. Arya and R. Zeyhger, *Phys. Rev.* B **24**, 1852 (1981).
67. P. H. Avouris and J. E. Demuth, *J. Chem. Phys.* **75**, 4783 (1981).
68. F. J. Adrian, *J. Chem. Phys.* **77**, 5302 (1982).
69. A. Otto, *J. Electron Spectrosc. Relat. Phenom.* **29**, 329 (1983).
70. J. Billman and A. Otto, *Solid State Commun.* **44**, 185 (1982).
71. T. E. Furtak and S. H. Macomber, *Chem. Phys. Lett.* **95**, 328 (1983).
72. M. L. A. Temperini, W. J. Barreto, and O. Sala, *Chem. Phys. Lett.* **99**, 148 (1983).
73. J. R. Lombardi, R. L. Birke, L. A. Sanchez, I. Bernard, and S. C. Sun, *Chem. Phys. Lett.* **104**, 240 (1984).
74. J. Billman and A. Otto, *Surface Sci.* **138**, 1 (1984).
75. A. Otto, J. Billmann, J. Eickmans, U. Erturk, and C. Pettenkofer, *Surface Sci.* **138**, 319 (1984).
76. J. Thietke, J. Billmann, and A. Otto, *Jerusalem Symp. Quantum Chem. Biochem.* **17** (Dyn. Surf.), 345 (1984).
77. F. R. Aussenegg and M. E. Lippitsch., *Chem. Phys. Lett.* **59**, 114 (1978).
78. M. E. Lippitsch, *Phys. Rev. B.* **29**, 3101 (1984).
79. A. G. Malshukov, *Solid State Commun.* **38**, 907 (1981).
80. A. G. Malshukov, *J. Phys. (Paris)* C10, 315 (1983).
81. A. C. Albrecht, *J. Chem. Phys.* **34**, 1476 (1961).
82. J. Tang and A. C. Albrecht, in: *Raman Spectroscopy, Theory and Practice* (H. A. Szymanski, ed.), Vol. 2, pp. 33-68, Plenum Press, New York (1970).
83. T. Watanabe, O. Kawannami, K. Honda, and P. Pettinger, *Chem. Phys. Lett.* **102**, 565 (1983).
84. J. R. Lombardi, R. L. Birke, T. Lu, and J. Xu, *J. Chem. Phys.* **84**, 4174 (1986).
85. H. S. Gold, *Appl. Spectrosc.* **33**, 649 (1979).
86. B. Simic-Galavaski, S. Zecevic, and E. Yeager, *J. Am. Chem. Soc.* **107**, 5625 (1985).
87. P. Gao, M. L. Patterson, M. A. Taddayoni, and M. T. Weaver, *Langmuir* **1**, 173 (1985).
88. P. Gao and M. J. Weaver, *J. Phys. Chem.* **89**, 5040 (1985).
89. C. C. Busby and J. A. Creighton, *J. Electroanal. Chem.* **140**, 379 (1982).
90. S. H. Macomber, T. E. Furtak, and T. M. Devine, *Chem. Phys. Lett.* **90**, 439 (1982).
91. F. Barz, J. G. Gordon, M. R. Philpott, and M. J. Weaver, *Chem. Phys. Lett.* **91**, 291 (1982).
92. T. T. Chen, K. U. Van Raben, J. F. Owen, R. K. Chang, and B. L. Laube, *Chem. Phys. Lett.* **91**, 494 (1982).
93. J. D. Jackson, *Classical Electrodynamics*, 2nd Edition, John Wiley and Sons, New York (1975), p. 150.
94. H. Seki and T. J. Chuang, *Chem. Phys. Lett.* **100**, 393 (1983).
95. J. I. Gersten and A. Nitzan, in: *Surface Enhanced Raman Scattering* (R. K. Chang and T. E. Furtak, eds.), pp. 89-107, Plenum Press, New York (1982).
96. P. B. Johnson and R. W. Christy, *Phys. Rev.* B **6**, 4370 (1972).
97. H.-J. Hagemann, W. Gudet, and C. Kunz, *J. Opt. Soc. Am.* **65**, 742 (1975).
98. M. Kerker, *J. Opt. Soc. Am.* B **2**, 1327 (1985).
99. M. Abramowitz and I. A. Stegun, *Handbook of Mathematical Functions*, Dover Publications, New York (1965), p. 752.
100. P. W. Barber, R. K. Chang, and H. Massoudi, *Phys. Rev. Lett.* **50**, 997 (1983).
101. P. W. Barber, R. K. Chang, and H. Massoudi, *Phys. Rev.* B **27**, 7251 (1983).
102. G. C. Schatz, *Acc. Chem. Res.* **17**, 370 (1984).
103. B. Pettinger, *Chem. Phys. Lett.* **78**, 404 (1981).
104. D. A. Weitz, S. Garoff, J. I. Gersten, and A. Nitzan, *J. Chem. Phys.* **78**, 5324 (1983).
105. M. E. Lippitsch, *Chem. Phys. Lett.* **74**, 125 (1980).
106. B. Pettinger, *Chem. Phys. Lett.* **110**, 576 (1984).

107. F. A. Cotton, *Chemical Applications of Group Theory*, 2nd Edition, Wiley Interscience, New York (1971).
108. E. B. Wilson, J. C. Decius, and P. C. Cross, *Molecular Vibrations*, Dover Publications, New York (1955), pp. 161.
109. G. R. Erdheim, R. L. Birke, and J. R. Lombardi, *Chem. Phys. Lett.* **69**, 495 (1980).
110. R. Dornhaus, M. R. Long, R. E. Benner, and R. K. Chang, *Surface Sci.* **93**, 240 (1980).
111. R. Dornhaus, *J. Electron Spectrosc. Relat. Phenom.* **30**, 1978 (1983).
112. J. K. Sass, H. Neff, M. Moskovits, and S. Holloway, *J. Phys. Chem.* **85**, 621 (1981).
113. M. Moskovits and J. S. Suh, *J. Phys. Chem.* **88**, 5526 (1984).
114. J. A. Creighton, *Surface Sci.* **124**, 209 (1983).
115. J. Xu, R. L. Birke, and J. R. Lombardi, *J. Am. Chem. Soc.* **109**, 5645 (1987).
116. R. L. Birke, I. Barnard, L. A. Sanchez, and J. R. Lombardi, *J. Electroanal. Chem.* **150**, 447 (1983).
117. L. A. Sanchex, R. L. Birke, and J. R. Lombardi, *J. Phys. Chem.* **88**, 176 (1984).
118. A. J. Bard and L. R. Faulkner, *Electrochemical Methods*, John Wiley and Sons, New York (1980), p. 521.
119. S. Farquharson, M. J. Weaver, P. A. Lay, R. H. Magnuson, and H. Taube, *J. Am. Chem. Soc.* **105**, 3351 (1983).
120. S. Farquharson, K. L. Guyer, P. A. Lay, R. H. Magnuson, and M. J. Weaver, *J. Am. Chem. Soc.* **106**, 5123 (1984).
121. M. A. Taddayoni, S. Farquharson, and M. J. Weaver, *J. Chem. Phys.* **80**, 1363 (1984).
122. T. Lu, R. L. Birke, and J. R. Lombardi, unpublished results (1985).
123. M. Fleischmann, P. R. Graves, and J. Robinson, *J. Electroanal. Chem.* **182**, 73 (1985).
124. D. E. Irish, D. Guzonas, and G. F. Atkinson, *Surface Sci.* **158**, 314 (1985).
125. J. S. Suh, D. P. Dilella, and M. Moskovits, *J. Phys. Chem.* **87**, 1540 (1983).
126. M. Takahashi, M. Fujita, and M. Ito, *Chem. Phys. Lett.* **109**, 122 (1984).
127. M. Moskovits and J. S. Suh, *J. Phys. Chem.* **88**, 1293 (1984).
128. T. M. Cotton, in: *Surface and Interfacial Aspects of Biomedical Polymers* (J. D. Andrade, ed.), Vol. II, p. 161, Plenum Press, New York (1985).
129. J. J. Kester, T. E. Furtak, and A. J. Bevolo, *J. Electrochem. Soc.* **129**, 1716 (1982).
130. J. Rubin, G. R. Gutz, O. Sala, and W. J. Orwille-Thomas, *J. Mol. Struct.* **100**, 571 (1983).
131. D. Thierry and C. Leygraf, *J. Electrochem. Soc.* **132**, 1009 (1985).
132. A. Wokaun, A. Baiker, S. K. Miller, and W. Fluhr, *J. Phys. Chem.* **89**, 1910 (1985).
133. M. Takahashi and M. Ito, *Chem. Phys. Lett.* **103**, 512 (1984).
134. M. Ohsawa and W. Suetaka, *J. Electron Spectrosc. Relat. Phenom.* **30**, 221 (1983).
135. M. Fleischmann, I. R. Hill, and G. Sundholm, *J. Electroanal. Chem.* **158**, 153 (1983).
136. J. C. Rubin, I. G. R. Gutz, and O. Sala, *J. Mol. Struct.* **101**, 1 (1983).
137. R. Dornhaus and R. K. Chang, *Solid State Commun.* **34**, 811 (1980).
138. M. Fleischmann and I. R. Hill, *J. Electroanal. Chem.* **146**, 367 (1983).
139. M. Fleischmann and I. R. Hill, in: *Comprehensive Treatise of Electrochemistry* (R. E. White, J. O'M. Bockris, B. E. Conway, and E. Yeager, eds.,), Vol. 8, pp. 373–342, Plenum Press, New York (1984).
140. H. Wetzel, B. Pettinger, and U. Wenning, *Chem. Phys. Lett.* **75**, 173 (1980).
141. M. Fleischmann, P. R. Graves, and J. Robinson, *J. Electroanal. Chem.* **182**, 57 (1985).
142. T. M. Cotton and M. Vavra, *Chem. Phys. Lett.* **106**, 491 (1984).
143. S. Farquharson, P. A. Lay, and M. J. Weaver, *Spectrochim. Acta A* **40A**, 907 (1984).
144. M. Kim and K. Itoh, *J. Electroanal. Chem.* **188**, 137 (1985).
145. M. A. Taddayoni, S. Farquharson, T.-T. Tomi, and M. J. Weaver, *J. Am. Chem. Soc.* **88**, 4701 (1984).
146. J. A. Chambers and R. P. Buck, *J. Electroanal. Chem.* **163**, 297 (1984).
147. H. Chang and K. C. Hwang, *J. Am. Chem. Soc.* **106**, 6586 (1984).
148. D. J. Rogers, S. D. Luck, D. E. Irish, D. A. Guzonas, and G. F. Atkinson, *J. Electroanal. Chem.* **167**, 237 (1984).
149. S. C. Sun, I. Bernard, R. L. Birke, and J. R. Lombardi, *J. Electroanal. Chem.* **196**, 359 (1985).

150. S. M. Heard, F. Grieser, and C. G. Barraclough, *Chem. Phys. Lett.* **95**, 154 (1983).
151. S. C. Sun, R. L. Birke, and J. R. Lombardi, unpublished work (1985).
152. H. R. Virdee and R. E. Hester, *J. Phys. Chem.* **88**, 451 (1984).
153. A. Regis and J. Corset, *J. Chim. Phys.* **78**, 687 (1981).
154. C. A. Melendres, P. C. Lee, and D. Meisel, *J. Electrochem. Soc.* **130**, 1523 (1983).
155. T. Lu, R. L. Birke, and J. R. Lombardi, *Langmuir* **2**, 305 (1986).
156. I. Pockrand, *Surface Enhanced Raman Vibrational Studies at Solid/Gas Interfaces*, Springer Tracts in Modern Physics, Vol. 104 (1984).
157. M. W. Howard and R. P. Cooney, *Chem. Phys. Lett.* **87**, 299 (1982).
158. M. L. Patterson and M. J. Weaver, *J. Phys. Chem.* **89**, 5046 (1985).
159. M. L. Patterson and M. J. Weaver, *J. Phys. Chem.* **89**, 1331 (1985).
160. L. M. Arbantes, M. Fleischmann, I. R. Hill, L. M. Peter, M. Mengoli, and G. Zotti, *J. Electroanal. Chem.* **164**, 177 (1984).
161. P. Podini and J. M. Schnur, *Chem. Phys. Lett.* **93**, 86 (1982).
162. C. Sandroff, H. E. King, Jr., and D. R. Herschbach, *J. Phys. Chem.* **88**, 5647 (1984).
163. H. Abe, K. Manzel, W. Schulze, M. Moskovits, and D. P. Dilella, *J. Chem. Phys.* **74**, 792 (1981).
164. W. Schulze, B. Breithaupt, K. P. Charle, and U. Kloss, *Ber. Bunsenges. Phys. Chem.* **88**, 308 (1984).
165. H. Abe, W. Schulze, and B. Tesche, *Chem. Phys.* **47**, 95 (1980).
166. K. Hutchinson, A. J. McQuillan, and R. E. Hester, *Chem. Phys. Lett.* **98**, 27 (1983).
167. A. M. Stacy and R. P. Van Duyne, *Chem. Phys. Lett.* **102**, 365 (1983).
168. V. V. Marinyuk, R. M. Lazorenko-Manevich, and M. Kolotyrkin, *Dokl Akad. Nauk. SSSR* **242**, 1382 (1978).
169. D. A. Guzonas, G. F. Atkinson, and D. E. Irish, *Chem. Phys. Lett.* **107**, 193 (1984).
170. L. A. Sanchez and T. G. Spiro, *J. Phys. Chem.* **89**, 763 (1985).
171. D. E. Irish, I. R. Hill, P. Archambault, and G. F. Atkinson, *J. Sol. Chem.* **14**, 221 (1985).
172. J. J. Kim and G.-S. Shin, *Chem. Phys. Lett.* **118**, 493 (1985).
173. J. E. Pemberton, *Chem. Phys. Lett.* **115**, 321 (1985).

7

ESR Spectroscopy of Electrode Processes

Richard G. Compton and Andrew M. Waller

1. Introduction

Electron spin resonance (ESR) is an attractive technique for the identification and study of species containing an odd number of electrons (radicals, radical cations and anions, and certain transition metal species). The experiment is based on the fact that whereas in the absence of a magnetic field, the two possible spin states, $+\frac{1}{2}, -\frac{1}{2}$, of an unpaired electron have identical energies, this degeneracy is lost when a field is applied. ESR spectroscopy involves the flipping of the spin between the two now different energy levels, an act which is brought about by the absorption of microwave radiation. As will be shown below, the presence of magnetic nuclei in the molecule is revealed as hyperfine structure in the ESR transition, which thus may be used to provide a positive identification of the odd-electron species. Since each magnetic nucleus contributes (at least in principle) to the hyperfine structure, a rather more intimate insight into molecular identity emerges than from, for example, UV–visible spectroscopy. It is this high information content of ESR spectroscopy, together with the sensitivity of the technique (radical concentrations on the order of 10^{-8} M may be observed with standard equipment[1]), that has made ESR the method of choice for investigating complex electrode reactions which proceed via radical intermediates. Other advantages are that the specificity towards paramagnetic species is advantageous in some cases, that the technique is nonperturbing (there is no activation or destruction of the sample), and that absolute concentrations may be measured (albeit with limited accuracy).

The first application of ESR in dynamic electrochemistry dates from 1958, when Ingram and coworkers[2] demonstrated the formation of radical anions in the polarographic reduction of the compounds anthracene, benzophenone, and anthraquinone in the solvent dimethylformamide. They filled tubes by

Richard G. Compton and Andrew M. Waller ● Physical Chemistry Laboratory, Oxford University, Oxford OX1 3QZ, United Kingdom.

withdrawing samples during electrolysis, froze them by chilling with liquid nitrogen, and then transferred the tubes to the ESR spectrometer, where spectra were recorded at low temperature. Broad, single-line spectra characteristic of solid-state samples of organic radicals were observed, indicating the existence of radicals as intermediates in the electroreduction processes. However, the lack of hyperfine splittings, characteristic of solution-phase spectra, precluded any conclusions being drawn as to the nature of the radicals involved.

It was left to the pioneering work of Maki and Geske[3] to first demonstrate the real possibilities of the technique. They obtained the first solution-phase ESR spectra from electrogenerated species and, in so doing, performed the first *in situ* electrolysis—that is to say that the electrolysis was carried out within the cavity of the ESR spectrometer. They used a two-electrode cell consisting of a 3-mm capillary tube with a platinum wire anode, and this was positioned along the axis of their cylindrical cavity so that the tip of the electrode was at the cavity center for maximum sensitivity. The first chemical system studied in this manner was the oxidation of $LiClO_4$ in acetonitrile. The spectrum displayed a hyperfine structure of four equally spaced lines of equal intensity, interpreted as due to a radical species containing a single chlorine nucleus (spin $\frac{3}{2}$) and initially thought to be the perchlorate radical.[3] It was later confirmed[4] to be $ClO_2^{\cdot}$, produced via the following route:

$$ClO_4^{-} \rightarrow ClO_4^{\cdot}$$

$$ClO_4^{\cdot} \rightarrow ClO_2^{\cdot} + O_2$$

The same workers then rapidly employed their *in situ* cell to observe the radical anions of nitrobenzene,[5] of *ortho-*, *meta-* and *para*-dinitrobenzenes,[6] and of a large number of substituted nitrobenzenes.[7] All of these anions were generated in nonaqueous media. While this work clearly demonstrated to electrochemists the involvement of the anion radicals as intermediates in the electrode reactions, it soon became clear to spectroscopists that *in situ* generation was a convenient way of making large quantities of relatively long-lived (seconds) radicals within an ESR cavity with a view to investigating their spectra. There are two particular advantages of electrogeneration over chemical oxidation or reduction. Firstly, one apparatus can find wide application without major experimental change since the "redox potential" of the working electrode is widely variable (within the limits of solvent decomposition), unlike chemical reagents. Furthermore, judicious choice of electrode material can lead to the technique being specific in its reactivity towards different substrates. Secondly, the possibility of interference of chemical reagents with the radical is reduced. Generation of the radical anion of nitrobenzene by reduction with alkali metals in solvents, such as tetrahydrofuran, may produce a spectrum complicated by ion-pairing of the radical anion with the alkali metal cation. This is not seen if the species is electrogenerated in acetonitrile with tetraalkylammonium salts as supporting electrolyte.[8]

The work of Adams and coworkers[9,10] led to an optimized version of the Maki and Geske *in situ* cell. This is shown in Figure 1 and consists of a platinum gauze electrode inside what is essentially a conventional ESR flat cell. The gauze provides an electrode of high area for the production of high concentrations of radicals with the cell. The cell was used in conjunction with a three-electrode system, the counter (auxiliary) electrode being located below the ESR cavity and the reference electrode above it. Initial experiments with the cell led to the study of the radical cations of various aromatic amines[9] and the radical anions of both aliphatic and aromatic nitro compounds. A measure of the success of Adams' design for producing copious quantities of radicals for spectroscopic purposes is that essentially the same cell is marketed still today by manufacturers of commercial ESR equipment (e.g., Bruker, Varian). Cells of this type have, over the years, been used to generate an enormous range of radicals and radical ions for spectroscopists, and modified versions of the cell have appeared that are suitable both for low-temperature work,[11] for example, in liquid ammonia,[12-14] and for high-temperature work,[15,16] for example, in molten salts.

Despite the success of the cell in generating relatively long-lived radicals, the cell is less than optimal for the electrochemist who, firstly, will wish to study his reactions under conditions where well-defined voltammetric measurements may be made and who, secondly, may require to study processes involving comparatively short-lived ($\ll$ seconds) paramagnetic species, with a

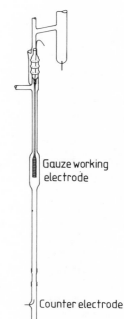

Gauze working electrode

Counter electrode

FIGURE 1. The *in situ* cell of Adams utilizes a platinum gauze of high area, with a counter electrode below the cavity and a reference electrode above it.

view not only to observing the radicals but also to establishing the kinetics and mechanisms of their decay. The limitations of the Adams design, in this context, are several. Firstly, it is reported[9] that the "resolution of recorded polarograms is very poor." This may be attributed partly to irregular and irreproducible natural convection within the cell (as may be evidenced by generating a highly colored radical ion in the cell,[17] which reveals swirling motions and convection patterns), and partly to the fact that the thinness of the cell produces a large change in electrical resistance over the surface of the electrode, which results in the distortion of current–voltage curves (so-called "ohmic drop"). A second limitation of the type of cell depicted in Figure 1 is that only radicals which live for at least a few seconds will be detectable. This is because there is no efficient or well-defined flow bringing fresh electroactive material to the electrode surface. Thus, as electrolysis proceeds, the current at the electrode will drop and hence so will the flux of radicals being generated. Only when the radical is sufficiently long-lived to survive for the period of time necessary to make the ESR measurements will the cell display adequate sensitivity. It is clear, however, that this problem may be alleviated and shorter-lived species observed by employing a hydrodynamic flow to bring unelectrolyzed material to the electrode, thereby sustaining the production of radicals. Lastly, it should be pointed out that meaningful kinetic measurements are also precluded with this cell geometry. It might be thought that such information could be readily obtained simply by establishing a radical population within the cell, then open-circuiting the electrode and monitoring a radical decay transient. However, the ESR cavity has a nonuniform sensitivity along its length and is spatially discriminating as a result (see below). The consequence of this is that in order to be able to deduce the radical kinetics, the radical concentration profile in both space and time must be known (i.e., calculable). As we have seen, because of the existence of natural convection of uncertain behavior, it must be recognized that this will not be feasible for the cell geometry under consideration.

Subsequent developments in electrochemical ESR methodology were aimed at improving the quality of the electrochemical behavior of the cell while hopefully enabling as large a number of radicals as possible to be generated. Additionally, it was endeavored to enable the lifetimes and kinetic modes of decomposition of the radical intermediates to be determined. These quests led to two alternative approaches, depending on whether the working electrode was located inside the ESR cavity or not—"in situ" or "ex situ" methods. The philosophy behind the ex situ approach was that the electrochemistry could be carried out under conditions free from the very limiting constraint of available sample space which necessarily arises when electrodes are located within the ESR cavity (see below). The ex situ methods rely on generating radicals under conditions approximating to conventional conditions and then transferring them to the cavity for detection, generally by means of a flow system. This approach, of course, imposes a restriction on the accessible radical lifetimes corresponding to the rate of transfer from the site of generation to the point of detection. In contrast, users of the *in situ* methods, at least until recently, chose to sacrifice the well-defined electrochemistry in favor of observing shorter-lived radicals than was feasible with the *ex situ* method. However, as revealed below in the section on Practice (Section 3), contemporary cell designs have now realized satisfactory electrochemical performances from *in situ* cells, and so the *in situ/ex situ* schism of the late 1960s and 1970s is to a large extent only of historical interest today. We will, nevertheless, briefly outline some of the techniques developed during this period, partly to show the intellectual development of the area and partly because for some applications these cells continue to find application at the present time. We will, in the remainder of this section, consider the two approaches separately and then, in Section 3, show how the better features of the earlier work are incorporated in modern cells.

1.1. External Generation Methods

Apart from the work of Ingram *et al.* outlined above, an early *ex situ* approach was that of Fraenkel and coworkers,[18-20] who generated radical anions at a mercury pool cathode inside a vacuum electrolytic cell. After electrolysis, the solution was tipped into a side-arm used for the ESR measurements. More subtly, Kastening[21] used an assembly in which the solution circulated between the electrode and the cavity. This was effected by the hydrostatic pressure produced by a large and rapidly rotating working electrode. At the time, this method was claimed to provide greater sensitivity than the then available *in situ* generation methods, such as those described above due to either Maki and Geske or Adams. The reason for this is that the existence of a flow of solution sustains the production of radicals during the course of the experiment.

In subsequent work by Kastening,[22,23] the solution passed from the working electrode through a capillary tube in the cavity under the effect of gravity; a constant head device kept a steady flow rate. A mixing chamber upstream of the cavity homogenized the solution before it reached the cavity and also permitted additional reagents to be added to the flowing radical stream from another solution reservoir. By varying the flow rate or by measuring the ESR signal at various distances along the capillary tube, the timescale of the radical decomposition kinetics was established. In this way, Kastening studied the lifetime of the nitrobenzene radical anion at various pH's. A similar apparatus was developed by Gerischer *et al.*[24]; in this case, a layer of sand was used to homogenize the solution before it entered the cavity.

The problem with external generation is, of course, as remarked previously, that the finite time between creating the radical at the electrode, mixing the solution, and pumping it to the ESR cavity restricts the technique to rather long-lived species. This problem was alleviated as much as possible given *ex situ* generation by Albery and coworkers[25-30] by the use of a tube electrode. The latter consists of an annular ring which forms part of the wall of a tube through which solution flows. The tube passes through the ESR cavity, which is located immediately downstream of the electrode, as depicted in Figure 2. The radicals consequently are generated as close as possible to the cavity without the electrode actually being inside the cavity. Unlike earlier external generation systems, the solution is not mixed and so the concentration of radicals in the solution stream entering the cavity is not homogeneous. However, since the flow was constrained to be laminar at all flow rates employed, it was possible to calculate the radical distributions as a function of flow rate, electrode current, and cell geometry, in much the same way as will be described in detail in Section 3 for an improved, related cell (the Compton–Coles channel electrode cell). This, unlike the Albery cell, is capable of *in situ* operation, while at the same time improving on the electrochemical features of the Albery cell. Theories were developed for stable radicals[26] and

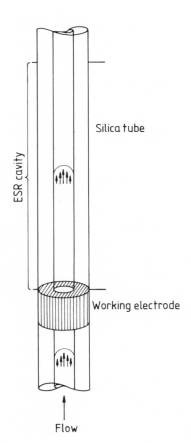

FIGURE 2. The *ex situ* flow cell of Albery is based on a tube electrode located immediately upstream of the ESR cavity.

for radicals decaying with first-[27] and second-order[28] kinetics, and these led to information about the mechanisms of radical decay. Chemical systems studied with this cell include the cations of various aromatic amines,[29,30] the radical anion of nitrobenzene,[28] and semiquinone species.[25,26]

1.2. Internal Generation Methods

In this section, we discuss some improvements on the Adams design of cell described above. Initially, Cauquis and coworkers[31-35] fabricated an experiment in which the solution flowed over a grid electrode in a flat cell. The flow was in a closed loop, and so for "stable" radicals, exhaustive electrolysis could be employed. Subsequent developments of the *in situ* method were aimed more at the elucidation of the rates of homogeneous reactions of the radicals. Thus, Dohrmann and coworkers[36-39] placed a foil electrode in the center of a flat cell and flowed solution over this. Kinetic information was obtained by interrupting the current and measuring ESR signal transients. Kastening *et al.*[140] described a similar apparatus which uses a number of strip electrodes in the center of the channel of a flow system. This permitted the use of large currents, the potential of each strip being adjusted to maximize the ESR signal. While both the Dohrmann and the Kastening designs are amenable, in principle, to the study of kinetic processes, it should be pointed out that to do this rigorously requires the solution of the appropriate convective-diffusion equation (see below). Hence, it would be necessary to have knowledge of the hydrodynamics within the two cells. In neither case is this simple, and uncertainties exist,[41] especially with the Kastening design. Such ambiguities are eliminated in the flow cells described in Section 3.

Bard, Goldberg, and coworkers[42-45] also have made kinetic measurements using an *in situ* stationary solution cell similar to that of Adams. As with that design, they found that ESR data had to be recorded rapidly before the onset of natural convection. The problem of ohmic drop was reduced by comparison with the Adams cell by locating the reference and counter electrodes inside the flat cell; this improved the electrochemistry, but problems arose due to the production of radicals at the counter electrode[45] for some chemical systems studied, since this was now located within an ESR-sensitive region. Initially, the cell was used to make simultaneous electrochemical ESR investigations of the relatively stable radicals formed from anthraquinone, azobenzene, and nitrobenzene in nonaqueous media. Subsequently, transient ESR measurements following a current pulse were utilized to study reactions of radicals. Reaction mechanisms involving first-order decomposition, radical ion dimerization, and radical parent dimerization were considered.[43] In particular, the mechanism of the cathodic dimerization of various olefins was examined in detail.[44]

Bond[46] recently has described an *in situ* cell with stationary solution which employs tiny volumes of solution (0.2 cm³) and a very small working electrode and which is capable of variable temperature operation. This permits the study of fairly stable radicals (~seconds) but under conditions where reasonable cyclic voltammograms can be recorded.

Finally, it only remains in this section, concerned with the historical development of the subject, to point out that a number of reviews have appeared dealing with the earlier work in the field.[47-54]

2. Theory

The purpose of this section is to provide an elementary introduction to those aspects of ESR that an electrochemist with no previous experience in the field would require in order to be able to carry out and interpret electrochemical ESR experiments. Those previously initiated in the theory and practice of ESR are advised to skip to Section 3.

2.1. Introductory Remarks

The origin of the ESR experiment lies in the spin possessed by an electron. Chemists will know that an electron "has a spin of one-half." By this we mean that the component of spin angular momentum, S_z, in a direction specified by an applied magnetic field (conventionally labeled the z-direction) can take on only two values— $+\frac{1}{2}\hbar$ and $-\frac{1}{2}\hbar$. The vector describing the spin angular momentum has a magnitude $[\frac{1}{2}(\frac{1}{2}+1)]^{1/2}\hbar$, and this vector is perceived as rotating ("precessing") around the z-direction at such an angle that its projection on this axis is consistent with one of the two values of S_z given above (Figure 3). It is useful to introduce the quantum number M_s to label the allowed eigenvalues of S_z. Clearly, $M_s = +\frac{1}{2}$ or $-\frac{1}{2}$ and gives the permitted values of the spin angular momentum in the z-direction in units of $\hbar$.

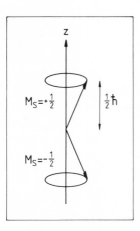

FIGURE 3. The precession of the spin angular momentum vector about a magnetic field applied in the z-direction. The vector has a magnitude $[\frac{1}{2}(\frac{1}{2}+1)]^{1/2}\hbar$ and the size of its component in the z-direction may be either $+\frac{1}{2}\hbar$ or $-\frac{1}{2}\hbar$.

The existence of spin angular momentum, together with the electrons' electrostatic charge, confers a magnetic moment, μ, on the electron which is proportional to the magnitude of the spin vector ($\mathbf{S}$):

$$\mu = -g\beta\mathbf{S} \tag{1}$$

where β is the electronic Bohr magneton, equal to $(e\hbar/2mc)$, with e and m the charge and the mass of the electron, respectively. The quantity g is a dimensionless constant known (imaginatively) as the electron g-factor. For a free electron, g has the value 2.0023.

The ideas embodied in Eq. (1) were well established long before magnetic resonance experiments were carried out, from the famous Stern–Gerlach experiment,[55,56] in which a beam of silver atoms was found to split into two separate beams upon passage through an inhomogeneous magnetic field. This was a direct reflection of the two only possible spin states of the unpaired electron within the atom. The existence of fine structure in atomic spectra provided further evidence for electron spin.

If we apply a steady magnetic field, $\mathbf{B}$, to the electron, then the energy of the interaction with the magnetic moment of the electron is represented by[1]

$$E = -\mu \cdot \mathbf{B} \tag{2}$$

If the field is applied in the z-direction, the scalar product in Eq. (2) can be simply written as follows:

$$E = g\beta S_z B_z \tag{3}$$

Now we have seen that S_z can have only the values $+\frac{1}{2}\hbar$ and $-\frac{1}{2}\hbar$, and so we see that the electron has just two possible energies in the presence of the field:

$$E = (\pm 1/2)g\beta B \tag{4}$$

Equation (4) tells us that whereas the two spin states are degenerate in the absence of an applied field, this degeneracy is lost when a field is switched on and the spin states adopt different energies, as depicted in Figure 4. The lowest-energy state is associated with the negative sign and has the magnetic moment parallel to the field. From Eq. (1), we see that this corresponds to the spin being antiparallel to the field. The difference in energy between the two spin states given in Eq. (4) can be related to the frequency of radiation, ν, necessary to induce transitions between the levels:

$$h\nu = g\beta B \tag{5}$$

If the relevant numerical values are substituted for the physical constants in Eq. (5), it is found that the resonance frequency of a free electron is about 9.4 GHz for an applied field of 3400 G (0.34 T). These are the approximate conditions most often used experimentally to observe ESR transitions. Technically, this is carried out, for reasons of stability, with a fixed frequency and a

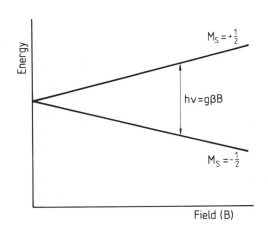

FIGURE 4. The effect of an applied, static magnetic field, B, on the $M_s = \frac{1}{2}$ levels of a free electron. Transitions between the levels may be induced by electromagnetic radiation of the correct frequency. For an applied field of 3400 G, this corresponds to microwave radiation (X-band) of 9.4 GHz.

swept field. The frequency quoted above is in the microwave region and corresponds to a wavelength of about 3 cm (X-band). Occasionally, measurements are made under other conditions: Q-band, which corresponds to 35 GHz, 12,500 G, and 0.86 cm, or S-band, which corresponds to 3.2 GHz, 1140 G, and 9.4 cm.

2.2. The g-Value

So far, our discussion has concerned a free electron. In reality, we are interested in looking at unpaired electrons within molecules. The first consequence of this is that in addition to the magnetism arising from the electron spin, there may also be a contribution from the electronic orbital motion. The reader will be familiar with elementary atomic structure and know that associated with atomic orbitals are angular momenta of magnitude $[l(l + 1)]^{1/2}\hbar$, where $l = 0$, 1, and 2 for s-, p-, and d-orbitals, respectively. It is apparent that a nonzero angular momentum is associated with orbital degeneracy. An orbital angular momentum $\mathbf{L}$ has associated with it a magnetic moment

$$\boldsymbol{\mu}_L = -\beta \mathbf{L} \tag{6}$$

For atoms and certain linear radicals, $\mathbf{S}$ and $\mathbf{L}$ combine to give a total angular momentum $\mathbf{J}$, and ESR examines transitions between the different J levels. In fact, however, in most molecular radicals the orbital contribution is lost ("quenched") because the existence of covalent bonds lifts the degeneracy of the orbitals, and providing the electron is in an orbital sufficiently removed in energy from other levels, the effective g-value will be close to the "spin-only" value quote above. The g-value is not exactly 2.0023, since spin–orbit coupling will mix in small contributions from orbitals of higher energy into the wave function describing the orbital occupied by the unpaired electron. We can therefore write for a radical containing one unpaired electron:

$$h\nu = g\beta B \tag{7}$$

where g now is a unique property of the molecule as a whole. In general, g is anisotropic, but in highly fluid solution, where there is rapid and random tumbling, an average g-value is observed. For the purposes of understanding the liquid-phase ESR spectra generated in electrochemical ESR experiments, this average value is sufficient. Notice, however, that for solid-state spectra this is not the case, and the g-factor must be regarded as a tensor.[57] Furthermore, the anisotropy of the g-tensor may contribute in some cases to the linewidth of the spectra from solution-phase radicals.

2.3. Hyperfine Splitting

We have seen that one consequence of putting a "free" electron into a molecule is to alter its g-value. Another is that the electron spin will be brought under the influence of any magnetic nuclei that there may be in the radical. The results of this are manifested in the occurrence of hyperfine splittings in the ESR spectrum. Consider an electron in a radical which sees a single magnetic nucleus of spin I. Just as a spinning electron is aligned by a magnetic field, so too is the nucleus. Quantum mechanics tells us that its spin angular momentum vector may take up $(2I + 1)$ orientations in the field, corresponding to angular momenta of $+I\hbar$, $+(I - 1)\hbar$, $+(I - 2)\hbar$, ..., $-(I - 1)\hbar$, $-I\hbar$ in the direction of the applied field (the z-direction). Again, the spin angular momentum vector (of magnitude $[I(I + 1)]^{1/2}\hbar$) is viewed as precessing around the z-direction, and we can label the eigenvalues of I_z, the nuclear spin angular momentum in the z-direction, $M_I = +I$, $+(I - 1)$, ..0, ..., $-(I - 1)$, $-I$. Just as spinning electrons have an associated magnetic moment so, too, do the spinning nuclei,

$$\boldsymbol{\mu}_N = g_N\beta_N\mathbf{I} \tag{8}$$

where now g_N is the nuclear g-factor and is a property of the particular nucleus in question, and β_N is the nuclear magneton. The latter is equal to $(e\hbar/2Mc)$, where e and M are the charge and the mass of the proton, respectively.

What is the effect of these oriented nuclei on the unpaired electron in the ESR experiment? The magnitude of the magnetic field experienced by the electron will depend on the field applied externally by the spectrometer, the size of the magnetic moment of the magnetic nucleus, and its *orientation*. Consequently, the value of the applied field necessary to being the ESR transition into resonance will depend on the orientation of the magnetic nuclei. For each spin state of the nucleus, a separate transition in the ESR spectrum will be seen. This is illustrated in Figure 5, which shows the ESR spectra of (electrogenerated) $ClO_2^{\cdot}$ and of Fremy's salt. In the first case, four lines are

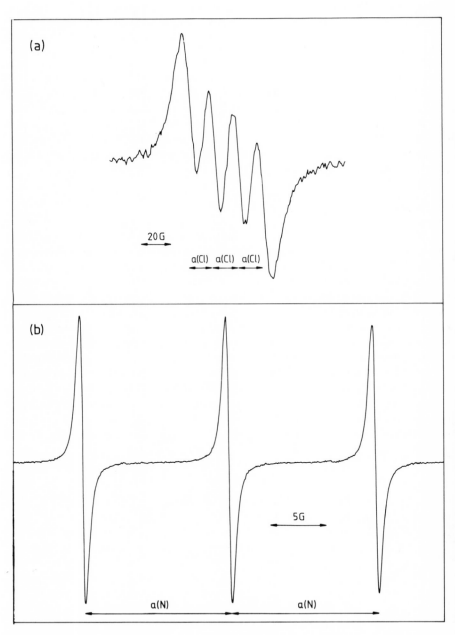

FIGURE 5. The ESR spectra of (a) $ClO_2^{\cdot}$ in acetonitrile and (b) Fremy's salt, $K_2NO(SO_3)_2$, in dimethylformamide.

seen corresponding to the spin of $\frac{3}{2}$ on the chlorine nucleus, while in the second case, three lines are seen due to the spin of 1 on the nitrogen (the molecules contain no other magnetic nuclei). The separations between the hyperfine lines give the hyperfine coupling constant $a(N)$ for the nucleus N in question, for that particular radical.

Implicit in the above is the notion that when the electron changes its spin, the nuclear spins are unaffected, i.e., that they remain in their original spin state. If this is the case, the selection rules for the ESR experiment can be stated as $\Delta M_I = 0$ and $\Delta M_s = 1$. This is a correct statement provided the magnitude of the hyperfine coupling is not too large, i.e., that the spectra are so-called first-order spectra. Otherwise, the spins are partially coupled and the behavior is rather more complicated.[57]

Quantitatively, under first-order conditions, the electronic energy levels of a one-electron, one-nucleus system are given by the following equation:

$$E = M_s(g\beta B + aM_I) \tag{9}$$

Figure 6 shows this equation applied to the radicals whose ESR spectra are shown in Figure 5.

One further feature is apparent from Figure 5, namely that the spectrometer records the spectrum as the first derivative of the absorption spectrum. This arises since the instrument employs a phase-sensitive method of detection.[58]

The different hyperfine lines within the same spectrum in Figure 5 have the same intensity. This implies that the different nuclear spin states are essentially equally populated. This arises because the energy of interaction of the nuclear spin with the applied magnetic field is much smaller than that of the electron spin with the field (ca. 1000×) which is essentially a result of the relative sizes of the electronic and nuclear magnetons. For example, for protons in a field of 3000 G, the magnetic energy is only about 3×10^{-4} cm^{-1}, whereas the thermal energy kT is about 200 cm^{-1} at room temperature. As a result, the populations of the two spin states differ by less than one part in 10^5, and so the two spin states will appear to be equally populated when examined by means of the hyperfine structure in the ESR experiment.

So far, we have limited our discussion to the hyperfine interaction arising from a single magnetic nucleus. The extension to the case where there is more than one nucleus is straightforward. Figure 7 contains the ESR spectrum of the radical anion of 2-nitropropane. The spectrum is composed of a triplet of doublets. The triplet arises from hyperfine interaction with the nitrogen and the doublet is due to the hydrogen on the central carbon atom in the molecule. The splittings due to the six hydrogens on the terminal carbon atoms are too small to resolve. The separations between the lines reveal the following coupling constants: $a(N) = 25.50$ G and $a(H) = 4.75$ G.

It often happens that several of the nuclei in a radical have identical coupling constants. This leads to a reduction in the number of lines from that which would have been found if the coupling constants were all different,

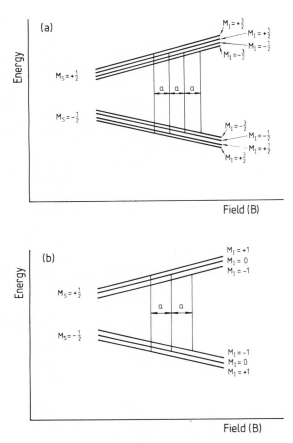

FIGURE 6. The variation with magnetic field of the $M_s = \frac{1}{2}$ levels in the presence of a single magnetic nucleus having (a) $I = \frac{3}{2}$ and (b) $I = 1$. Note that the allowed transitions have $\Delta M_I = 0$ and that the allowed spin I produces $(2I + 1)$ hyperfine lines.

since now some of the different transitions occur at the same field. The lines no longer have the same intensities since they correspond to different degeneracies. To illustrate this, let us consider a radical with coupling to two protons 1 and 2. Equation (9) becomes

$$E = M_s(g\beta B + a_1 M_1 + a_2 M_2) \qquad (10)$$

where a_1 and a_2 are the two coupling constants and M_1 and M_2 are the nuclear spin quantum numbers. Figure 8a shows that when the two splitting constants are different in size, four lines are seen and the spectrum appears as a "doublet of doublets." When, however, $a_1 = a_2$, we find that two lines coincide and a

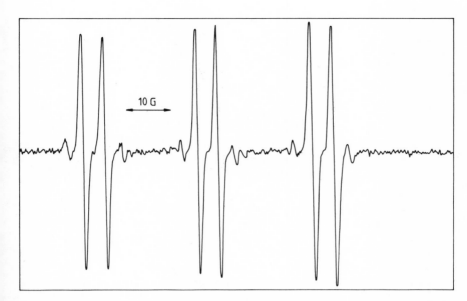

FIGURE 7. The ESR spectrum of the electrogenerated radical anion of 2-nitropropane, $(CH_3)_2CHNO_2$, in aqueous solution, showing a triplet of doublets. The former arise from hyperfine interaction with ^{14}N and the latter with ^1H.

triplet of intensities $1:2:1$ results, as shown in Figure 8b. In general, it is found that an electron interacting with n equivalent protons produces $(1 + n)$ lines whose relative intensities may be predicted from the coefficients of the binomial expansion of $(1 + x)^n$. These can most readily be recalled with the

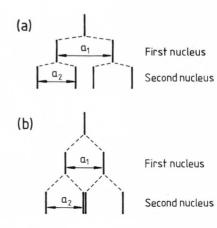

FIGURE 8. The hyperfine splitting patterns due to the interaction of an electron spin with two protons with (a) different hyperfine coupling constants and (b) identical hyperfine coupling constants.

aid of Pascal's triangle:

$n = 0$ 1

    1   1   1

    2   1   2   1

    3   1   3   3   1

    4   1   4   6   4   1

    5   1   5   10   10   5   1

To illustrate this, Figure 9 shows the ESR spectrum of the electrogenerated anion radical of p-benzoquinone, which has four equivalent protons. The spectrum is clearly a $1:4:6:4:1$ quintet, as would be predicted from the above ideas.

It is instructive to consider the mechanism by which the hyperfine interaction arises. It might be anticipated that a dipole–dipole interaction between the magnetic moments of the nucleus and the electron is involved. This is indeed the case in the solid state, where this leads to an anisotropic hyperfine interaction, which has to be represented by a tensor. However, in solution, provided that there is sufficiently rapid tumbling of the radical, the dipolar

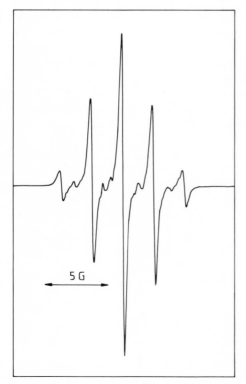

5 G

FIGURE 9. The ESR spectrum of the radical anion of p-benzoquinone generated by *in situ* electrolysis in dimethylformamide. The spectrum is a $1:4:6:4:1$ quintet due to the four equivalent hydrogens in the molecule.

contribution is averaged to zero. What remains is an isotropic hyperfine interaction which results from a nonclassical interaction, known as the Fermi or contact interaction since it depends on the presence of a finite unpaired spin density at the position of the nucleus. Thus, the coupling constant a is proportional to the square of the electronic wave function at the nucleus:

$$a = (8\pi/3)g\beta g_N \beta_N |\psi(0)|^2 \tag{11}$$

Now only s-orbitals have a finite probability density at the nucleus; p-, d-, and f-orbitals all have nodes at this point. Consequently, the contact interaction can only occur when the electron occupies a molecular orbital with some s-character. It is therefore not immediately obvious how hyperfine splittings arise in the cases of, for example, either the p-benzoquinone radical anion (Figure 9), where the electron occupies a π-orbital which has a nodal plane coincident with the plane of the aromatic ring, or the ethyl radical ($CH_3CH_2\cdot$), which shows splittings of 26.87 G from the methyl (or β-) protons—a larger value than is seen from the α-protons (22.38 G)!

Let us examine first the coupling mechanism to α-protons. To do this, consider an isolated $>\dot{C}-H$ unit with one electron in the π-orbital on the C-atom. Of course, this orbital is perpendicular to the plane of the three trigonal σ-bonds. Proceeding qualitatively, we note that the electrons in the $C-H$ bond, on account of their mutual repulsion, will tend to keep apart, and this may be represented pictorially, as in Figure 10. Two "structures" are possible, depending on whether the electron spin nearest the C-atom is parallel or antiparallel to the spin of the electron in the π-orbital. But for the presence of exchange forces[57] between the π and σ systems, the two structures shown in Figure 10 would have identical energies. In fact, the structure with the parallel spins at the C-atom has a slightly lower energy. The reason is analogous to the stabilization of states of highest multiplicity in atoms (Hund's rule). The result is that the spins in the $C-H$ bond are slightly polarized so that the spin of the electron in the σ-orbital on the C-atom takes on slightly more of the character of the spin in the π-orbital. This means that the electron at the H-atom must have a small excess of the opposite spin, and this, of course, resides in the $1s$ orbital used to form the $C-H$ bond. Therefore, the result is a finite spin density at the H-nucleus and, hence, a hyperfine splitting is seen.

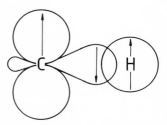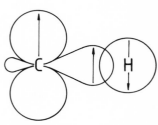

FIGURE 10. Canonical structures of the $C-H$ fragment (see text).

If the arguments in the foregoing paragraph are pursued quantitatively,[57] we are led to the conclusion that the hyperfine splitting constant should be proportional to π-electron density, ρ, on the carbon atom. This is expressed mathematically by McConnell's relationship:[59]

$$a(H) = Q\rho \tag{12}$$

where Q is a proportionality constant and takes a value of about -23 G. Notice that the equation predicts a *negative* coupling constant. This is because, as we have seen, the unpaired spin density at the H-atom is of opposite sign to that in the π-orbital on the C-atom. This prediction of a negative coupling constant for α-protons has no impact on the (isotropic) solution-phase spectra but has been confirmed experimentally both by NMR spectroscopy and by solid-state ESR data. Equation (12) is well illustrated if we compare the proton hyperfine coupling in the methyl radical (-23.04 G) with that in the benzene radical anion, $C_6H_6^{\overline{\cdot}}$ (-3.75 G). The splittings are close to being in the ratio of $6:1$ that would be predicted from McConnell's relationship on the basis of the relative amounts of electron density in the π-orbitals on the C-atom(s). Similar results are found for other cyclic polyene radicals.[1]

Having seen how hyperfine coupling to α-protons occurs, we now consider β-protons. It can be readily appreciated by a straightforward extension of the ideas in the above paragraph that exchange forces are capable of transmitting spin density through more than one bond, and, indeed, this is thought to contribute to some extent to couplings to β-protons. A more dominant effect, however, is due to hyperconjugation. This mechanism involves the direct overlap of the orbital containing the unpaired electron with the C—H σ-bond(s) formed by the β-hydrogens, as depicted in Figure 11. The magnitude of this interaction will depend on the relative orientation of the C—H bond to the orbital containing the unpaired electron. If the angle between the two

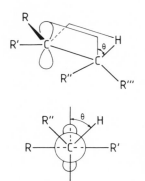

FIGURE 11. The magnitude of the hyperfine coupling between an electron and a β-proton depends on the dihedral angle, θ, between the C—H bond and the p-orbital containing the unpaired electron.

is θ, then quantitatively,

$$a(H) = A + B \cos^2 \theta \tag{13}$$

where A is close to zero and B is about 46 G. In the ethyl radical, the methyl group is freely rotating and so an average value is seen. Since the averaged value of $\cos^2 \theta$ is $\frac{1}{2}$, we would predict a value of ca. 23 G for the coupling to these protons, in excellent agreement with the experimental value quoted above.

2.4. Linewidths

Hitherto, we have discussed the appearance of ESR spectra in terms of, firstly, the g-factor, which governs the field at which resonance occurs, and, secondly, the effect of magnetic nuclei, which produce hyperfine splittings. The final aspect of the observed spectrum that merits attention is the question of lineshapes and linewidths. One contribution to the lineshape will, of course, arise from unresolved hyperfine splittings, which can give rise to a whole range of possible shapes, although if a large number of these are present, a Gaussian lineshape is encountered. In the absence of such complications, lines in solution almost always show the so-called Lorentzian lineshape defined by:

$$g(\omega) = (T_2/\pi)[1 + T_2^2(\omega - \omega_0)^2]^{-1} \tag{14}$$

where T_2 is the spin–spin relaxation time, and the lineshape has been defined using an angular frequency scale $\omega = \nu/2\pi$. The term ω_0 corresponds to the resonant frequency of the ESR transitions, and Eq. (14) relates to an unsaturated line. If high microwave powers are used, then this may not be the case, and the rate at which the system can attain thermal equilibrium then becomes important. This is controlled by T_1, the spin–lattice relaxation time, which thus controls the degree of saturation. Spin–lattice relaxation relies on the existence of processes restoring a thermal population of spins that is being perturbed by microwave-induced transitions. It therefore requires the flipping of electron spins and must originate in time-dependent magnetic or electric fields at the electron, which themselves must arise from random thermal motions in the solution. Furthermore, to be effective, these motions must have the appropriate timescales in order to be able to induce the required transitions. In particular, interactions at around 10^{10} Hz will be most effective, while higher and lower frequencies will be much less so.

The processes which contribute to T_1 also influence T_2, which, as we have seen, controls the linewidth. They do so by modifying the instantaneous local magnetic field at the electron. However, it is clear that the random forces which modulate the electron spin energy levels at low frequencies (below 10^{10} Hz) will contribute to T_2, whereas they have no influence on T_1. The source of these fluctuating magnetic fields lies in the anisotropic magnetic interactions within the molecule due to the anisotropy in the g tensor and to the dipolar hyperfine coupling with magnetic nuclei. Both of these effects are fairly weak in organic radicals, which consequently have narrow linewidths (≥ 50 mG) and well-resolved spectra. This conclusion assumes that the radical concentration is not too high; otherwise, spin exchange occurs as a result of radical–radical collisions, and broadening of the ESR lines is seen. The process involves the interchange of the spins of the colliding radicals: $(+\frac{1}{2}, -\frac{1}{2})$ becomes $(-\frac{1}{2}, +\frac{1}{2})$. At very high concentrations, the spin "sees" an average environment as it moves from radical to radical, and so the hyperfine splitting is lost and a single line results. This is the reason that solid-state ESR spectra of powders of stable free radicals, such as diphenyl picryl hydrazyl, show a single line generally a few gauss wide. Chemical kinetic processes can also influence linewidths. For example, if an aromatic radical anion is formed in the presence of its parent aromatic, then the electron can "hop" from one aromatic molecule to the next. This has the result of broadening the spectrum. The broadening can be used to calculate the rate constant for the chemical exchange process.

2.5. The ESR Spectrometer

Having discussed in outline the theoretical basis of ESR, it remains to describe briefly the essential features of an ESR spectrometer. Figure 12 is a schematic of the basic features. It can be seen that the applied field from the electromagnet is augmented by a further, smaller field modulated at 100 kHz. This is to facilitate detection of the spectrum by phase-sensitive detection (see above). Microwaves are generated in a klystron and travel through a waveguide to the ESR cavity, which contains the sample under investigation. Standing electromagnetic waves are set up within the ESR cavity, as shown in Figure 13 for the case of the type of cavity most frequently encountered; namely a

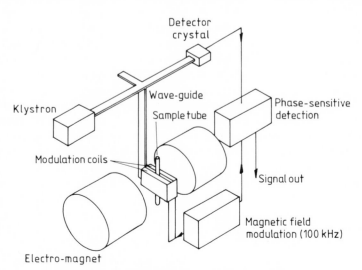

FIGURE 12. The basic features of an ESR spectrometer.

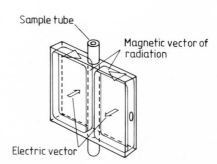

Sample tube

Magnetic vector of radiation

Electric vector

FIGURE 13. A TE_{102} ESR cavity.

rectangular TE_{102} cavity. This results in the spatial separation of the electric and magnetic components so that the magnetic part, which is responsible for inducing the ESR transition, has a maximum intensity at the cavity center and is thus concentrated on the sample. In particular, the distribution of the microwave magnetic field within the ESR cavity is such that the sensitivity profile along the vertical axis of the cavity obeys a $\sin^2$ relationship; there being zero sensitivity at the cavity edges and a maximum sensitivity in the cavity center. The cavity is thus spatially discriminating—the sensitivity varies with distance into the cavity. Notice that the electric component of the microwave field vanishes at the cavity center. This is useful since some samples, particularly aqueous samples, show appreciable dielectric loss at microwave frequencies. These losses would preclude the measurement of spectra but for the fact that one can contain samples within a thin flat cell in which the solution is held on the nodal plane and thus escapes interaction with the electric component of the microwave field. The need to minimize interaction with the electric field dictates that the thickness of the flat cell does not exceed about 0.5 mm. Cells are generally fabricated from silica, since glass normally contains ESR-visible paramagnetic impurities. As well as the rectangular cavity described above, cylindrical cavities and microwave helices[60] have found application in electrochemical ESR. The former are discussed in the following section.

The ideas sketched in this section are treated in depth in the numerous textbooks on ESR.[1,57,58,61-63] In particular, the book by Symons[61] provides an excellent and stimulating introduction for chemists. All aspects of ESR instrumentation are thoroughly covered in the work by Poole.[64]

3. Practice

This section is concerned with describing in detail contemporary methodology for electrochemical ESR. Experimental designs for two aims will be considered: the detection of short-lived electrogenerated intermediates and the investigation of the kinetics and mechanism of their decay.

3.1. The Allendoerfer Cell

Probably the most sensitive *in situ* cell for obtaining ESR spectra from short-lived species is that due to Allendoerfer and coworkers.[65-67] Instead of the rectangular TE_{102} cavity described in Section 2, this cell is based on a cylindrical cavity (TE_{011}). The basis of the design is shown in Figure 14. The cavity is modified by the presence of a cylindrical metallic conductor located along the axis of the cavity. This serves two purposes. Firstly, it acts as a wall of the ESR cavity, thus transforming it into a coaxial cylindrical cavity. Secondly, the conductor is used as the working electrode for the production of radicals. The enhanced sensitivity of this design arises from the significantly larger electrode area that is possible compared to cells which are built around the "flat cell in a TE_{102} cavity" design detailed in Section 1. In practice, the metal cylinder takes the form of a wire finely wound into the form of a shallow-pitched helix (of surface area ~ 22 cm^2), which fits snugly against the inner wall of a 6-mm-i.d. quartz test tube which contains the solution under investigation. In this configuration, the microwaves only "see" the portion of the solution between the helix and the quartz tube, and there is no penetration inside the helix (Figure 15). As a result, any material located within the helix

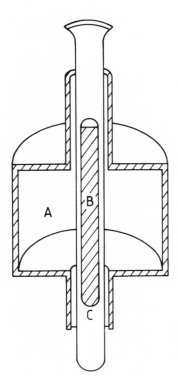

FIGURE 14. A schematic diagram of the Allendoerfer cell cavity design. The TE_{011} cylindrical cavity (A) is shown in cutaway section. A metal cylinder (B) is located along the axis of the cavity within a silica sample tube (C).

QUARTZ

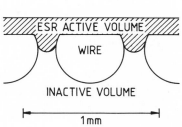

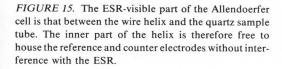

FIGURE 15. The ESR-visible part of the Allendoerfer cell is that between the wire helix and the quartz sample tube. The inner part of the helix is therefore free to house the reference and counter electrodes without interference with the ESR.

does not influence the ESR experiment. It follows that this space is a convenient location for the counter and reference electrodes. The counter electrode is a platinum rod, positioned along the axis of the cavity so that the current flow is radial, and this geometry ensures that the current density over the electrode surface is uniform. A Luggin capillary links the working electrode to the reference electrode. This arrangement enables electrochemical measurements to be made that are free of the distortions due to ohmic drop. Allendoerfer *et al.* observed cyclic voltammograms that were "indistinguishable from those run in a normal polarographic cell at a Hanging Mercury Drop Electrode, except for the 1000-fold increase in current due to the increased electrode area." They deduced that the uncompensated resistance of the cell was less than one ohm, even when highly resistive nonaqueous solvents were used. It was estimated theoretically that if high concentrations of electroactive materials were employed (up to $10^{-1} M$), then radicals as short-lived as 10^{-5} s would be open to study. This estimate is in agreement with that of Dohrmann and Vetter,[37] who arrived at the same figure for the shortest-lived radical open to study by *in situ* electrochemical ESR. The question of how far these estimates are met in practice will be discussed below.

The Allendoerfer cell described above has been modified by Ohya-Nishiguchi and coworkers for operation at low temperatures.[68-70] Their design was successfully used to record the ESR spectra of 20 aromatic radical ions at temperature as low as $-90°C$.[68-70]

It was established in Section 1 that the presence of an efficient hydrodynamic flow is an essential feature of any cell seeking application to extremely short-lived species. This is because of the need to sustain a steady supply of electroactive material to the electrode surface and so ensure a steady flux of radicals. For this reason, Carroll[67] has adapted the Allendoerfer cell to provide the capability for flow. The resulting changes can be seen from Figure 16. A series of baffles are used to prevent the solution from flowing in the central ESR-inactive part of the cell. It was found that the electrogenerated radical

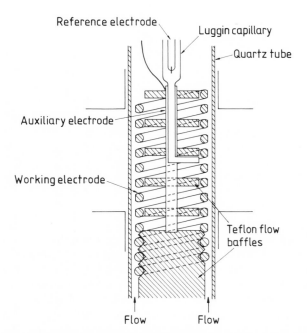

FIGURE 16. The Allendoerfer–Carroll cell as modified to provide the capability for solution flow.

anion of nitromethane in aqueous conditions could be observed using this flow cell.[67] This species was shown to have a lifetime of 10 ms and probably represents the shortest-lived species yet detected experimentally by electrochemical ESR. Encouraging calculations were presented[68] to show that the figure of 10^{-5} seconds, mentioned above, was likely to be obtainable in practice with this cell.

We next consider cells designed with the intention of studying the kinetics and mechanism of radical decay, as well as identifying the presence of particular radicals through their ESR spectra. We have noted (see above) that for this end, it is desirable not only to have a hydrodynamic flow over the electrode surface, but also that this flow be well defined and calculable so that the distribution of radicals in space and time may be calculated by solving the relevant convective-diffusion equation. We suggested above that this process could be expected to be difficult for the flow cells of Dohrmann and of Kastening because of the uncertain hydrodynamics of those cells. Likewise, the nature of the flow in Allendoerfer's cell cannot be confidently predicted, since it involves a complex flow between the inside of a silica tube and the surface of a coiled helix. We now describe the flow cell due to Compton and Coles,[71] which was shown to display predictable and calculable hydrodynamics.

3.2. The Compton–Coles Cell

Figure 17 illustrates the Compton and Coles design. The cell is essentially a demountable channel electrode constructed in synthetic silica. The electrode is a rectangular foil, for example, of platinum, cemented onto a silica cover plate and polished flat. Electrical connection is made to the rear of the electrode via a hole through the cover plate using 0.12-mm copper wire and conductive silver paint. The cover plate is cemented to the channel unit using a low-melting wax. The assembled unit is supported within a standard rectangular TE_{102} ESR cavity by means of PTFE spacers inside a silica tube, which runs right through the ESR cavity and which is held by nylon collets. The cell position within the cavity may be finely adjusted by movement of this supporting tube, and the tube also protects the cavity should any leakage occur. Generally, the electrode is positioned centrally in the cavity so that it is at the position of highest ESR sensitivity.

The flow cell is incorporated into a flow system capable of delivering a variable flow rate, usually in the range 10^{-1} to 10^{-4} cm^3 s^{-1}. This can be achieved either by gravity feed using various capillaries for flow range setting or by means of a mechanically driven syringe delivery system. The practicalities of constructing both sorts of flow system have been discussed by Brett and Oliveira-Brett.[72] A platinum gauze counter electrode is placed downstream of the working electrode, just outside the cavity, and a calomel reference electrode is located upstream.

When assembled, the channel has cross-sectional dimensions of 0.4 mm × 6 mm and is 30 mm long. The upstream edge of the working electrode is

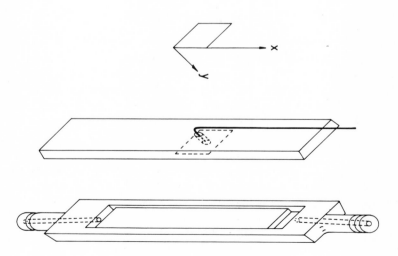

FIGURE 17. The Compton–Coles *in situ* flow cell.

normally located 15 mm from the inlet port of the channel. At the flow rates used, the flow is laminar—the Reynolds number Re can be calculated to be less than 100, even at the fastest flow rates, and provided that there is a sufficiently long lead-in length, a parabolic velocity profile will develop across the short dimension of the channel. If the depth of the channel is $2b$, then this lead-in length is given by $0.1 \times b \times Re$.[73] It can be seen that the 15-mm length available is more than adequate under the usual experimental conditions. The suggested hydrodynamics were fully confirmed experimentally.[71]

Compatibility of the cell with the microwave field inside the ESR cavity is achieved by placing the cell at the center of the TE_{102} cavity with the plane of the channel parallel to the direction of the electric field (refer to Figure 13). Fine adjustment of the cell position both vertically and horizontally, by movement of the silica tube retaining the cell, is essential to obtain maximum ESR sensitivity. By means of measurements of the cavity Q-factor (sensitivity varies as Q) under the least favorable conditions with high loss water inside the cell, it was shown that with sufficient care, ESR sensitivities close to those obtainable in conventional aqueous-sample ESR methods could be achieved. With the cell located as described, it was possible to illuminate the electrode using the normal irradiation port in the end wall of the ESR cavity, the cell position being such that the working surface of the electrode faces the port. Thus, the cell is well suited to the study of photoelectrochemical phenomena at metallic electrodes.

The distribution of concentration $c(x, t)$ of a species flowing in the channel can be described by the following (time-dependent) convective-diffusion equation:

$$\frac{\partial c}{\partial t} = D\frac{\partial^2 c}{\partial y^2} - V_x\frac{\partial c}{\partial x} - kc^n \tag{15}$$

where D is the diffusion coefficient ($cm^2\,s^{-1}$) of the species, x the distance along the channel (measured from the upstream edge of the electrode; Figure 17), y the distance normal to the electrode (measured from the center of the cell; Figure 17), k_n the rate constant for the decomposition of the species by nth order kinetics, and V_x the velocity of the solution in the x-direction. The last quantity we have seen to be parabolic:

$$V_x = \bar{U}[1 - (y^2/b^2)]; \qquad V_y = V_z = 0 \tag{16}$$

where $\bar{U}$ is the mean solution velocity ($cm\,s^{-1}$). Equation (15) shows that the dominant contributions to mass transport in the cell arise from diffusion normal to and convection parallel to the electrode. Diffusion in the direction of the flow can be shown to be negligible.[74] Equations (15) and (16) are solvable in general by numerical methods[28,74-77] and in some cases by analytical procedures,[26,27,78-81] for example, based on Laplace transformation. Analytical methods generally rely on an approximation originally invoked by Leveque in $1928^{(82)}$ in respect of the equivalent heat transfer problem, namely

$$V_x \simeq 2\bar{U}[1-(y/b)] \qquad \text{for } y \simeq \pm b \qquad (17)$$

This result effectively approximates the parabolic velocity profile by a linear one near the electrode surface and is valid provided diffusion is slow compared to convection, i.e., that concentration changes are confined close to the electrode surface. Using this approximation, Levich[73] calculated analytically the transport-limited current at a channel electrode of the type illustrated in Figure 17. He deduced that,

$$I_{\text{LIM}} = 1.165nFcD^{2/3}(\bar{U}/b)^{1/3}wx_E^{2/3} \qquad (18)$$

where w (cm) is the width and x_E (cm) the length of the electrode. The concentration in this equation is measured in units of mol cm^{-3} and n is the number of electrons transferred in the electrode reaction. Current–voltage curves measured for a wide range of electroactive species in the Compton–Coles cell[71,78,83,84] were found to give the flow rate dependence predicted by Eq. (18). Measured diffusion coefficients were in excellent agreement with either literature values or those found by other electrochemical techniques, such as rotating-disk voltammetry.

The dependence of the steady-state ESR signal (S) on the current (I) and the flow rate was investigated theoretically for the same cell, using approximations akin to those made to derive the Levich equation, in order to be able to solve Eq. (15). For a stable electrogenerated radical, it was deduced[71] that the following equation held:

$$S/I \propto \bar{U}^{-2/3} \qquad (19)$$

and this was shown to be true for radicals produced under both electrochemically reversible and irreversible conditions. Again excellent agreement was found when Eq. (19) was tested against experiment, using species such as the radical cation of N,N,N',N'-tetramethylphenylenediamine or the radical anion semifluorescein.[71] Typical data for the latter system are shown in Figure 18. Given the successful analysis of the transport-limited current data in terms of the Levich theory and the ESR–current–flow rate data using Eq. (19), it may be reasonably concluded that the hydrodynamics in the Compton–Coles cell are as described and quantified by Eq. (17). With this crucial point established, subsequent work was directed toward the study of unstable species.

Two different strategies exist for the determination of radical kinetics. Either the steady-state ESR signal can be measured as a function of the electrode current and the solution flow rate, or else a transient ESR signal can be recorded after the working electrode has been open-circuited once a steady-state has been established. In the latter method, one has a direct measure of

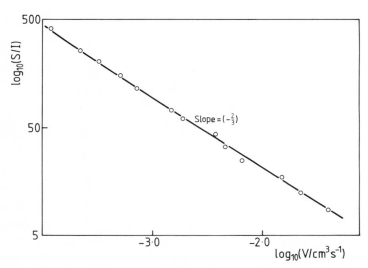

FIGURE 18. Typical ESR signal (S)–current (I)–volume flow rate (V) data obtained with the cell shown in Figure 17 and analyzed according to Eq. (19). The line drawn has the expected gradient of $(-\frac{2}{3})$.

the radical lifetime rather than relying on using variable flow rate to probe this indirectly. However, the former method uses steady-state signals, which may be recorded with suitable filtering and will be more sensitive and less subject to noise.

Let us consider the steady-state method first and illustrate this with reference to a radical reacting via first-order kinetics. The results of calculations for this case[85] are most conveniently expressed in terms of the "ESR detection efficiency," M_K, given by

$$M_K = S\bar{U}^{2/3}/I \tag{20}$$

where K represents the normalized (dimensionless) first-order rate constant for the radical decay and is defined by the following equation:

$$K = k_1(b^2 l^2 / 9 \bar{U}^2 D)^{1/3} \tag{21}$$

in which l is the length of the ESR cavity. For a stable radical, M_0 is constant. Figure 19 shows how the ratio M_K for an electrode positioned in the center of the ESR cavity varies with the parameter K. The dependence is shown for several electrode lengths in the range of 2 to 5 mm. Analysis of the experimental data proceeds along the following lines. Firstly, values of S and I are found for different flow rates $\bar{U}$. It is desirable to employ as wide a range of flow rates as possible. Values of the parameter M_K then are calculated from these values using Eq. (20). Figure 19 (or, the appropriate equivalent figure for either a different-sized electrode or a noncentrally located electrode) is used

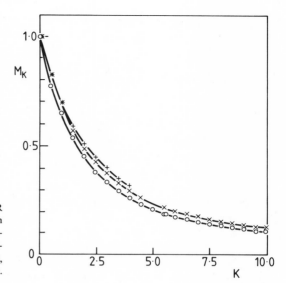

FIGURE 19. The variation of the ESR detection efficiency, M_K (eq. 20), with the normalized first-order rate constant, K (eq. 21), for different electrode lengths (+, 2 mm; ×, 3 mm; ○, 4 mm) located centrally in the cavity.

to deduce values of K corresponding to the different flow rates $\bar{U}$. Finally, Eq. (21) tells us that we can deduce the sought rate constant k_1 simply by plotting a graph of K against $\bar{U}^{-2/3}$ and using the slope of the graph together with the known geometry of the cell and the diffusion coefficient of the reacting species.

In this way, it was shown that the anion radical of 2-nitropropane formed during the reduction of the parent compound in aqueous solution (pH 10.2) at a mercury-plated foil electrode undergoes a first-order decay with a rate constant of $0.36\,\text{s}^{-1}$. A combination of the *in situ* electrochemical ESR data and voltammetric measurements made simultaneously at the channel electrode produced the spectrum shown in Figure 7. They, additionally, permitted the deduction that the reduction of 2-nitropropane under the specified conditions proceeded via an ECE mechanism in which the chemical step was rate-determining proton uptake by the radical anion. In the buffered media employed, this was seen as a pseudo–first-order step. The electrochemical reaction sequence can therefore be written:

Electrode $\qquad (CH_3)_2CHNO_2 \xrightarrow{+e} [(CH_3)_3CHNO_2]^{\overline{\cdot}}$

Solution $\qquad [(CH_3)_2CHNO_2]^{\overline{\cdot}} \xrightarrow[k\sim0.30\,s^{-1}]{+H^+} (CH_3)_2CH-N{\overset{\textstyle OH}{\underset{\textstyle O^{\cdot}}{\Big<}}}$

Electrode $\qquad (CH_3)_2CH-N{\overset{\textstyle OH}{\underset{\textstyle O^{\cdot}}{\Big<}}} \xrightarrow{+e} (CH_3)_2CH-N{\overset{\textstyle OH}{\underset{\textstyle O^-}{\Big<}}}$

$$\text{Solution} \quad (CH_3)_2CH-N\overset{\displaystyle OH}{\underset{\displaystyle O^-}{\Big\langle}} \quad \xrightarrow{\;-OH^-\;} \quad (CH_3)_2CH-N{=}O$$

We next consider the deduction of radical kinetics through ESR transient signals.[85,86] These are best obtained by potentiostating the working electrode at a potential corresponding to the diffusion-limited current plateau region of the electroactive substrate (for maximum signal) and then opening a switch so that the electrode is open-circuited. In this way, it is ensured that there is no contribution to the radical decay due to say reoxidation of an radical anion produced by an electroreduction (or the converse for an initial electrooxidation). This would be likely if the transient were obtained by stepping between two potentiostated values should the formation of the radical be reversible or quasi-reversible.

Figure 20 contains a typical ESR transient for the 2-nitropropane system already discussed. It may be seen that the decay is approximately exponential, as might be predicted for first-order decay. In fact, the curve is *not* a perfect exponential because there is a contribution to the decay from convection of the radical out of the cavity. Obviously, this is most noticeable at fast flow rates and for slowly decaying radicals. Theory has been presented which describes the transient shape under any conditions.[86] Conveniently, however, it was found that for all but the most stable radicals ($k_1 < 5\,s^{-1}$), a simple method of analysis was applicable.[85] This involves treating the measured transients as if they were exponentials and deducing the "effective" first-order rate constants at the flow rates employed. These may then be extrapolated

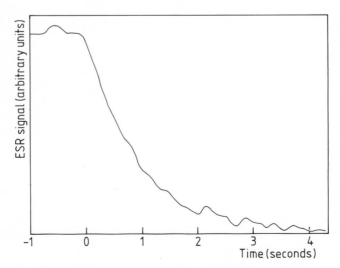

FIGURE 20. A typical ESR transient for the 2-nitropropane radical anion, produced by open-circuiting the working electrode.

back to zero flow rate (by plotting the apparent rate constant against $\bar{U}^{2/3}$) to give the true value of the rate constant. Results for the 2-nitropropane system, such as those in Figure 20, were found to be in quantitative agreement with the conclusions deduced from the steady-state measurements outlined above.

The Compton–Coles cell has been used for quantitative study of EC,[85] ECE,[85,87] and DISP[87] processes. An equivalent cell design has been used by Albery *et al.*[88] to investigate other kinetic schemes. This cell uses the tubular electrode geometry rather than the channel geometry described above, i.e., the cross section of the electrode is circular rather than rectangular. However, provided the tube diameter is not too small (a condition invariably met by practical electrodes which are built with as large a diameter as possible—subject to avoiding appreciable dielectric loss in the cavity—for reasons of sensitivity), then the convective-diffusion equations for the two designs are essentially equivalent, subject to trivial geometrical transformations. Albery has developed the steady-state theory for M_K for the case where electrogenerated radicals decay by second-order kinetics.[28] Because Eq. (15) for this situation is in this case nonlinear, numerical simulations were needed for its solution. The results of the theory could be described by a M_K versus K plot, as described above [see Eqs. (20) and (21)], and an analogous treatment of the experimental data used to find k_2. The theory was tested for an *ex situ* tubular electrode located at the upstream edge of the cavity and was found to agree with experiment. The theory is, with minor changes, equally applicable to the *in situ* Compton–Coles cell described here. Although the tubular and channel geometries are equivalent in terms of their theoretical description, in general the latter are to be preferred for several reasons. Firstly, the channel cells are demountable, and so the electrode surface can be examined and polished readily. This is not the case for the tubular design. Secondly, there are doubts as to whether all parts of the electrode surface are ESR active because of the effect of the 100-kHz magnetic field modulation inducing eddy currents in the electrode surface (see below). In the channel electrode design, the modulation is always parallel to the electrode surface and such effects cannot arise. Finally, we note that the channel geometry suffers from rather less ohmic drop and also enables irradiation of the electrode surface.

3.3. The Compton–Waller Cell

What sort of radical lifetimes are open to study with the *in situ* channel electrode cell? The answer to this question is obviously radical dependent to some extent—the more hyperfine splittings and/or the greater the linewidth, the less sensitivity one has. However, as a rough guide, one can probably say that radicals with lifetimes greater than tens to hundreds of milliseconds are accessible. This is rather larger than for the Allendoerfer cell and is attributable simply to the relative areas of the working electrodes in the two cells. It is to some extent the price one pays for winning the kinetic information. However, a considerable improvement in this respect was realized in a cell design due to Compton and Waller,[89] which utilizes the better features of both the Allendoerfer cell and the channel electrode cell. In particular, a coaxial cylindrical cavity is utilized, allowing a large-area electrode; at the same time, a well-characterized hydrodynamic flow is retained and so the investigation of electrode reaction mechanisms is possible.

This improved cell design is shown in Figure 21. In essence, it comprises a TE_{011} cylindrical cavity which is converted into a coaxial mode by the addition of a smooth polished copper rod positioned centrally in the cavity.[89] The diameter of the rod is close to 9 mm. The rod itself is located within a precision-bore silica tube (labeled G in Figure 21), so that an annulus (typically ~0.1 mm in width) is available through which the solution flows. The central 4 mm of the copper rod is insulated from the rest of the rod and this central portion (C in Figure 21) acts as the working electrode as well as part of the

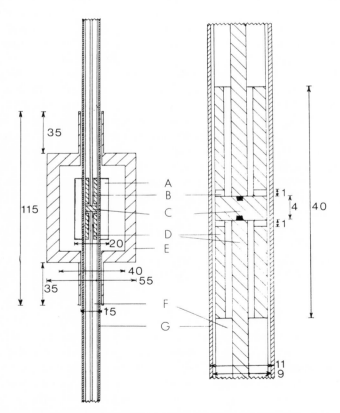

FIGURE 21. The Compton–Waller cell. A, PTFE annulus; B, PTFE insulation; C, electrode (mercury-plated copper); D, copper; E, TE_{011} cylindrical cavity; F, PTFE sheath; G, precision-bore silica tubing. The numbers represent the dimensions in millimeters.

wall of the coaxial cavity. The electrode is plated with mercury to permit a negative maximum potential range. The cell is connected to a flow system which delivers electrolyte at a controlled variable rate. A reference electrode is placed upstream in the flow system and a counter electrode downstream (to avoid possible counter electrode products being swept into the cavity). Because there is quite a long length of resistive solution between the working and the counter electrode, it is necessary to have the latter capable of delivering quite a high voltage—up to a few kilovolts may be required for nonaqueous solutions!

The ESR performance of this cell is analogous to that of the Allendoerfer cell. Firstly, it was found that the resonant frequency of the cavity was shifted above that of the empty cavity because of the reduction in effective size of the cavity produced by adding the copper rod. This shift may be sufficient to take the resonant frequency outside the tuning range of the klystron (9–10 GHz). The frequency can be returned to the workable range by partly filling the cavity with PTFE (A in Figure 21). Next, the sensitivity of the cavity, as it varied along and around the electrode surface, was checked by moving a minute crystal of a stable free radical in the gap between the rod and the outer silica tube. The $\sin^2$ sensitivity function expected[64] as the crystal was moved axially was obtained. When the crystal was moved round the copper rod at a fixed distance into the cavity, however, a $\cos^2$ behavior was observed. This could not be rationalized in terms of the microwave field within the cavity (which should have cylindrical symmetry). Instead, it may be attributed to a perturbation of the component of the magnetic field which is modulated at 100 kHz. This is caused[67] by eddy currents induced in the copper rod by the oscillating field, which set up a magnetic field which opposes the applied field. The effect is maximum when the applied field is perpendicular to the copper surface and zero when the applied field is parallel to the surface. This produces the observed $\cos^2$ variation, which has also been reported by Carroll[67] in the Allendoerfer cell.

Next, we examine the nature of the flow in the annular gap between the rod and the silica tube. Again, at the flow rates used, the flow will be laminar and, after a lead-in of insignificant length, a parabolic velocity profile will develop across the width of the annular gap. Thus, the hydrodynamics are equivalent to those of an ordinary channel electrode, and it would be expected that the diffusion-limited current would be given by the Levich equation (Eq. 18), where now w represents the circumference of the copper rod. This was found to be the case experimentally[89] at all but the very slowest flow rates ($>10^{-3}$ cm^3 s^{-1}), where small deviations were found because the cell was beginning to act like a thin-layer cell and the onset of exhausting electrolysis of the solution passing over the electrode surface was seen. However, the equivalence of the hydrodynamics with that of the channel electrode cell over virtually all of the flow range means that all the theory for that cell is immediately applicable to the Compton–Waller cell. This provides the capability for investigating radical kinetics with an enhanced sensitivity over the earlier channel electrode cell. The magnitude of the enhancement is readily perceived to be simply in proportion to the relative electrode areas in the two cells. This is because in the limit of very fast kinetics, the radicals are constrained to occupy a thin "skin" on the electrode surface. The number of radicals in the cavity therefore is directly related to the area of the electrode. From the relative dimensions of the working electrodes in Figures 15, 17, and 21, it can be seen that the Compton–Waller cell lies intermediate between the channel cell (tens to hundreds of milliseconds) and the Allendoerfer cell (possibly down to 10^{-5} seconds) in terms of sensitivity.

3.4. Some Practical Hints

Having described the various contemporary cells we recommend for electrochemical ESR, we will conclude this section by making a few practical points. Firstly, it is normal electrochemical practice to exclude oxygen from solutions under study. This is because of its electroactivity at a wide range of electrode materials in both aqueous and nonaqueous solvents. The desirability of the exclusion of oxygen is even more acute in joint electrochemical/ESR work since (a) oxygen is a notorious radical scavenger and (b) even if it does not react with a radical, it is likely to significantly broaden the ESR linewidth, with resulting loss of sensitivity and possible reduction of resolution. Secondly, the prospective experimentalist should be aware that the high intrinsic sensitivity of ESR means that rigorous standards of solvent, background electrolyte, and substrate purification and preparation are called for, since impurities may easily give detectable signals. Likewise, materials used for the construction of cells should be checked for possible paramagnetic impurities. Silica should always be used rather than glass. More importantly, materials such as iron or steel placed in a magnetic field will prove at best embarrassing, or at worst expensively damaging. Finally, when electrochemistry-related signals are observed, it is important to ensure by quantitative measurements that the signals represent the major reaction pathway and are not due to a very minor side reaction in the electrolytic process. This necessitates making absolute measurements of the number of spins present in the cavity and correlating this with the current generating the radicals. The practicalities of quantitative ESR have been thoroughly discussed by Goldberg.[90] In the case of cells compatible with rectangular ESR cavities, the use of a double cavity can be recommended. Thus, a TE_{104} cavity accommodates both the cell and a standardizing sample, and the two signals can be measured without the need to exchange samples, which considerably improves the accuracy of measurement.

4. Applications

This section describes typical applications of the methodology described in the previous section. Examples will be included of radical identification, the use of spin trapping to detect radicals too short-lived for direct observation, the determination of radical decay mechanisms and kinetics, and finally the deduction of dynamic processes from lineshapes.

4.1. Radical Identification

We consider first an example of radical identification with the aim of deducing the nature of the electrode reaction. This concerns the reduction of 1,2-dicyanobenzene in dimethylformamide (DMF) at a mercury electrode. This has been studied by Gennaro and coworkers,[91] who observed two waves during the first scan of their cyclic voltammetric experiments. The first wave at −1.32 V (vs. Ag/AgCl electrode in chloride-saturated DMF + acetonitrile) was attributed to the formation of the radical anion of dicyanobenzene. The second wave at −2.35 V was assigned to an ECE process in which the anion formed during the first wave gained a further electron, forming a dianion which, it was suggested, could undergo reaction with the solvent, producing benzonitrile. Since benzonitrile is known to be reducible at the potential of the second wave, a further electron transfer takes place, forming the benzonitrile radical anion. Experiments by Waller[92] using the channel electrode cell described in the previous section produced the spectrum shown in Figure 22a when the electrode was potentiostated at the second wave. This was found to be indistinguishable (in terms of both hyperfine couplings and g-value) from that obtained by the direct reduction of benzonitrile itself at these potentials. Potentiostating on the first wave gave the spectrum in Figure 22b, which was attributed to the 1,2-dicyanobenzene radical anion. Thus, the mechanism inferred indirectly from cyclic voltammetry was shown to be vindicated by the *in situ* electrochemical ESR study.

As a second example of the use of radical identification in determining electrode reaction mechanisms, we consider the oxidation of triphenylacetic acid in acetonitrile. This was first studied by Kondrikov et al.,[93,94] who obtained an ESR spectrum by using a modified version of the Adams design of *in situ* cell (see above) in which the counter and reference electrodes were contained inside the ESR-sensitive part of the cell. The spectrum was attributed to the triphenylacetoxyl radical, Ph_3CCO_2·. The authors suggested that there was restricted rotation about the carbon–carbon bond in this species and that this led to the third phenyl group not contributing hyperfine splittings to the spectrum. The four *ortho*-protons in the two contributing phenyl groups were assigned a hyperfine coupling constant of 2.2 G and the two *para*-protons a value of 3.5 G, although these numbers do not seem consistent with the published spectrum,[93,94] which appears to be essentially a 1:4:6:4:1 quintet (four equivalent hydrogens; $a \sim 2$ G) with some further much smaller, and difficult to resolve, splittings.

Different observations were made by Goodwin and coworkers,[95] using the *in situ* cell of Bard and Goldberg described in Section 1. Their work was

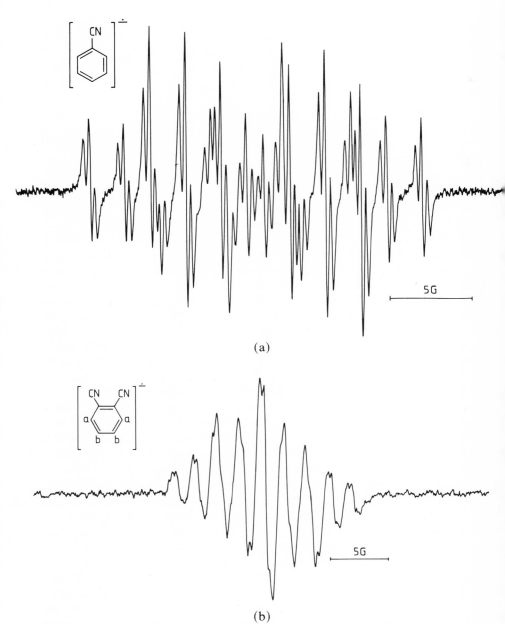

FIGURE 22. The ESR spectra obtained by *in situ* electrolysis of 1,2-dicyanobenzene (a) at −2.35 V, where the spectrum was shown to be that of the benzonitrile radical anion [hyperfine coupling constants: a (N) = 2.17 G, $a(o\text{-H})$ = 3.72 G, $a(m\text{-H})$ = 0.35 G, $a(p\text{-H})$ = 8.50 G] and (b) at −1.32 V, where the spectrum was shown to be that of the 1,2-dicyanobenzene radical anion [$a(\text{N})$ = 1.79 G, $a(\text{H}_a)$ = 0.39 G, $a(\text{H}_b)$ = 4.14 G].

prompted by the notion that the triphenylacetoxyl radical would be likely to decompose to triphenylmethyl and CO_2 at a rate likely to be nearly synchronous with electron transfer. They found that no spectrum could be seen on the direct oxidation of triphenylacetic acid. However, if the electrode was first held at an oxidizing potential (2.0 V versus Ag pseudo reference electrode) corresponding to the oxidation of the acid, then stepped negative to a value of about 0.35 V, a spectrum attributable to the triphenylmethyl radical was observed. These results were interpreted in terms of the formation of a carbonium ion in a two-electron oxidation at the more positive potential, followed by its one-electron reduction at the more negative potential:

$$Ph_3CCO_2H \xrightarrow{-2e} Ph_3C^+ + CO_2 + H^+ \xrightarrow{+e} Ph_3C^\cdot$$

It was surmised that the Russian workers might have been seeing signals arising from species generated at the counter electrode, which was situated in an ESR-visible part of the cell. No further insight into the nature of these signals was obtained by Goodwin *et al.*, the spectrum of the triphenylmethyl radical being utterly different from that reported by Kondrikov *et al.*

Compton *et al.*[96] showed that signals resembling those originally obtained by Kondrikov *et al.* could be found if moist acetonitrile was used and an oxidizing–reducing potential sequence similar to that of Goodwin *et al.* was employed, but in which the reducing step was to negative of -1.80 V versus SCE. The nature of the spectrum obtained was found to be critically dependent on the amount of water present. At high water concentrations, a pure $1:4:6:4:1$ quintet was obtained and the spectral parameters were found to be identical to those of the (electrogenerated) anion radical of benzoquinone. The spectra observed at lower water levels were shown to be due to a mixture of two radicals—the benzoquinone anion radical and the benzophenone anion radical—and the measured spectra could be simulated from the hyperfine splittings and *g*-values of these anion radicals (measured under conditions where they were produced in isolation by reduction of the parent compounds) by assuming varying relative amounts of the two species. The production of the two radical anions was rationalized by the following kinetic scheme:

Oxidation

$$Ph_3CCO_2H \xrightarrow{-2e} Ph_3C^+ + CO_2 + H^+$$

$$Ph_3C^+ + H_2O \longrightarrow Ph_3COH + H^+$$

$$Ph_3COH \xrightarrow{-2e} Ph_3CO^+ + H^+$$

$$Ph_3CO^+ \longrightarrow Ph_2\overset{+}{C}OPh$$

$$Ph_2\overset{+}{C}OPh + H_2O \longrightarrow Ph_2C(OH)OPh + H^+$$

$$Ph_2C(OH)OPh \longrightarrow Ph_2C{=}O + PhOH$$

$$PhOH + [Ox] \longrightarrow O{=}\emptyset{=}O$$

Reduction

$$Ph_3C^+ \xrightarrow{+e} Ph_3C^{\cdot}$$

$$O{=}\phi{=}O \xrightarrow{+e} \text{benzoquinone radical anion}$$

$$Ph_2C{=}O \xrightarrow{+e} \text{benzophenone radical anion}$$

This scheme invokes the well-established rearrangement of Ph_3CO^+ cations to form $Ph_2\overset{+}{C}OPh$, which is known to react in the presence of water to form phenol and benzophenone. The formation of benzoquinone requires that the phenol formed should become oxidized in some way. This is not unreasonable, since it was shown that at the potentials needed to oxidize triphenylacetic acid, phenol was electrochemically oxidized to form the required benzoquinone. The dependence of the ESR spectrum on the water concentration can be understood because the radical anion of benzophenone is known to react to form Ph_2CHOH in the presence of water. Further evidence in support of the suggested mechanism came from the observation that triphenylmethanol, Ph_3COH, under similar oxidizing–reducing conditions, gave similar ESR spectra to those discussed above.[97] Also, the potential required to be able to observe the signals in both cases corresponded with the reduction potential of benzophenone. Stepping to less negative potentials results in either the benzoquinone signal alone (between -0.55 and -1.70 V) or the spectrum of the triphenylmethyl radical (between $+0.35$ and ca. 0 V). Given that an oxidizing–reducing sequence is essential to see the ESR spectra under well-defined experimental conditions, it must be concluded that in the original experiments of Kondrikov, this was being effected by electrolysis products from the counter electrode migrating to the working electrode, which was being potentiostated at oxidizing potentials. Redox reactions between reduction products from the counter electrode must have produced the observed radicals.

The case history of triphenylacetic acid provides a good example of the mechanistic detail that can be obtained from coupling electrochemistry with ESR. At the same time, note should be taken of the possible pitfalls both in cell design and in the interpretation of spectra.

4.2. Spin Trapping

We continue our discussion of the identification of radical intermediates in electrolytic reactions by describing the technique of spin trapping, which is particularly valuable when the radicals are not sufficiently long-lived to be observable directly. Spin trapping, which was introduced by Janzen and Blackburn in 1968,[98,99] makes use of a diamagnetic compound (the spin trap) to react with the reactive radical to give a relatively stable, ESR-observable free radical. Optimally, the original radical can be identified from the ESR parameters of the spin adduct. Spin trapping therefore extends the capabilities of ESR in general. A further benefit of spin traps in the specific area of joint electrochemical/ESR work is that they can be employed under *ex situ* conditions using conventional electrochemical instrumentation, and hence the need for relatively complex *in situ* methods is avoided.

To illustrate the possibilities of spin trapping in electrochemical ESR, let us consider the oxidation of various organoborides, BR_4^-, in acetonitrile as studied by Blount and coworkers.[100] Direct oxidation of BBu_4^- under *in situ* conditions revealed no ESR signals. However, in the presence of the spin trap phenyl *N-tert*-butylnitrone (PBN), signals were recorded, and these were attributed to the adduct of PBN with butyl radicals:

$$BBu_4^- \xrightarrow{-e} \textit{n-Bu} \cdot \longrightarrow \text{products}$$

$$\downarrow \text{+PBN}$$

$$\begin{array}{cc} Bu & O^{\bullet} \\ | & | \\ PhCH{-}N{-}Bu' \end{array} \quad \text{(spin adduct)}$$

Conversely, tetraphenylboride, BPh_4^-, gave no spectrum, and although biphenyl was observed among the reaction products, it was concluded that this could not be formed via the intermediacy of free phenyl radicals.

Various applications of spin trapping in both electrode oxidations and reductions have been reported.[100-102] The widely used spin traps fall into two classes of compounds, the nitrones and the nitroso compounds. The former are typified by PBN, which has the following structure:

$$\begin{array}{c} O^- \\ | \\ Ph{-}CH{=}\underset{+}{N}{-}Bu' \end{array}$$

The advantage of nitrones is that they show a wide potential range over which they are inactive electrochemically. For PBN in acetonitrile at platinum electrodes, this stretches from +1.4 V versus SCE to −2.4 V. (These limits are obviously solvent and electrode dependent; thus, for example, PBN is inert cathodically in water at platinum only to −1.9 V.) The potential range may be readily extended in either direction by substitution of the phenyl group.[103] The less attractive feature of nitrones is that the trapped radical is rather distant from the location of the spin in the spin adduct. Thus, spectral parameters do not change much as the structure of the trapped species varies, and as a consequence, unambiguous radical identification is difficult without independent synthesis of the presumed adduct.

The other class of spin traps are the nitroso compounds:

$$R{-}N{=}O \xrightarrow{+R'} \begin{array}{c} R{-}N{-}R' \\ | \\ O^{\bullet} \end{array}$$

In this case, the structure of the radical is much more easily identified since the resulting unpaired electron is close to the site of trapping. The price for this is that the nitroso spin-traps show a more limited potential "window" of electroinactivity. For example, the trap nitroso-*tert*-butane (NtB), which finds widespread application in ordinary solution-phase mechanistic ESR studies, is reduced at -0.98 V at mercury in dimethylformamide although at platinum this stretches to -2.06 V (in acetonitrile). Nitrosobenzene has been used as a spin trap, and here as with PBN, appropriate substitution of the aromatic ring can often shift the decomposition potential to within desired ranges.[104] A further complication with the nitroso compounds is that they participate in a monomer–dimer equilibrium in solution. Only the monomeric form acts as a radical scavenger, and this may be present as a rather low percentage of the total amount of spin trap added.[105] A knowledge of the kinetics and equilibria of dissociation of these compounds is therefore desirable when they are to find application. These data are available for a limited number of nitroso compounds in Ref. 105. Nitroso compounds also have the disadvantage of being thermally and photochemically unstable,[99,106] and this is widely appreciated to be a major liability in their application.

4.3. The Kinetics and Mechanisms of Electrode Reactions

We next consider an example of the determination of the mechanism and kinetics of an electrode reaction. This concerns the apparent two-electron reduction of the molecule fluorescein (F) to leuco-fluorescein (L) in buffered aqueous solution, pH range 9–10. Experiments by Compton *et al.*[87] using the Compton–Coles cell revealed strong ESR signals (Figure 23) attributable to semifluorescein (S·) where:

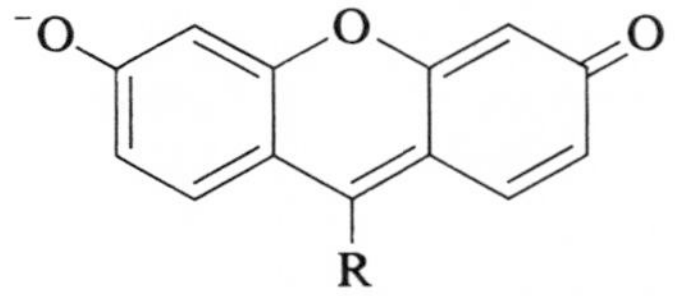

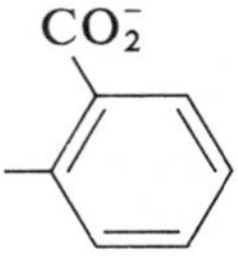

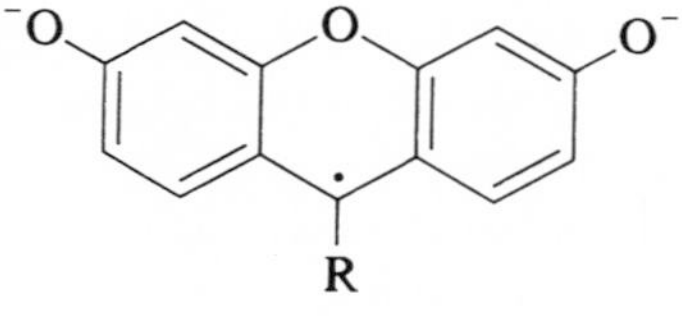

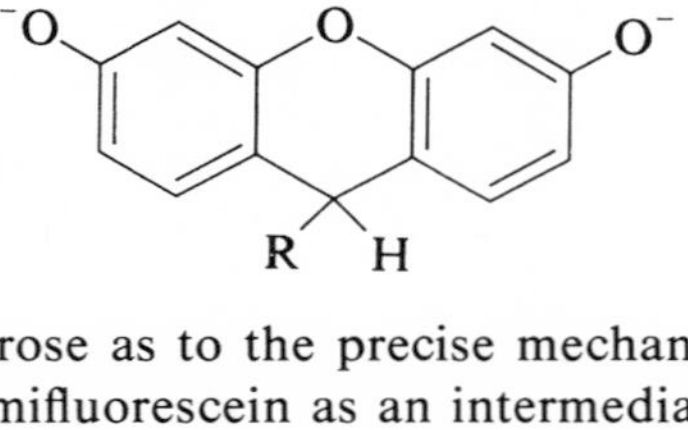

The question arose as to the precise mechanism of the overall process. The existence of semifluorescein as an intermediate implies two discrete electron

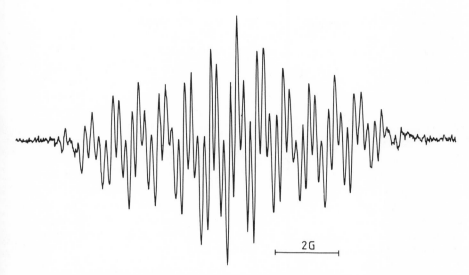

FIGURE 23. The ESR spectrum of electrogenerated semifluorescein radicals.

transfer steps, and since the final product, leucofluorescein, has an additional proton to the starting material and the intermediate, there must be a protonation step after the first electron transfer. Figure 24 shows the transport-limited current at the channel cell, as a function of the electrolyte flow rate. The drawn solid lines correspond to the behavior that would be predicted on the basis of Eq. (18) and the measured diffusion coefficient of F for a simple one- or two-electron reduction process. It is apparent that at fast flow rates, the data points tend towards one-electron behavior, while at slow flow rates, they tend towards two-electron behavior. This is characteristic of an ECE-type process, i.e., one in which the chemical step is sandwiched between the two one-electron transfer steps. The behavior in Figure 24 arises because at fast flow rates, the product of the first electron transfer is swept off the electrode surface before there is time for the chemical step to take place. One-electron behavior is therefore seen. Conversely, at slow flow rates, there is ample time for the chemical step to proceed and so the second electron transfer takes place.

Although for a long period, reactions of the above type were written as involving two heterogeneous electron transfers, it is now recognized that the second electron transfer may occur homogeneously via disproportionation; a so-called DISP process.[107,108] Hence, for the reduction of fluorescein, we can write the following general kinetic scheme:

$$F + e \longrightarrow S^\cdot \qquad (a)$$

$$S^\cdot + H^+ \longrightarrow SH^{+\cdot} \qquad (b)$$

$$SH^{+\cdot} + e \longrightarrow L \qquad (c)$$

$$SH^{+\cdot} + S \longrightarrow L + F \qquad (d)$$

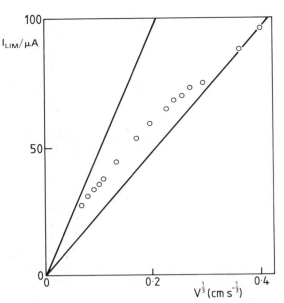

FIGURE 24. The transport-limited current (I_{LIM}) for the reduction of fluorescein at the Compton–Coles channel electrode cell. The solid lines show the predicted flow rate (V) ($cm^3 s^{-1}$) behavior for simple one- and two-electron reductions.

If the reaction proceeds via steps (a), (b), and (c), then we have an ECE process. The sequence (a), (b), (d) corresponds to a DISP reaction. Within the latter scheme there are two further possibilities, depending on whether step (b) or (d) is rate determining. In the former case, we have a DISP1 process, since it is a (pseudo-) first-order reaction (in buffer), while in the latter case, we have a DISP2 process, since it is second-order. Conventional electrochemical methods readily recognize DISP2 processes but, with only a few exceptions (double potential-step chronoamperometry[108] and possibly microelectrodes[109]), they cannot be used to discriminate between ECE and DISP1. It emerges that a combination of ESR transient and electrochemical data from the channel electrode cell can make this distinction.

Figure 25 illustrates a typical ESR transient obtained by potentiostating the electrode at a potential corresponding to the transport-limited current and setting the magnetic field at a value corresponding to a peak in the spectrum shown in Figure 23, then monitoring the ESR signal as the working electrode is open-circuited. Analysis of this transient according to the procedure outlined in Section 3 showed the decay to be first-order with a rate constant of 1.05 s^{-1} (at pH 10.05). This eliminated DISP2 as a mechanistic possibility and left either ECE or DISP1.

Attention was then turned to the electrochemical behavior shown in Figure 24. Because of the well-defined and known hydrodynamics of the cell, the dependence of the transport-limited current on the flow rate can be calculated for both ECE and DISP1 processes[87] using analytical theory and the Leveque approximation. The predicted behavior is governed by a normalized rate

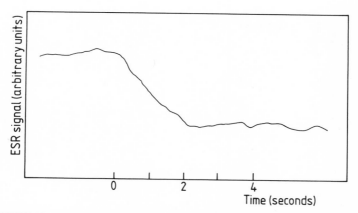

FIGURE 25. A typical ESR transient for the decay of the semifluorescein radical.

constant, K', defined by,

$$K' = k'(b^2 x_E^2/9\bar{U}^2 D)^{1/3} \tag{22}$$

where k' is the pseudo–first-order rate constant for reaction (b) and the other symbols have been defined in Section 3. Figure 26 reveals how the effective number of electrons transferred depends on the parameter K' for the two mechanisms. Different behavior for ECE and DISP1 is predicted. Analysis of data such as those in Figure 24 produced different values of k' according to the choice of mechanism. Only the values deduced assuming the DISP1

FIGURE 26. The effective number of electrons transferred at the channel electrode in ECE (———) and DISP1 (· · · · ·) processes, respectively.

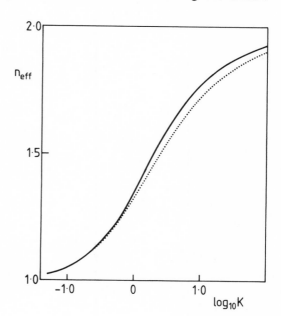

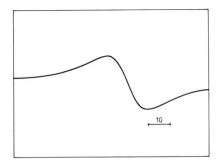

FIGURE 27. The ESR signal obtained from the oxidation of electrodes coated with the polymer poly(N-vinylcarbazole). A single-line spectrum near $g = 2$ is evident as expected for a typical "powder" spectrum of an organic radical at high concentration.

mechanism were found to be consistent with the ESR transient data. Thus, by considering the electrochemical data alongside the ESR kinetic data, the mechanism of the reduction of fluorescein to leucofluorescein was unambiguously shown to be a DISP1 process. This was subsequently confirmed by Compton *et al.*[110] using an independent method.

The above example shows several virtues of electrochemical ESR. Firstly, the existence of a radical intermediate is indicated, then its chemical identity (including state of protonation) is revealed and its lifetime and the order of its kinetic decay measured. Finally, because the channel electrode cell behaves as a satisfactory hydrodynamic electrode, the full mechanism of the electrolytic reaction is elucidated.

4.4. Dynamic Processes and ESR Lineshapes

As a final example of typical areas of application of the technique, we consider the information that can be found from the lineshapes of ESR lines and their intensities. In particular, we examine the contributions of electrochemical ESR to the field of polymer-coated electrodes and conducting polymers.

The *in situ* cells described in Section 3 have been shown to have sufficient sensitivity to detect radicals present in low concentrations in thin polymer films on electrode surfaces. For example, Figure 27 shows a signal obtained by Compton, Davis, and coworkers[111,112] from an oxidized coat of poly(N-vinylcarbazole) on a platinum electrode. It is seen that a single-line spectrum is produced, centered near $g = 2$ with a peak-to-peak linewidth of about 3.8 G. This is as expected for a typical "powder" spectrum of a concentrated organic radical. Strong exchange interactions between the radicals have washed out any hyperfine structure and narrowed the effects of the dipolar interactions in the solid. Quantitative ESR measurements showed that the radical was only a minor component of the film (3–10%). The observation that the dominant product of the oxidation was diamagnetic, together with potential-step electrochemical data, showed that the oxidation process involved the dimerization of the initially formed radical cations of the pendant carbazole groups and the further oxidation of the resulting bicarbazolyl group in a two-electron step, as shown in the scheme below:

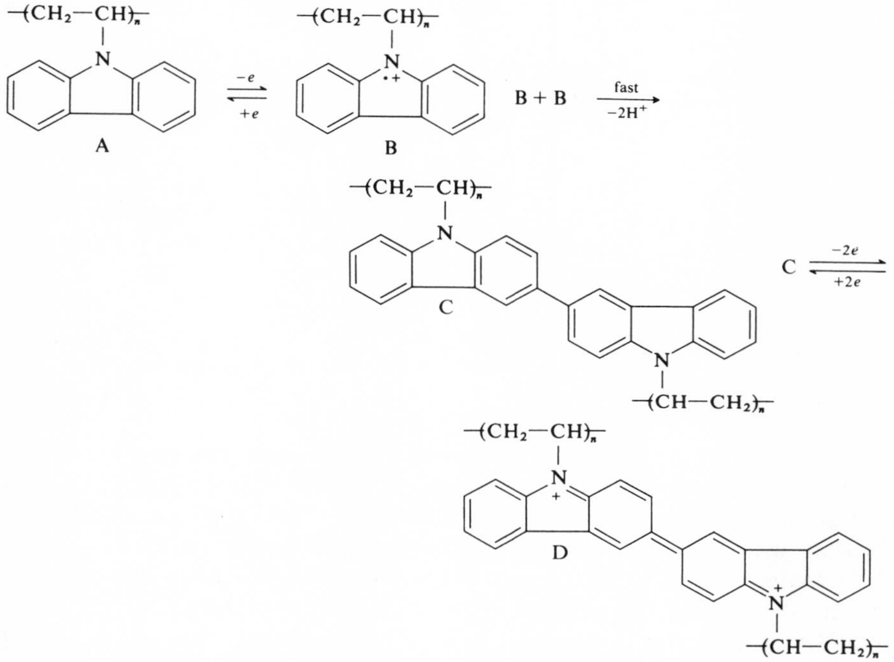

$-(CH_2-CH)_n-$

A

$\xrightarrow[+e]{-e}$

$-(CH_2-CH)_n-$

B

$B + B \xrightarrow[-2H^+]{fast}$

$-(CH_2-CH)_n-$

C

$C \xrightarrow[+2e]{-2e}$

$-(CH-CH_2)_n-$

$-(CH_2-CH)_n-$

D

$-(CH-CH_2)_n-$

It is clear from the signal-to-noise quality of the spectrum in Figure 27, which was obtained from a film of about 10^{-7}-m thickness of an electrode of area 50 mm^2, that much thinner films are open to study. In particular, in the case of electrochemically grown films of polypyrrole, Jones and Albery[113] have been able to see spectra when the electrode is coated with just four monolayers of polymer. This example is especially interesting in that the lineshape differs from that seen usually, as typified by Figure 27. As shown in Figure 28, the signal seen from oxidized polypyrrole is asymmetric and was shown to have a "Dysonian" lineshape.[114] The latter is indicative of metal-like conductivity in the polymer film and arises because microwaves incident on a metallic conductor are absorbed within a short (submicron) depth of the surface—the "skin depth" of the conductor. If the electron mobility is so great that it can diffuse in and out of this region on the timescale of the ESR experiment, then a Dysonian lineshape results.[64] This is a useful fingerprint for metallic conductivity. The effect is most dramatically seen in the case of the salt TTF$^+$TCNQ$^-$,[115] which is a compound in which the flat aromatic ions are lined up in stacks. Crystals of the solid show an anisotropic conductivity which is metallic in the direction of the stacks but near-insulating at right angles to them. The observed ESR spectrum shows either Dysonian or ordinary behavior, depending on whether the magnetic component of the microwave field in the cavity is perpendicular or parallel to the axis of the stacks and samples metallic conductivity or not.

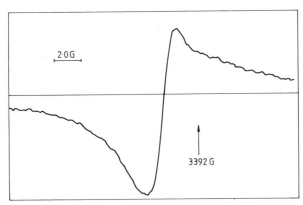

FIGURE 28. ESR signal from a polypyrrole coat showing a Dysonian lineshape.

A final application of ESR to the study of polymer films on electrodes is the use of spin probes to examine rotational motion within the films with the aim of understanding aspects of their morphology. Spin probes are stable free radicals whose spectra, in viscous media, show the effects of incomplete rotational averaging of the (anisotropic) g- and hyperfine coupling constant tensors. Analysis of the spectrum allows deduction of details of the molecular rotation. In particular, the isotropy, or anisotropy, of the motion and rotational correlation times emerges. Kaifer and Bard[116] have studied the behavior of various cationic spin probes incorporated into Nafion. The latter is a perfluorinated ion-exchange polymer which can electrostatically bind cationic species at its sulfonate groups. The material has found appreciable application in the design of chemically modified electrodes for diverse applications.[117] The results showed that the motion of a given cation within Nafion is determined by the nature of the cation itself. Simple inorganic cations such as Ti^{3+}, VO^{2+}, and Mn^{2+} retain the fast-tumbling spectra characteristic of their behavior in aqueous solution, suggesting that inorganic cations reside in aqueous-like environments within the Nafion matrix. Organic cations show more complex behavior, ranging from free tumbling to being strongly anchored by the polyelectrolyte. It was thought that the more delocalized the charge over the cation, the less likely was strong anchoring.

Similar experiments by Compton and Waller[118] on spin probes in plasticized poly(vinyl chloride) coats revealed a dependence of the rotational correlation time on the molecular volume of the spin probe, the temperature, and the loading level and nature of the plasticizer.

4.5. Adsorbed Radicals

The foregoing shows that the presence of very small amounts of radicals on electrodes can be observed by ESR. This naturally raises the question of

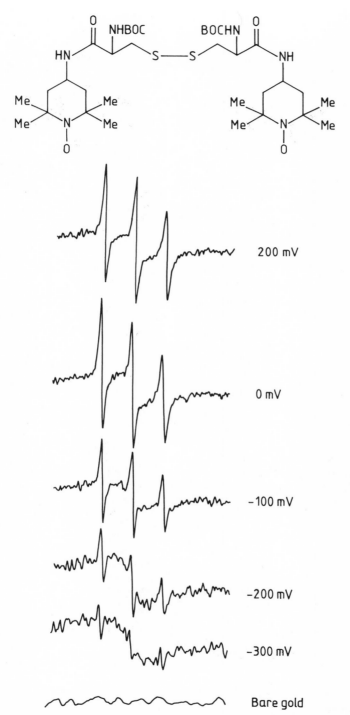

FIGURE 29. The spin label used by Hill and coworkers for the preparation of their "spin-labeled" electrode and the resulting potential-dependent ESR spectra.

whether a monolayer of adsorbed radicals could be detected by *in situ* electrochemical ESR. While this question has not been definitively answered experimentally, cause for optimism is provided by Hill and coworkers who have reported[119] the synthesis of a "spin-labeled electrode" by the adsorption of the spin label (shown in Figure 29) onto the surface of a gold electrode. As shown, the observed spectrum of the electrode in aqueous solution had two components, one with hyperfine splittings characteristic of a solution-phase radical superimposed on a broad spectrum attributed to adsorbed nitroxide. These observations were interpreted in terms of an equilibrium between surface-bound and free spin label. This equilibrium was found to be potential dependent. This work leads one to anticipate the exciting prospect of the application of ESR in the study of adsorption on electrode surfaces.

Acknowledgments

The authors wish to thank the following for stimulating and helpful discussions: Dr. R. A. Allendoerfer, Dr. B. A. Coles, Dr. F. J. Davis, and Mr. M. J. Day.

References

 1. A. Carrington and A. D. McLachlan, *Introduction to Magnetic Resonance*, Chapman and Hall, London (1979).
 2. D. E. G. Austen, P. H. Given, D. J. E. Ingram, and M. E. Peover, *Nature* **182,** 1784 (1958).
 3. A. H. Maki and D. H. Geske, *J. Chem. Phys.* **30,** 1356 (1959).
 4. M. C. R. Symons and M. M. Maguire, *J. Chem. Res.,* 330 (1981).
 5. A. H. Maki and D. H. Geske, *J. Am. Chem. Soc.* **82,** 2671 (1960).
 6. A. H. Maki and D. H. Geske, *J. Chem. Phys.* **33,** 825 (1960).
 7. A. H. Maki and D. H. Geske, *J. Am. Chem. Soc.* **83,** 1852 (1961).
 8. T. Kitagana, T. Layloff, and R. N. Adams, *Anal. Chem.* **36,** 925 (1964).
 9. L. H. Piette, R. Ludwig, and R. N. Adams, *Anal. Chem.* **34,** 916 (1962).
10. L. H. Piette, R. Ludgwig, and R. N. Adams, *Anal. Chem.* **34,** 1587 (1962).
11. D. Levy and R. J. Myers, *J. Chem. Phys.* **41,** 1062 (1962).
12. D. Levy and R. J. Myers, *J. Chem. Phys.* **42,** 3731 (1965).
13. D. Levy and R. J. Myers, *J. Chem. Phys.* **43,** 3063 (1965).
14. D. Levy and R. J. Myers, *J. Chem. Phys.* **44,** 4177 (1966).
15. G. D. Luer and D. E. Bartak, *J. Org. Chem.* **47,** 1238 (1972).
16. V. E. Norvell, K. Tanemati, G. Mamanta, and L. N Klatt, *J. Electrochem. Soc.* **128,** 1254 (1981).
17. R. N. Adams, *J. Electroanal. Chem.* **8,** 151 (1964).
18. G. K. Fraenkel and P. H. Rieger, *J. Chem. Phys.* **39,** 609 (1963).
19. G. K. Fraenkel, P. H. Rieger, I. Bernal, and V. H. Reinmuth, *J. Am. Chem. Soc.* **85,** 683 (1963).
20. G. K. Fraenkel and J. Bolton, *J. Chem. Phys.* **40,** 3307 (1964).
21. B. Kastening, *Z. Anal. Chem.* **224,** 196 (1967).
22. B. Kastening, *Ber. Bunsenges. Phys. Chem.* **72,** 20 (1968).
23. B. Kastening and S. Vavricka, *Ber. Bunsenges. Phys. Chem.* **72,** 27 (1968).
24. H. Gerischer, D. Kolb, and W. Wirths, *Ber. Bunsenges. Phys. Chem.* **73,** 148 (1969).
25. W. J. Albery, A. M. Couper, B. A. Coles, and A. M. Garnett, *J. Chem. Soc. Chem. Commun.* 198 (1974).

26. W. J. Albery, A. M. Couper, and B. A. Coles, *J. Electroanal. Chem.* **65**, 901 (1975).
27. W. J. Albery, R. G. Compton, B. A. Coles, A. T. Chadwick, and J. A. Lenkait, *J. Chem. Soc. Faraday Trans. 1*, **76**, 139 (1980).
28. W. J. Albery, A. T. Chadwick, B. A. Coles, and N. A. Hampson, *J. Electroanal. Chem.* **75**, 229 (1977).
29. W. J. Albery and R. G. Compton, *J. Chem. Soc. Faraday Trans. 1*, **78**, 1561 (1982).
30. W. J. Albery, R. G. Compton, and I. S. Kerr, *J. Chem. Soc. Perkin Trans. 2*, 825 (1981).
31. G. Cauquis, J. P. Billon, and J. Cambrisson, *Bull. Soc. Chim., Fr.* 2062 (1960).
32. G. Cauquis and M. Genies, *Bull. Soc. Chim., Fr.* 3220 (1967).
33. G. Cauquis, J. P. Billon, and J. Raisson, *Bull. Soc. Chim., Fr.* 199 (1967).
34. G. Cauquis, M. Genies, H. Lemaire, A. Rassat, and J. Ravet, *J. Chem. Phys.* **47**, 4642 (1967).
35. G. Cauquis, C. Barry, and M. Maivey, *Bull. Soc. Chim., Fr.* 2510 (1968).
36. J. K. Dohrmann, F. Galluser, and H. Wittchen, *Faraday Discuss. Chem. Soc.* **56**, 350 (1973).
37. J. K. Dohrmann and K. J. Vetter, *J. Electroanal. Chem.* **20**, 23 (1969).
38. J. K. Dohrmann and F. Galluser, *Ber. Bunsenges. Phys. Chem.* **75**, 432 (1971).
39. J. K. Dohrmann, *Ber. Bunsenges. Phys. Chem.* **74**, 575 (1970).
40. B. Kastening, J. Divicek, and B. Costica-Mihelcic, *Faraday Disc. Chem. Soc.* **56**, 341 (1973).
41. M. Fleischmann and R. E. W. Jansson, *Faraday Discuss. Chem. Soc.* **56**, 373 (1973).
42. A. J. Bard and I. B. Goldberg, *J. Phys. Chem.* **75**, 3281 (1971).
43. A. J. Bard and I. B. Goldberg, *J. Phys. Chem.* **78**, 290 (1974).
44. A. J. Bard, I. B. Goldberg, D. Boyer, and R. Hirasawa, *J. Phys. Chem.* **78**, 295 (1974).
45. A. J. Bard, P. P. Gordi, J. Gilbert, and C. John, *J. Electroanal. Chem.* **59**, 163 (1975).
46. R. N. Bagchi, A. M. Bond, and R. Colton, *J. Electroanal. Chem.* **199**, 297 (1986).
47. R. N. Adams, *J. Electroanal. Chem.* **8**, 151 (1964).
48. B. Kastening, *Chem. Ing. Tech.* **42**, 190 (1970).
49. B. Kastening, *Adv. Anal. Chem. Instrum.* **10**, 421 (1974).
50. T. M. McKinney, in: *Electroanalytical Chemistry* (A. J. Bard, ed.), Vol. 10, pp. 1–117, Marcel Dekker, New York (1977).
51. B. Kastening, *Progr. Polarogr.* **3**, 195 (1972).
52. J. Robinson, in: *Electrochemistry: Royal Society of Chemistry Specialist Periodical Reports* (D. Pletcher, ed.), Vol. 9, p. 156, Royal Society of Chemistry, London (1984).
53. I. B. Goldberg and T. M. McKinney, in: *Laboratory Techniques in Electroanalytical Chemistry* (P. T. Kissinger and W. R. Heineman, eds.), p. 675, Marcel Dekker, New York (1984).
54. I. B. Goldberg and A. J. Bard, in: *Magnetic Resonance in Chemistry and Biology* (J. N. Herak and K. J. Adanini, eds.), Chapter 10, Marcel Dekker, New York (1975).
55. O. Stern, *Z. Physik* **7**, 249 (1921).
56. W. Gerlach and O. Stern, *Ann. Phys. Leipzig* **74**, 673 (1924).
57. N. M. Atherton, *Electron Spin Resonance*, Ellis Horwood, Chichester (1973).
58. P. B. Ayscough, *Electron Spin Resonance in Chemistry*, Methuen, London (1967).
59. H. M. McConnell and D. B. Chesnut, *J. Chem. Phys.* **28**, 778 (1958).
60. P. Boyer and J. Dericbourg, *C. R. Acad. Sci. Paris*, 429 (1967).
61. M. C. R. Symons, *Electron Spin Resonance Spectroscopy*, Van Nostrand Reinhold Co., London (1978).
62. K. A. McLauchlan, *Magnetic Resonance*, Oxford University Press, Oxford (1972).
63. J. E. Wertz and J. R. Bolton, *Electron Spin Resonance*, McGraw-Hill, New York (1972).
64. C. P. Poole, *Electron Spin Resonance: A Comprehensive Treatise on Experimental Techniques*, 2nd Edition, John Wiley and Sons, New York (1983).
65. R. D. Allendoerfer, G. A. Martinchek, and S. Bruckenstein, *Anal. Chem.* **47**, 890 (1973).
66. R. D. Allendoerfer and J. B. Carroll, *J. Mag. Res.* **37**, 497 (1980).
67. J. B. Carroll, Ph.D. thesis, State University of New York at Buffalo (1983).
68. H. Ohya-Nishiguchi, *Bull. Chem. Soc. Jap.* **52**, 2064 (1979).
69. F. Gerson, H. Ohya-Nishiguchi, and G. Wydler, *Angew. Chem. Int. Ed. Eng.* **15**, 552 (1976).
70. J. Bruken, F. Gerson, and H. Ohya-Nishiguchi, *Helv. Chim. Acta* **60**, 1220 (1977).
71. R. G. Compton and B. A.Coles, *J. Electroanal. Chem.* **144**, 87 (1983).

72. C. M. A. Brett and A. M. Oliveira-Brett, in: *Comprehensive Chemical Kinetics* (C. H. Bamford and R. G. Compton, eds.), Vol 26, Elsevier, Amsterdam (1986).
73. V. G. Levich, *Physicochemical Hydrodynamics*, Prentice-Hall, Englewood Cliffs, New Jersey (1982).
74. J. B. Flanagan and L. Marcoux, *J. Phys. Chem.* **78**, 718 (1974).
75. J. L. Anderson and S. Moldoveanu, *J. Electroanal. Chem.* **175**, 67 (1984).
76. J. L. Anderson and S. Moldoveanu, *J. Electroanal. Chem.* **179**, 107 (1984).
77. J. L. Anderson and S. Moldoveanu, *J. Electroanal. Chem.* **179**, 119 (1984).
78. B. A. Coles and R. G. Compton, *J. Electroanal. Chem.* **127**, 37 (1981).
79. S. G. Weber and W. C. Purdy, *Anal. Chim. Acta*, **99**, 77 (1978).
80. K. Tokuda, K. Aoki, and H. Matsuda, *J. Electroanal. Chem.* **127**, 211 (1977).
81. H. Matsuda, *J. Electroanal. Chem.* **15**, 325 (1967).
82. M. A. Leveque, *Annales des Mines Mémoires Ser. 12* **13**, 201 (1928).
83. R. G. Compton and P. J. Daly, *J. Electroanal. Chem.* **178**, 45 (1984).
84. R. G. Compton and P. R. Unwin, *J. Electroanal. Chem.* **206**, 57 (1986).
85. R. G. Compton, D. J. Page, and G. R. Sealy, *J. Electroanal. Chem.* **161**, 129 (1984).
86. R. G. Compton, D. J. Page, and G. R. Sealy, *J. Electroanal. Chem.* **163**, 65 (1984).
87. R. G. Compton, P. J. Daly, P. R. Unwin, and A. M. Waller, *J. Electroanal. Chem.* **191**, 15 (1985).
88. W. J. Albery, C. C. Jones, and R. G. Compton, *J. Am. Chem. Soc.* **106**, 469 (1984).
89. R. G. Compton and A. W. Waller, *J. Electroanal. Chem.* **195**, 289 (1985).
90. I. B. Goldberg, in: *Electron Spin Resonance: Royal Society of Chemistry Specialist Periodical Reports* (P. B. Asycough, ed.), Vol 6, pp. 1–22, Royal Society of Chemistry, London (1981).
91. A. Gennaro, F. Maran, A. Maye, and E. Vianello, *J. Electroanal. Chem.* **185**, 353 (1985).
92. A. M. Waller, unpublished work.
93. N. B. Kondrikov, V. V. Orlov, V. I. Ermakov, and M. Ya. Fioshin, *Electrokhimiya* **8**, 920 (1972).
94. N. B. Kondrikov, V. V. Orlov, V. I. Ermakov, and M. Ya. Fioshin, *Russ. J. Phys. Chem.* **47**, 368 (1973).
95. R. D. Goodwin, J. C. Gilbert, and A. J. Bard, *J. Electroanal. Chem.* **59**, 163 (1975).
96. R. G. Compton, B. A. Coles, and M. J. Day, *J. Electroanal. Chem.* **200**, 205 (1986).
97. R. G. Compton, B. A. Coles, and M. J. Day, *J. Chem. Res. (S)*, 260 (1986).
98. E. G. Janzen and B. J. Blackburn, *J. Am. Chem. Soc.* **90**, 5909 (1968).
99. E. G. Janzen, *Acc. Chem. Res.* **4**, 31 (1977).
100. H. Blount, E. E. Bancroft, and E. G. Janzen, *J. Am. Chem. Soc.* **101**, 3692 (1979).
101. B. W. Gara and B. P. Roberts, *J. Chem. Soc. Perkin Trans. 2*, 150 (1978).
102. P. Martigny, G. Mahon, J. Simonet, and G. J. Mousset, *J. Electroanal. Chem.* **121**, 349 (1981).
103. G. L. McIntire, H. N. Blount, H. J. Stronks, R. V. Shetty, and E. G. Janzen, *J. Phys. Chem.* **84**, 916 (1980).
104. G. Gronchi, P. Courbis, P. Tordo, G. Monsset, and J. Simonet, *J. Phys. Chem.* **87**, 1343 (1983).
105. M. Culais, P. Tordo, and G. Gronchi, *J. Phys. Chem.* **90**, 1403 (1986).
106. S. Terabe, K. Kuruma, and R. Konaka, *J. Chem. Soc. Perkin Trans. 2*, 1252 (1973).
107. C. Amatore and J. M. Saveant, *J. Electroanal. Chem.* **85**, 27 (1977).
108. C. Amatore, M. Gareil, and J. M. Saveant, *J. Electroanal. Chem.* **147**, 1 (1983).
109. M. Fleischmann, F. Lasserre, and J. Robinson, *J. Electroanal. Chem.* **177**, 115 (1984).
110. R. G. Compton, R. G. Harland, P. R. Unwin, and A. M. Waller, *J. Chem. Soc. Faraday Trans. 1*, **83**, 1261 (1987).
111. R. G. Compton, F. J. Davis, and S. C. Grant, *J. Appl. Electrochem.* **16**, 239 (1986).
112. F. J. Davis, H. Block, and R. G. Compton, *J. Chem. Soc. Chem. Commun.*, 890 (1984).
113. C. C. Jones and W. J. Albery, private communication.
114. W. J. Albery and C. C. Jones, *Faraday Discuss. Chem. Soc.* **78**, 193 (1984).
115. S. K. Khanna, E. Ehrenfreund, A. F. Garito, and A. J. Heeger, *Phys. Rev. B* **10**, 2205 (1974).
116. A. E. Kaifer and A. J. Bard, *J. Phys. Chem.* **90**, 898 (1986).
117. R. W. Murray, *Ann. Rev. Mater. Sci.* **14**, 145 (1984).
118. R. G. Compton and A. M. Waller, unpublished work.
119. K. di Gleria, H. A. O. Hill, D. J. Page, and D. G. Tew, *J. Chem. Soc. Chem. Commun.*, 460 (1986).

8

Mössbauer Spectroscopy

Daniel A. Scherson

1. Introduction

Mössbauer spectroscopy is a technique that relies on the phenomenon of recoilless emission and absorption of γ-rays for the investigation of nuclear quantum states. The energy associated with such quantum states is modified by the interactions of the nucleus with the surrounding electric and magnetic fields. Hence, the analysis of information derived from such measurements may be expected to afford considerable insight into the structural, electronic, and magnetic properties of a variety of condensed-phase materials. Although restricted to only a few elements, this methodology has found wide application in a number of research areas, including chemistry, biology, and metallurgy, and has recently emerged as a powerful tool in the investigation of systems of interest to physical electrochemistry.

This chapter will provide a brief summary of fundamental theoretical and experimental aspects of Mössbauer spectroscopy that may aid in the design of *in situ* electrochemical experiments and in the interpretation of spectral data. In addition, it will present a number of illustrations that may be regarded as model systems for further applications of this technique to the investigation of a wider class of interfacial phenomena.

2. Theoretical Aspects

Nuclear transitions involve energies which are orders of magnitude larger than those associated with the excitation of vibrational or electronic states. It thus becomes necessary to account for the effects of mechanical recoil in the emission and absorption of such high-energy radiation in order to interpret the results of resonance-type measurements.

Daniel A. Scherson • Case Center for Electrochemical Sciences and the Department of Chemistry, Case Western Reserve University, Cleveland, Ohio 44106.

2.1. Recoil Energy, Resonance, and Doppler Effect

Consider a gas-phase experiment in which quanta emitted by excited atoms of an element are used to excite atoms of the same element in their ground state. For the sake of simplicity, only a single transition with energy ϵ_0 and average lifetime τ will be assumed to be involved.

It can be shown based on the laws of energy and momentum conservation that the average energy of the emitted quantum, ϵ_0', will be in general different from ϵ_0. This is due to three contributing factors: mechanical recoil, and a linear and a second-order Doppler effect, which account for the motion of the atom prior to the emission. A directly analogous phenomenon also can be expected to occur in the case of absorption of radiation by gas atoms in the ground state. For emitting and absorbing atoms initially at rest (or moving with the same velocity), the displacement of the average emission energy with respect to the average absorption energy is given to a good degree of approximation by ϵ_0^2/mc^2 (m is the mass of the atom), a quantity that corresponds to twice the recoil energy ϵ_R. Hence, for an ensemble of excited- and ground-state atoms moving with precisely the same velocity, the degree of overlap between the emission and absorption bands, which is directly related to the extent of resonance, would be essentially negligible for values of $2\epsilon_R$ much larger than $\Gamma = \hbar/\tau$, the natural width of the transition band. This is represented pictorially in Figure 1. Such a situation is expected to occur, for instance, in the case of nuclear transitions involving the ground and first excited states of ^{57}Fe ($\epsilon_0 = 14.4$ keV), for which $\tau \sim 1.44 \times 10^{-7}$ s, and hence $\Gamma \sim 4.56 \times 10^{-9}$ eV.

A collection of gas particles, however, exhibits a statistical distribution of velocities, a phenomenon that will give rise to a broadening in the emission and absorption energies. The characteristic width of this so-called Doppler broadening may be shown to be given by $D = 2(\epsilon_R kT)^{1/2}$, where k is the Boltzmann constant. It follows that the resonance probability will be determined by the relative values of Γ/ϵ_R and Γ/D.

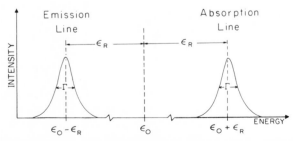

FIGURE 1. Effects of recoil on the transition lines associated with the absorption and emission of high-energy radiation by isolated atoms. The dashed curve in the center of the figure corresponds to a recoil energy $\epsilon_R = 0$. For $\epsilon_R \neq 0$, the separation between the transition energies is $2\epsilon_R$. No overlap occurs for $2\epsilon_R \gg \Gamma$ and thus no resonance absorption would be observed. (Adapted from Figure 2.4 in Ref. 2b.)

From a more general perspective, two possible means of enhancing resonance absorption for spectroscopic techniques involving high-energy radiation can be envisioned: (i) to raise the temperature of the emitter so as to increase the Doppler broadening and (ii) to compensate for the recoil energy loss by imparting to the emitter an additional momentum, either by mechanical means or by relying on a preceding nuclear event. Although experiments based on these principles were proven in many cases successful, the general interest in this area prior to 1957 was rather restricted. In that year, R. Mössbauer made two observations of far-reaching consequences, while investigating the nuclear absorption of γ-rays emitted by ^{191}Ir by metallic iridium.[1] First, he found that the extent of resonance increased as the temperature decreased, which is in direct contrast with what would be expected based on the arguments put forward above, and second, he observed that the resonance absorption could be totally destroyed if the source of the radiation was moved with respect to the absorber with velocities of the order of a few centimeters per second. These findings led Mössbauer to conclude that the phenomenon had to involve nuclear events in which there were no recoil losses both in the emission and in the absorption of radiation. For this outstanding discovery, Mössbauer was awarded the Nobel prize in 1961.

The following sections will provide a brief introduction to some of the principles upon which Mössbauer spectroscopy is based. The discussion will be restricted to Fe and Sn as to date only these nuclides have been the subject of *in situ* investigations involving electrochemical systems. More extensive treatments of theoretical and experimental aspects of Mössbauer spectroscopy may be found in a number of excellent specialized books and monographs.[2]

2.2. Phonons, Mössbauer Effect, and Recoilless Fraction

The recoil energy associated with the emission or absorption of radiation by an atom bound to a solid of macroscopic dimensions will be distributed among the available translational and vibrational quantum states. This may involve (i) a transfer of momentum to the host matrix as a whole, a process that is expected to make a negligible contribution in view of the large mass involved, (ii) a site displacement, for which the threshold energy is in the range of 10 to 50 eV, and (iii) a thermal excitation of the lattice for lower energies. If the free-atom recoil energy is of the order of 10 meV, however, and thus of the same magnitude as the separation between the collective vibrational energy levels in the solid, or phonons, there is a finite probability that the emission or absorption will occur without exchange of energy between the atom involved and the lattice. In honor of its discoverer, such a phenomenon has become known as the Mössbauer effect.

An explicit expression for the fraction of such recoilless or zero-phonon processes, commonly denoted as f, can be derived from the specific model used to represent the solid. The Debye model, for instance, predicts an increase in the magnitude of f when either the γ-ray energy or the temperature decreases, and also when the Debye temperature increases. The latter is a measure of the bond strength between the emitting or absorbing atom and the lattice. It is precisely through a spectral analysis of such recoil-free events that information regarding electrical and magnetic interactions involving the nucleus and its environment, of direct interest to problems of physicochemical relevance, can be obtained.

2.3. Electric Hyperfine Interactions

The energy associated with the electrostatic interaction of a nucleus and the surrounding charges may be separated into two contributions: the electric monopole interaction, an effect that generates a shift in the nuclear energy levels, and the electric quadrupole interaction, which lifts the degeneracy of the energy levels. These serve as a basis for defining two important parameters in Mössbauer spectroscopy, the isomer shift and the quadrupole splitting, which can be related to certain aspects of structure and bonding in a variety of materials.

2.3.1. Isomer Shift

Within the nonrelativistic approximation, only electrons in s-type orbitals have a nonzero probability density at the nucleus and can thus interact with the charge density therein, leading to a shift in the energy of the nuclear states. Nuclear excitations are accompanied by changes in the nuclear charge density distribution, and therefore the actual magnitude of the energy shift may not be expected to be the same for the ground and excited states. In particular, the energy difference between the excited and ground states for a nucleus in the absence, E_0, and the presence, E_s, of this specific electrostatic interaction, denoted as ΔE, is given by:

$$\Delta E = E_s - E_0 = \tfrac{2}{3}\pi Z e^2 |\psi(0)|^2 [\langle r^2 \rangle_e - \langle r^2 \rangle_g] \tag{1}$$

$e|\psi(0)|^2$ in this expression is the electronic charge density at the nucleus, $\langle r^2 \rangle_i$ the expectation value of the square of the nuclear radius in either the ground ($i = g$) or excited ($i = e$) state, and Ze the nuclear charge. The term $[\langle r^2 \rangle_e - \langle r^2 \rangle_g]$ involves intrinsic nuclear parameters, whereas $|\psi(0)|^2$ depends on the specific environment surrounding the nuclei and the atom or ion as a whole. Hence, the quantity ΔE in general will vary for different compounds or host lattices. If one of such materials is regarded as the source S, and the other as the absorber A, the difference between $(\Delta E)_A$ and $(\Delta E)_S$ defines an experimentally accessible parameter known as the isomer shift, δ, given explicitly by:

$$\delta = (\Delta E)_A - (\Delta E)_S = \tfrac{2}{5}\pi Z e^2 (|\psi(0)|_A^2 - |\psi(0)|_S^2) \cdot (R_e^2 - R_g^2) \tag{2}$$

The expectation value of the nuclear radius in this equation has been replaced by R, a quantity that represents the radius of a spherically symmetric nucleus

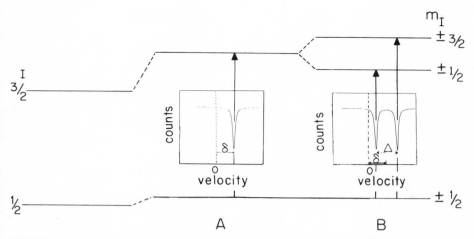

FIGURE 2. Effects on the ground and first excited nuclear energy levels of ^{57}Fe due to (A) the isomer shift and (B) the quadrupole splitting, showing the Mössbauer absorption transitions and the expected spectra. The isomer shift, δ, is indicated with reference to some arbitrary standard. I and m_I are the nuclear spin and magnetic spin quantum numbers, respectively. (Adapted from Figure 1.5 in Ref. 2c.)

with a uniform charge density, in either the excited or the ground state. According to Eq. (2), δ is a relative quantity, and therefore the standard or reference with respect to which the isomer shift is being measured or reported must always be specified. A schematic representation of the nuclear level displacement for ^{57}Fe due to the isomer shift, including the associated Mössbauer spectrum and the corresponding absorption transitions, is shown in part A of Figure 2.

The electronic density at the nucleus may be separated into contributions due to inner-core filled orbitals and to partially filled valence orbitals, and thus the isomer shift may be expected to be sensitive to the specific nature of the chemical bonds involving the Mössbauer-active species and adjacent atoms in the molecule or lattice. Furthermore, the value of δ may be modified not only by varying the s-orbital occupation but also by a shielding effect associated with orbitals of other types. The removal of a d-electron from an iron ion, for instance, will lead to a contraction of the occupied s-orbitals and thus to a shift of δ in the negative direction. A correlation between the isomer shift and oxidation states, which may be useful in the analysis of certain spectral features associated with iron in molecules and other materials, is shown in Figure 3.

2.3.2. Quadrupole Splitting

The interactions between the nuclear quadrupole moment eQ, which is a measure of the extent of deviation of the nuclear charge distribution from

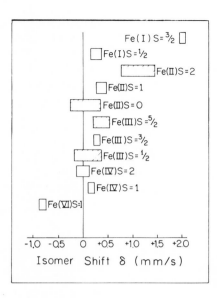

FIGURE 3. Oxidation state/spin quantum state–isomer shift correlation diagram for iron in a variety of compounds. The values of δ are referred to the α-Fe standard. (Adapted from Figure 3.2 in Ref. 2b.)

spherical symmetry, and the gradient of the electric field associated with the presence of electrons and ions surrounding a specific nucleus may give rise to a lifting of the degeneracy of the nuclear states. This is expected to occur when both $Q \neq 0$, which is true only if the nuclear state spin quantum number I is greater than $\frac{1}{2}$, and the charges around the nucleus are not distributed in cubic symmetry.

The off-diagonal components of the matrix which represents the electric field gradient 3×3 second-rank tensor can be made zero by performing an appropriate coordinate transformation. This makes it possible to specify the electric field gradient by two independent quantities: V_{zz}, the diagonal element with the largest absolute value, and $\eta = (V_{xx} - V_{yy})/V_{zz}$, a non-negative quantity known as the asymmetry parameter. V_{zz} and η have contributions due to ions in the lattice and to the valence electrons, which can be evaluated from molecular or crystal structural parameters and molecular orbital calculations, respectively.

In the case of ^{57}Fe, the electric quadrupole interaction will split the first excited state $(I = \frac{3}{2})$ into two substates without shifting the baricenter of the state, giving rise to two absorption lines in the spectra. The energy difference between these states, represented by the distance between the two lines, is called the quadrupole splitting Δ (B, Figure 2). If the electric field gradient is axially symmetric, for instance, η becomes zero, and $\Delta = eQV_{zz}/2$. Since Q is an intrinsic property of the Mössbauer-active nuclide, the differences in the value of Δ may be attributed solely to changes in the electric field gradient tensor, which in turn will be a function of the molecular and electronic structure. An interesting illustration of such phenomena is provided by certain planar organometallic molecules capable of forming axially coordinated adducts. In the case of iron phthalocyanine, a widely studied transition metal macrocycle, the presence of pyridine, imidazole, or picoline in the axial positions results in a decrease in the value of Δ over the unsubstituted species, the magnitude of which appears to correlate with the strength of the Lewis base character of the ligand.[3]

2.4. Magnetic Hyperfine Interaction

The nuclear magnetic dipole moment can interact with a magnetic field, inducing a splitting of a nuclear state with spin quantum number I into $2I + 1$ equally spaced, nondegenerate substates. The energies of these substates are given by:

$$E(m_I) = -g_N \beta_N H m_I \qquad (3)$$

where g_N is the nuclear Landé factor, β_N the nuclear magneton, H the strength of the magnetic field, and m_I the nuclear magnetic spin quantum number. In the case of ^{57}Fe, the magnetic field splits the ground state $(I = \frac{1}{2})$ and first excited state $(I = \frac{3}{2})$ into two and four sublevels, respectively. According to the magnetic dipole selection rules, only transitions for which $\Delta I = 1$ and $\Delta m = 0, \pm 1$ will be allowed, yielding in the case of $V_{zz} = 0$ a symmetric six-line spectrum such as that shown in Figure 4. The line intensities of these spectral

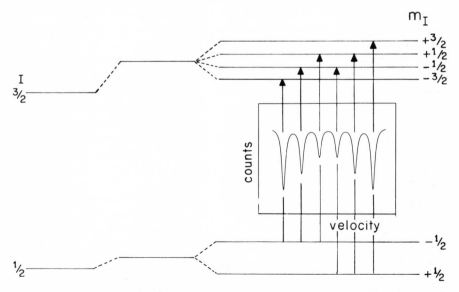

FIGURE 4. Effects of the magnetic splitting on the nuclear energy levels of ^{57}Fe, showing the Mössbauer absorption transitions and the resulting spectrum in the absence of quadrupole splitting. The magnitude of the splitting of the lines is proportional to the total magnetic field at the nucleus. (Adapted from Figure 1.6 in Ref. 2c.)

features are not the same due to differences in the transition probabilities, which also have different angular distributions. An asymmetry in the magnetically split spectrum will be introduced if $V_{zz} \neq 0$, from which the actual sign of V_{zz} and thus the electric field gradient can be determined.

3. Experimental Aspects

A detailed description of the principles of operation of conventional Mössbauer spectrometers may be found in a number of excellent monographs.[2] Although some of these concepts will be reviewed here, most of the attention will be focused on those aspects which are of special importance to the design of *in situ* experiments.

3.1. Instrumentation and Modes of Operation

Based on the theoretical formalism outlined in the previous section, the electric and magnetic environments surrounding a specific nucleus modify the energy of the nuclear states. A Mössbauer spectrometer is an instrument that enables the relative velocity of a source and an absorber to be varied while simultaneously monitoring the absorption or emission of recoil-free γ-rays. This makes it possible to compensate for the minute transition energy differences between a given Mössbauer-active nucleus in a source and an absorber and thus determine the energies at which the resonance conditions are met. In its most common form, the apparatus consists of a source mounted on a velocity transducer, a detector, and ancillary components that allow a direct correlation between the detector response, expressed most often in terms of the number of counts, and the velocity of the source. Recent advances in microelectronics technology have made it possible to reduce appreciably the size and cost of commercially available standard-type units capable of fulfilling essentially all the requirements associated with *in situ* spectroscopy experiments.

Most Mössbauer spectroscopy experiments are conducted either in the transmission mode, in which a source of well-defined characteristics is used to examine the spectral properties of an unknown absorber, or in the emission mode, in which the source of radiation becomes the sample under investigation and a known or standard absorber is employed to determine the transition energy differences. In both cases, any of a number of γ-ray detectors is utilized to record the amount of radiation transmitted through the absorber.

An alternate way of acquiring Mössbauer spectra consists in measuring events associated with the deexcitation of nuclei in the sample following the absorption of γ-rays. Such a process involves the emission of γ-rays, X-rays, or electrons, which are most often detected in a backscattering geometry. The probability associated with different relaxation pathways for ^{57}Fe in the $I = \frac{3}{2}$ state is shown in Table I. One of the main advantages of this approach is that the spectral background is small compared to that associated with transmission measurements, as the measured photons or electrons originate solely from Mössbauer excited nuclei. In addition, it makes it possible to record spectra of materials too thick to be examined in the transmission configuration. It is interesting to note that due to their small mean free path in a solid, only electrons emitted from the outermost layers of the material may be expected to escape and then be detected. Because of its inherent surface sensitivity, this specific technique known as conversion electron Mössbauer spectroscopy is of utmost importance to the study of interfaces. A comparison of the different backscattering detection techniques is given in Table II.

Emission Mössbauer spectroscopy is several orders of magnitude more sensitive than conventional measurements in the transmission mode and thus offers the opportunity of detecting species at very low concentrations. Despite this advantage, however, there are certain problems associated with this methodology that require special consideration. Specifically, the emission process in the case of cobalt, as shown in Figure 5, involves the capture of an inner-shell electron by the nucleus, generating a highly excited iron nucleus and a hole. The hole can then move to a higher level and emit an X-ray, a process that would not change the charge state of the original Co nucleus. An alternate channel for deexcitation involves the filling of the core hole by an electron from a higher level, followed by the emission of another electron in order to conserve energy. Such events are referred to as Auger transitions and may yield a variety of charged states for the iron nuclide. The predominant mechanism for deexcitation will be determined primarily by the relative lifetimes of different excited states, which in turn will depend on the properties of the host lattice. These two deexcitation modes, collectively known as *after effects*, are especially important in materials with rather low electronic conductivity. In such materials, highly ionized states may have long enough lifetimes to modify the chemical environment surrounding the Mössbauer-active nuclide.

TABLE I. Probability of Relaxation Products from
Excited ^{57}Fe $(I = \frac{3}{2})$

Type of emission	No. out of 100 absorptions
14.4-keV γ-ray	9
6.3-kev X-ray	27
7.3-keV $K\,e^-$	81
13.6-keV $L\,e^-$	9
14.3-keV $M\,e^-$	1
6.4-keV KLL Auger e^-	63

TABLE II. Backscattering Detection in Mössbauer Spectroscopy Measurements

Advantages	Disadvantages
• Background suppression improves signal-to-background ratio.	• Restricted solid angle for detection reduces considerably overall sensitivity.
• Equivalent signal-to-noise ratio can be achieved theoretically in two orders of magnitude less time than in transmission.	*X-Ray*
X-Ray	• Detector window materials attenuate signal.
• Applicable for measurements in the low-temperature range.	• Half-absorption length of water for keV radiation only a fraction of a mm. Special cell would be required for *in situ* measurements through front of the electrode.
• *In situ* scattering experiments from back of electrode feasible.	
Conversion electron	*Conversion electron*
• Enhanced sensitivity.	• Measurements at low-temperature require vacuum chamber. Use of Channeltron, for example, offsets higher sensitivity due to more restricted solid angle for detection.
• Surface specificity.	
• Ideally suited for room temperature measurements with gas flow counter.	• Half-absorption length of keV electrons by water is of the order of a fraction of a μm.
	• Although quasi *in situ* experiments are feasible, *in situ* measurements seriously impaired.

Another mechanism by which the measured Fe charged state may be different from that of the original parent Co involves an actual electron transfer reaction in which other host lattice species serve as either donor or acceptor sites. The difficulties in the interpretation of emission Mössbauer spectra introduced by chemical and after effects could be diminished to some extent by conducting detailed studies of the behavior of known solids.

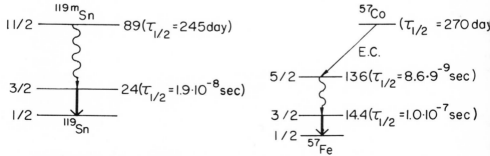

FIGURE 5. Decay schemes of ^{119m}Sn and ^{57}Co. The nuclear transitions associated with Mössbauer spectroscopy measurements are shown by the heavy arrows. Also indicated are the transition lifetimes, $\tau_{1/2}$, for the states involved. Internal conversion processes have not been included. Energies are given in keV. E.C. refers to electron capture. (Adapted from Ref. 2a.)

3.2. Sources, Data Acquisition, and Data Analysis

The most common type of source for ^{57}Fe Mössbauer spectroscopy consists of elemental ^{57}Co incorporated into a host metal lattice such as rhodium or copper. In the case of ^{119}Sn measurements, ^{119m}Sn-enriched $CaSnO_3$ or $BaSnO_3$ is used as a source. Schematic diagrams of the radioactive decay schemes for these two isotopes are shown in Figure 5. In addition to these transitions, internal conversion processes may give rise to emission of radiation of other energies. For example, in the case of ^{57}Fe, the $I = \frac{3}{2}$ state may decay via the ejection of a K-shell 7.3-keV electron, and the hole created be filled by an L-shell electron, leading to the emission of either a 6.4-keV electron (Auger process) or X-ray in order to conserve energy.

In most applications, the source is moved at constant acceleration between two prescribed velocities in a periodic fashion. This generates a spectrum consisting of mirror images about the minimum or maximum velocity. The data collected in a given cycle are added electronically, usually with a multi-channel analyzer, to the cumulative sum of previous cycles until the signal-to-noise ratio becomes high enough to perform a reliable data analysis. Prior to an actual experiment, it is customary, and indeed advisable, to record the spectrum of a standard in order to verify the proper operation of the instrument and to calibrate the velocity scale. The separation between the outermost lines of the six-line spectrum for metallic iron, which amounts to 10.167 mm s^{-1}, is most often used for this purpose. Alternatively, the actual velocities can be measured directly by employing interferometric techniques.

The Mössbauer spectral lineshapes in the case of thin absorbers can be represented fairly accurately by Lorentzian-type functions. The values of the isomer shift, quadrupole splitting, linewidth, strength of the effective magnetic field at the nucleus, and other parameters of interest can be determined by a statistical treatment of the data. A number of computer routines have become available which enable such analyses to be performed in a rather straightforward fashion.

3.3. In Situ Mössbauer Spectroscopy†

A key factor in the design of electrochemical cells for *in situ* transmission Mössbauer measurements is to decrease the attenuation of the γ-ray beam, so as to reduce the time required for spectra acquisition. This may be accomplished by selecting low-absorption materials for windows and electrode supports and by minimizing the amount of electrolyte in the beam path. Radiation in the keV range penetrates rather deeply into matter, and therefore small amounts of rather high-Z elements can be tolerated without seriously comprising the overall cell transmission. As a means of illustration, the half-absorption length for 14.4-keV X-rays in water is about 3.5 mm, which is approximately

† For recent reviews, see Ref. 4.

12.5 times larger than for 6.4-keV X-rays. From an experimental viewpoint, the decrease in intensity of the 14.4-keV γ-rays for aqueous electrolytes is compensated to a certain extent by the higher absorbance of the 6.4-keV X-rays as it makes the use of an external filter unnecessary.

Highly conducting thin films of gold vapor deposited on Melinex[6] and self-supported thin glassy carbon disks have been used extensively as substrates for *in situ* transmission Mössbauer spectroscopy studies involving electroactive layers. In the case of small particle dispersions, very satisfactory results have been obtained with Teflon-bonded high-area carbon electrodes of the type used in fuel cell applications.[7] These electrodes are prepared by adding the dispersion in dry form to an aqueous suspension of emulsified Teflon under ultrasonic agitation. The slurry then is filtered and the resulting paste thoroughly homogenized by repeated spreading with a spatula onto a Teflon sheet. After the excess liquid is eliminated, the actual electrode is formed under pressure in a die and the emulsifier later removed by heat treatment under an inert atmosphere. The dry, circularly shaped electrode is placed once again in the die and a current collector, consisting most often of a flexible, open metal grid, is attached to one of the faces by compression. A more detailed description of the overall procedure involved in the fabrication of this type of electrode may be found in the original literature.[7]

A reduction of the electrolyte volume in the beam path has been most often realized for *in situ* Mössbauer applications by using a geometry similar to that of conventional spectroelectrochemical thin-layer cells. It is customary, in this configuration, to place the counter electrode away from the working electrode to avoid blocking the radiation. This, however, generates a path of high ionic resistance, which increases the time response to potential changes. Although this factor may not be important in many applications, it introduces uncertainties in the interpretation of results involving systems in which the rate of change of the potential drop across the interface modifies the nature of the resulting products, such as passive film formation on iron. One way of

---➤

FIGURE 6. Spectroelectrochemical cells for *in situ* Mössbauer measurements. (a) The working electrode is mounted on a rod that can be moved vertically without disturbing the overall cell operation. During potential changes, it is placed in front of the fixed counter electrode in the upper part of the cell and returned to the measuring area after the current decreases to a small value. A. Working electrode, B. counter electrode, C. collapsible cell, D. reference electrode, E. gas bubbler, F, Teflon holder for working electrode, G. stopper with press-fitted nickel wire connected to counter electrode, H. aluminum frame, I. Viton O-ring, J. Teflon cell top, K. aluminum plates, L. brass screws, M. nickel screen annulus (current collector), N. Teflon ring for working electrode mounting and mechanical attachment to F through pin (not shown in the figure). (Reproduced from Ref. 7.) *Inset*: Schematic diagram of the front and side views of electrochemical cell, showing sliding mechanism for working electrode and geometrical configuration during Mössbauer measurements. [Adapted from D. Scherson, S. B. Yao, E. B. Yeager, J. Eldridge, M. E. Kordesch, and R. W. Hoffman, *Appl. Surface Sci.* **10**, 325 (1982).] (b) Cell specifically designed for studies involving passive film formation on iron. The working electrode is fixed and the counter electrode is mounted on a Lucite plunger. A. Passivated Fe working electrode, B. electrolyte, C. Au on Melinex counter electrode. (Adapted from Ref. 8.)

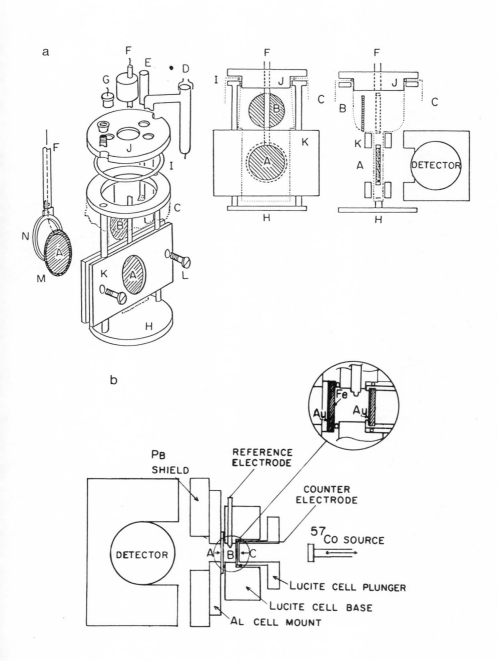

circumventing this problem is by using a cell with flexible walls made out of a γ-ray-transparent material such as polyethylene (Figure 6a).[7] In this design, the working electrode is placed in front of the counter electrode in the upper part of the cell during changes in polarization to improve the current distribution. After the current reaches a small value, the working electrode is moved downwards under potential control and placed in the γ-ray beam path. Two aluminum plates of the form shown in the figure are then used to compress lightly the cell walls so as to decrease the amount of solution in the γ-ray beam path without blocking the radiation. Alternatively, a semitransparent counter electrode can be placed permanently in front of the working electrode as illustrated in Figure 6b.[8]

An area of special interest to the general field of electrocatalysis is the *in situ* spectroscopic examination of electrodes during actual operation, as considerable insight may be gained into mechanistic aspects of redox processes. At least two *in situ* Mössbauer experiments of this type, both involving oxygen cathodes, have been reported in the literature.[9,10] A geometry that appears particularly suited for the application of Mössbauer spectroscopy to studies of this type is that of a conventional fuel cell. This device consists of oxygen-fed and hydrogen-fed Teflon-bonded high-area carbon electrodes placed at a close distance in front of one another. A sheet of Teflon is often attached to the electrode side facing the gas compartment to prevent electrolyte leakage. The feasibility of conducting *in situ* MES measurements on an operating fuel cell was demonstrated by using a heat-treated iron-macrocycle-based electrode as the oxygen cathode.[10] This specific fuel cell consisted of two separate sets of elements which, upon assembly, formed the cathodic and anodic cell compartments. A schematic diagram of the fully assembled fuel cell as well as the geometric arrangement for Mössbauer measurements in the transmission mode are given in Figure 7. Among the many advantages of this configuration are a uniform current distribution and the establishment of steady-state conditions. The latter is a very important factor since it might be possible to sustain a finite concentration of reaction intermediates and thus provide improved conditions for their detection and study.

3.4. Quasi In Situ Mössbauer Spectroscopy

In situ spectroelectrochemical techniques may be regarded as a type of methodology in which spectroscopic information about the electrode, the electrode/electrolyte interface, and/or the electrolyte solution is sought under conditions in which the potential across the electrode/electrolyte interface is controlled during the data acquisition. There are some instances, however, in which because of intrinsic physical limitations, experiments cannot be conducted in a conventional *in situ* fashion. Two techniques that appear to fall in such a category, referred to hereafter as quasi *in situ*, will be presented in the following sections.

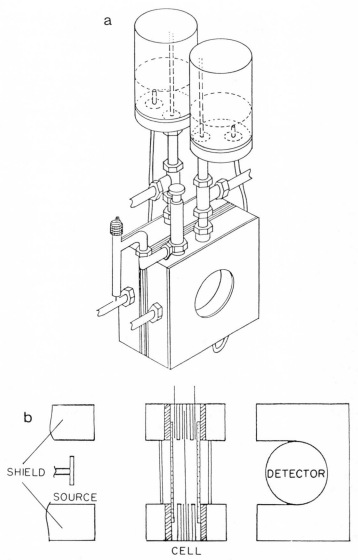

FIGURE 7. Diagram of fuel cell and geometric arrangement for *in situ* Mössbauer measurements in the transmission mode.[10]

3.4.1. Quasi In Situ Conversion Electron Mössbauer Spectroscopy

As was mentioned in Section 3.1, conversion electron Mössbauer spectroscopy (CEMS) provides an advantageous means of studying surface structure as it affords much greater sensitivity than measurements in the transmission mode. Electrons, however, cannot penetrate through detector windows or thick layers of electrolyte, and thus CEMS cannot be readily applied to the *in situ* study of electrochemical interfaces. Recently, Kordesch *et al.*[11] developed a quasi *in situ* technique that makes it possible to detect conversion electrons using a continuously emersed electrode similar in design to that used by Rath and Kolb in their work function studies.[12] A schematic diagram of the complete electrochemical cell–Mössbauer spectrometer system is shown in Figure 8. The electrode is a disk with its lower half in the solution under potential control and its upper half in the conversion electron detector surrounded by the counting gas. The disk is mounted on a motor that continuously rotates the polarized surface. This carries with it only a very thin layer of solution into the γ-ray beam while returning the previously measured area into the electrolyte. It may be noted that the counting gas components, He and CH_4, are practically inert from an electrochemical standpoint and thus are not expected to interfere with the intrinsic behavior of the interface. Although essentially identical results have been obtained in some cases for spectra recorded *in situ* both in transmission and by the conversion electron method described above, additional experiments will be required to determine whether this approach consistently reproduces the spectroscopic behavior found under conventional *in situ* conditions.

3.4.2. Low-Temperature Quenching

Considerable insight into the nature of a variety of materials can be obtained by a detailed examination of Mössbauer spectral features as a function of temperature. Unfortunately, most electrolytes undergo freezing at rather high temperatures, making it essentially impossible to conduct proper *in situ* measurements in a wide temperature range. An approach that could offer interesting possibilities is the fast low-temperature quenching of polarized electrodes using, for instance, liquid nitrogen. It should be mentioned in this regard that the feasibility of performing traditional electrochemical measurements, such as cyclic voltammetry, at very low temperatures has been demonstrated by Stimming, Schmickler, and coworkers[13]; their results could provide a framework for attempting meaningful Mössbauer experiments of the type suggested above.

3.5. Limitations of the Technique

Besides the fact that the number of Mössbauer-active nuclides for which measurements can be conveniently made is rather small, there are other factors which tend to restrict the type of electrochemical system that can be examined with this spectroscopic technique. In the case of ^{57}Fe in the transmission mode, for instance, the number of scatterers required to obtain adequate statistics in a reasonable period of time, which often amounts to several hours or even days, is of the order of 10^{17}. Hence, this technique lacks the time resolution

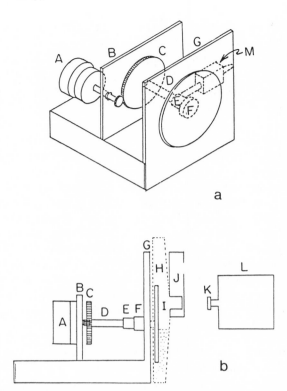

a

b

FIGURE 8. (a) Electrochemical cell and ancillary components for quasi *in situ* conversion electron Mössbauer measurements.[11] The counter and reference electrodes are not shown in this figure. (b) Schematic diagram of rotating system. A. Motor, B. aluminum support, C. reduction gear, D. phenolic shaft, E. brass contact, F. Teflon bushing, G. aluminum support, H. electrochemical cell, I. working electrode (disk), J. conversion electron counter, K. Mössbauer source, L. Mössbauer Doppler velocity transducer, M. carbon brush assembly.

of other spectroscopic methods. Also to be considered is the fact that the natural abundance of the Mössbauer active isotopes is 8.58% for ^{119}Sn and only 2.19% for ^{57}Fe. Therefore, the possibility of conducting transmission Mössbauer measurements in this specific mode involving species at monolayer coverages on smooth surfaces, even with fully enriched compounds, appears very unlikely. It may be mentioned in this regard that, to date, no conclusive evidence has been obtained regarding the detection of molecules adsorbed at such low coverages on even high-area materials in electrochemical environments.

Emission Mössbauer spectroscopy affords a much more sensitive means of acquiring spectral information as only 10^{12} atoms are required usually to obtain an adequate signal-to-noise ratio in a reasonable period of time. Besides the problems associated with the interpretation of emission spectra discussed in the previous section, a further constraint is that the appropriate radioactive isotope must be incorporated in the compound under study. This involves the development of synthetic pathways which optimize the utilization of rather expensive isotopes without introducing a high degree of dilution. In addition, an appropriate handling of radioactive substances is required. Although these may not represent serious obstacles, such experiments indeed demand careful planning.

4. Model Systems

A number of examples will be provided in this section of the use of *in situ* Mössbauer spectroscopy as applied to the study of electrochemical phenomena, involving transmission, emission, and quasi *in situ* conversion electron modes. It is expected that these examples may serve as a guide for the design of experiments involving a much wider variety of interfacial systems. Except where otherwise indicated, the isomer shifts, δ, are referred to the α-Fe standard, and δ, the quadrupole splittings Δ, and widths Γ are given in mm s^{-1}.

4.1. Electrochemical Properties of Iron and Its Oxides

A detailed understanding of iron passivation and corrosion is of crucial importance to the development of new iron-based materials and coatings with optimized chemical and structural characteristics to withstand prolonged exposure to a large variety of aggressive gaseous and liquid environments. Because of its specificity, Mössbauer spectroscopy appears to provide an ideal means for studying key aspects of these phenomena. Hence, it is not surprising that most of the literature in the area of *in situ* application of this technique has been devoted to their investigation.

This subsection has been divided into two parts. The first will present results of two rather recent contributions from which considerable insight into the electrochemical behavior of the iron oxyhydroxide system has been obtained, whereas the second part will address studies of the passive film in borate buffer media.

4.1.1. The Iron Oxyhydroxide System

One of the earlier Mössbauer studies of the electrochemical properties or iron was that of Geronov *et al.*,[14] who investigated the spectral changes induced by the charge and discharge of high-area iron–carbon polymer-bonded electrodes in strongly alkaline media. Despite the fact that the experiments were not conducted under strict potential control, as the circuit was opened during data acquisition, these authors made a number of interesting observations regarding the behavior of iron electrodes in 5 M KOH. In particular, electrodes in the fully charged state (-0.9 V versus Hg/HgO, OH$^-$) were found to exhibit features associated with metallic iron and $Fe(OH)_2$, whereas two additional peaks, attributed to β-FeOOH, were observed upon discharge of these electrodes under galvanostatic conditions (-0.5 V versus Hg/HgO, OH$^-$). Similar experiments conducted in the presence of LiOH in the same solution led to a conversion of the β-FeOOH into bulk magnetite (γ-Fe$_3$O$_4$) as evidenced by the appearance of the characteristic strong-field Zeeman-split six-line spectrum.

The first *in situ* Mössbauer investigation involving the behavior of iron oxides in electrochemical environments was the result of a fortuitous incident in which a specimen containing iron phthalocyanine, FePc, dispersed on a high-area carbon was accidentally decomposed during a rather mild heat treatment to yield a very fine dispersion of a ferric oxide.[15] FePc is a highly conjugated transition metal macrocycle which has been found to exhibit high activity for the electrochemical reduction of dioxygen when supported on high-area carbon.[7] The studies in question were aimed at characterizing the interactions of FePc with the carbon substrate and their role in the overall catalytic process.

The actual samples were prepared by adding Vulcan XC-72, a high-area carbon of about 250 m^2 g^{-1}, to a solution of FePc in pyridine under ultrasonic agitation. The carbon suspension was later transferred to a distillation apparatus and heated until all the excess solvent was removed. The dry dispersion, containing about 10% w/w FePc/XC-72, was then placed into a small crucible and heated in a furnace under a flowing inert gas atmosphere at 280°C so as to eliminate pyridine axially coordinated to the macrocycle. After two hours, the heating was interrupted and the sample allowed to cool without disconnecting the stream of inert gas.

Figure 9 shows the cyclic voltammetry of an FePc/XC-72 dispersion, prepared in such fashion, in the form of a thin porous Teflon-bonded coating electrode in a 1 *M* NaOH solution. A description of the methodology involved in the preparation of this type of electrode may be found in Ref. 15. As can be clearly seen, the voltammetry of this specimen exhibits two sharply defined peaks separated by about 330 mV. The potentials associated with these features are essentially identical to those found by other workers for the reduction and oxidation of films of iron oxyhydroxide formed on a number of host surfaces, including iron and carbon.[16]

A 10% w/w highly enriched ^{57}FePc/XC-72 dispersion, prepared according to the same methodology as that described above, was used in the Mössbauer measurements. The ^{57}FePc was synthesized by heating a mixture of hydrogen-reduced, highly divided metallic ^{57}Fe with dicyanobenzene in an evacuated, sealed ampoule for over a day. It was then extracted with acetone and subsequently purified by vacuum sublimation under reduced pressure. The electrode employed in the *in situ* Mössbauer measurements was the same as that involved in the operating fuel cell experiments using the *in situ* spectroelectrochemical cell shown in Figure 6a.

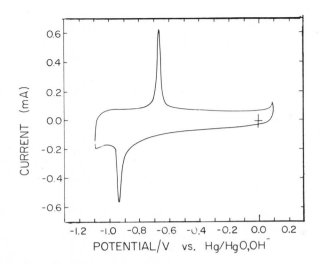

FIGURE 9. Cyclic voltammetry of 7% w/w iron phthalocyanine dispersed on Vulcan XC-72 carbon, after a heat treatment at 300°C in a flowing inert atmosphere. The measurement was conducted with the material in the form of a thin porous Teflon-bonded coating electrode in 1 M NaOH at 25°C. Sweep rate: 5 mV s^{-1}.

The *in situ* Mössbauer spectrum obtained at 0.0 V is given as curve A of Figure 10. The parameters associated with this doublet (Table III) are similar to those reported by various groups for high-spin ferric oxyhydroxides (Table IV). Also, they appear in agreement with those observed for certain magnetically ordered oxides for which the characteristic six-line spectrum collapses into a doublet as the particles become smaller in size. This phenomenon, known as superparamagnetism,[2] is attributed to the flipping of the magnetic moment of each microcrystal between easy directions. This occurs in a shorter time than either the Larmor precessional period of the nucleus or the lifetime of the excited $\frac{3}{2}$ state of the ^{57}Fe nucleus, or both, when the temperature is sufficiently high. Such behavior has been observed, for example, by Hassett *et al.*[17] for magnetite dispersed in lignosulfonate. Low-temperature quenching experiments could be highly valuable in attempts to establish on a firm basis the actual nature of this thermally generated material.

It may be noted that the quadrupole splitting of the heat-treated FePc/XC-72 electrode measured *ex situ* prior to the electrochemical experiments was larger than that found *in situ*. Smaller values for Δ have been reported for certain ferric hydroxide gels and for small particles of FeOOH (Table IV), and thus the effect observed for this specimen is most probably related to the incorporation of water into the oxide structure. Based on this information, the material observed *in situ* at this potential will be referred to hereafter as FeOOH (hydrated), without implying any specific stoichiometry.

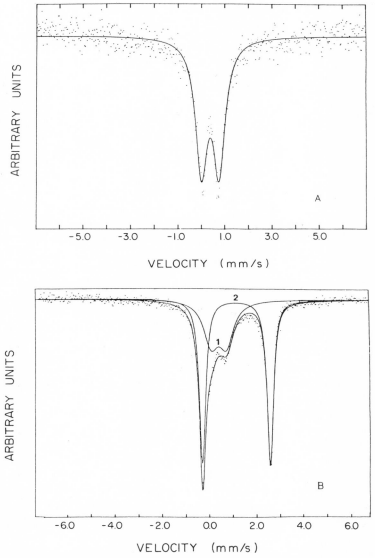

FIGURE 10. *In situ* Mössbauer spectra for FePc dispersed on Vulcan XC-72 carbon subjected to the same heat treatment as specified in the caption to Figure 9, obtained at (A) 0.0 V and (B) −1.05 V versus Hg/HgO, OH⁻. Other conditions given in caption to Figure 9.

The *in situ* spectrum obtained at −1.05 V, shown as curve B of Figure 10, yielded a doublet with an isomer shift and quadrupole splitting in excellent agreement with those of crystalline $Fe(OH)_2$ (Table III). This provides rather definite evidence that the redox process associated with the voltammetric peaks

TABLE III. *In Situ* Mössbauer Parameters for Small Particles of a Hydrated Ferric Oxyhydroxide, FeOOH (hydrated), Dispersed on High-Area Vulcan XC-72 Carbon[a]

Potential (V versus Hg/HgO, OH$^-$)	Isomer shift (δ/mm s^{-1} versus α-Fe)	Quadrupole splitting (Δ/mm s^{-1})	Width (Γ/mm s^{-1})	Figure
0.0	0.37	0.76	0.62	10A
−0.4	0.37	0.74	0.65	
−0.75	0.37	0.66	0.57	
−0.85	0.35	0.63	0.60	
−1.05	$\begin{cases} 0.41 \\ 1.14 \end{cases}$	$\begin{cases} 0.65 \\ 2.85 \end{cases}$	$\begin{cases} 0.77 \\ 0.32 \end{cases}$	10B

[a] This material was prepared by the heat treatment of iron phthalocyanine dispersed on the carbon in an inert atmosphere at 280°C.

TABLE IV. *Ex Situ* Mössbauer Parameters of Various Iron Oxides and Oxyhydroxides at Room Temperature[a]

Specimen	Isomer shift (δ/mm s^{-1} versus α-Fe)	Quadrupole splitting (Δ/mm s^{-1})	Width (Γ/mm s^{-1})	H_{eff} (kOe)
α-FeOOH (goethite)	0.44	0.16	0.86	367
α-Fe$_2$O$_3$[b] (diam. <10 nm)	0.32	0.98		
β-FeOOH	0.38 (61.4)	0.53	0.26	
	0.39 (38.6)	0.88	0.30	
γ-FeOOH (lepidocrocite)	0.38	0.59	0.27	
α-Fe$_2$O$_3$ (hematite)	0.38	0.24	0.29	523
γ-Fe$_2$O$_3$ (maghemite)	0.43	0.06	0.45	506
Fe$_3$O$_4$ nonstoichiometric	0.39 (49.6)	0.11	0.53	506
	0.78 (50.4)	0.28	0.38	465
γ-Fe$_3$O$_4$ stoichiometric (magnetite)	0.37 (4.8)[c]	0.59	0.43	
	0.34 (34.4)	0.12	0.29	491
	0.72 (59.8)	0.10	0.31	461
γ-Fe$_3$O$_4$[d] (diam. <5 nm)	0.37	0.89		
Fe(OH)$_2$[e]	1.18	2.92		
Ferric oxide[f] (small particles)	0.33	0.70		
Ferric oxide[g] (hydrated)	0.35	0.62		
FeOOH[h] (small particles)	0.39	0.62		

[a] Data from S. Music, I. Czako-Nagy, S. Popovic, A. Vertes, and M. Tonkovic, *Croat. Chem. Acta* **59**, 833 (1986), except where otherwise indicated.
[b] W. Kundig, H. Bommel, G. Constabaris, and H. Lindquist, *Phys. Rev.* **142**, 327 (1966).
[c] This feature is due to FeOOH.
[d] S. Aharoni and M. Litt, *J. Appl. Phys.* **42**, 352 (1971).
[e] A. M. Pritchard and B. T. Mould, *Corros. Sci.* **11**, 1 (1971).
[f] D. G. Rethwisch and J. A. Dumesic, *J. Phys. Chem.* **90**, 1863 (1986).
[g] P. P. Bakare, M. P. Gupta, and A. P. B. Sinha, *Indian J. Pure Appl. Phys.* **18**, 473 (1980).
[h] P. O. Vozniuk and V. N. Dubinin, *Sov. Phys.—Solid State* **15**, 1265 (1973).

is given by:

$$\text{FeOOH (hydrated)} + e^- + H^+ \rightarrow \text{Fe(OH)}_2 \text{ (crystalline)} \quad (4)$$

It may thus be concluded that the specific methodology involved in the dispersion of FePc on high-area carbon leads to the thermal decomposition of the macrocycle at temperatures much below those expected for the bulk material, generating small particles of hydrated FeOOH upon exposure to an alkaline solution.

Subsequently, Fierro *et al.*[18] have reported a series of *in situ* Mössbauer experiments aimed at investigating the electrochemical behavior of the iron oxyhydroxide system in strongly alkaline media. A hydrated form of a ferric oxyhydroxide precipitated by chemical means on a high-area carbon was used in these experiments. This material was prepared by first dissolving a mixture of highly enriched metallic ^{57}Fe and an appropriate amount of natural iron in concentrated nitric acid to achieve about one-third isotope enrichment in the final product. This solution was then added to an ultrasonically agitated water suspension of Shawinigan black, a high-area carbon of about 60 m^2 g^{-1}, and the iron was subsequently precipitated by the addition of 4 M KOH. A Teflon-bonded electrode was prepared with this material following the same procedure as that described in Section 3.3, except that no heat treatment was performed to remove the Teflon emulsifier. The ^{57}Fe/XC-72 w/w ratio was in this case 50% and thus much higher than in the heat-treated FePc experiments described earlier. The electrochemical cell for the *in situ* Mössbauer measurements is shown in Figure 11.

The *ex situ* Mössbauer spectrum for the partially dried electrode yielded a doublet with $\delta = 0.34$ and $\Delta = 0.70$ mm s^{-1}. A decrease in the value of Δ was found in the *in situ* spectrum of the same electrode immersed in 4 M KOH at -0.3 V versus Hg/HgO, OH$^-$ (Figure 12a), in direct analogy with the behavior observed for the heat-treated FePc. It is thus conceivable that this material is the same as that found after the thermal decomposition of FePc dispersed on carbon, reported by other workers, and that the variations in the value of Δ are simply due to differences in the degree of hydration of the lattice.

No significant changes in the spectra were found when the electrode was polarized sequentially at -0.5 and -0.7 V, by scanning the potential to these values at 10 mV s^{-1}. This is not surprising since the cyclic voltammetry for an identical, although nonenriched, iron/carbon mixture, shown in the inset of Figure 12, indicated no significant faradaic currents over this voltage region for the sweep in the negative direction. In a subsequent measurement at a potential of -0.9 V, the resonant absorption of the doublet underwent a marked drop. This may be due to an increase in the solubility of the oxide and thus in a loss of solid in the electrode and/or to a modification in the recoilless fraction of the solid induced by the hydration of the lattice.

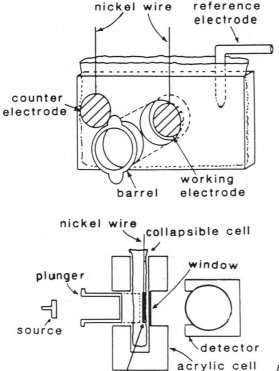

FIGURE 11. Electrochemical cell for *in situ* Mössbauer spectroscopy measurements.

The electrode was then swept further negative to -1.1 V, a potential more negative than the onset of the faradaic current in the voltammogram, yielding after about two hours of measurement, a strong, clearly defined doublet (Figure 12b) with parameters in excellent agreement with those of $Fe(OH)_2$ (Table IV). The potential was then *stepped* to -0.3 V. In contrast to the doublet obtained originally at this voltage, a magnetically split six-line spectrum was obtained in this case (Figure 13a). The Zeeman effect and the value of δ are consistent with those of a magnetically ordered ferric oxide species (Tables IV and V). Unfortunately, the strength of the internal field, H_{eff}, cannot be used as a definite identifying parameter as the calculated value seems significantly smaller than that expected for a bulk iron oxide, a behavior often attributed to superparamagnetism (see above). Furthermore, the asymmetric broadening of the peaks may be ascribed to a distribution of effective magnetic fields, providing evidence for the presence of an ensemble of small particles of varying sizes. From a statistical viewpoint, this is accounted for by a Gaussian distribution of Lorentzians, a feature that is built into the computer routine

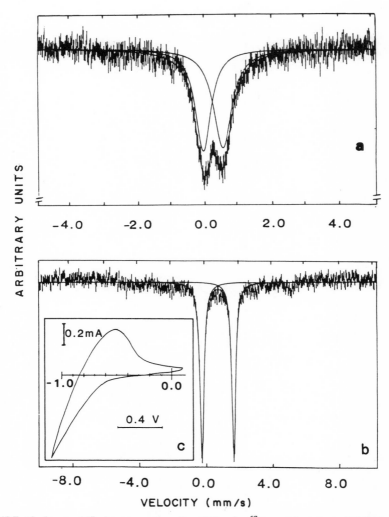

FIGURE 12. *In situ* Mössbauer spectra of a 50% w/w ^{57}Fe-enriched hydrated ferric oxide precipitated on Shawinigan black high-area carbon in the form of a Teflon-bonded electrode in 4 M KOH at (a) −0.3 V and (b) −1.1 V versus Hg/HgO, OH⁻. *Inset*: Cyclic voltammogram of the same, although nonenriched, material as in (a) in the form of a thin porous Teflon-bonded coating electrode deposited on an ordinary pyrolytic graphite electrode, in 4 M KOH. Scan rate: 10 mV s⁻¹.

used to fit the data shown in Figure 13a. It may be noted that Hassett *et al.*[17] have reported a strikingly similar six-line spectrum for small particles of magnetite dispersed in a lignosulfonate matrix.

In a subsequent measurement, the potential was swept in the negative direction to −1.1 V, yielding once more a Mössbauer spectrum characteristic

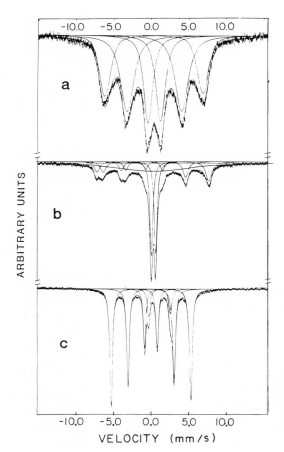

FIGURE 13. *In situ* Mössbauer spectra of the same electrode as in figure 12a at −0.3 V, (a) after a potential *step* and (b) after a potential *sweep*, from −1.1 V. The curve in (c) was obtained at a potential of −1.2 V.

of $Fe(OH)_2$, and later *swept* at 2 mV s^{-1}, rather than *stepped* positive, to −0.3 V. As shown in Figure 13b, the resulting spectrum was different from either that associated with the original material or that obtained after a voltage step. The apparent splitting observed for two of the absorption lines located at negative velocities is typical of magnetite (Fe_3O_4) in bulk form at room temperature (Table IV). This is due to the superposition of spectra arising from ferric cations in tetrahedral sites and ferrous and ferric cations in octahedral sites. The broad background centered at 0.24 mm s^{-1} may be the result of several effects including particle size and structural disorder among magnetite crystals, which would distort the Mössbauer spectrum. The sharp doublet in the center of the spectrum, as judged by the parameter values given in Table V, can be attributed to the same hydrated ferric oxyhydroxide observed originally.

The results of these experiments may be explained in terms of the influence on the nature of the particles generated of the specific way in which the ferrous

TABLE V. *In situ* Mössbauer Parameters for Iron Oxides and Oxyhydroxides Dispersed on High-Area Shawinigan Black Carbon Electrode

Potential (V versus Hg/HgO, OH$^-$)	Isomer shift (δ/mm s^{-1} versus α-Fe)	Quadrupole splitting (Δ/mm s^{-1})	H_{eff} (kOe)	Figure
−0.3 (initial)	0.33 (0.34)^a	0.58 (0.70)		12a
−1.1	1.10	2.89		12b
−0.3 (step)	0.37	0.04	406	13a
−0.3 (sweep)	0.33	0.57		13b
	0.28		463	
	0.56	0.15	437	
	0.24			
−1.2	0.00		330	13c
	1.15	2.90		

^{a} Values in parentheses are those obtained for the same electrode dry.

oxide is electrochemically oxidized. In particular, a potential step is expected to promote the formation of a multitude of nuclei large enough to exhibit a Zeeman splitting but on the average smaller than the size required to yield a spectrum characteristic of bulk magnetite. When the oxidation is performed by sweeping the potential, however, a few magnetite nuclei are generated which grow to a size sufficiently large to show bulklike behavior.

Considerable insight into the nature of these species could be obtained by conducting quenching experiments of the type described in Section 3.4.2, which would enable a careful examination of the behavior of the Mössbauer spectral features as a function of the temperature.

At the end of these measurements, the electrode was polarized by sweeping the potential to −1.2 V, yielding a six-line spectrum corresponding to metallic iron with some contribution from Fe(OH)$_2$ (Figure 13c). The potential was then scanned up to −0.3 V and a spectrum essentially identical to that recorded at −1.2 V was observed. This result clearly indicates that the iron metal particles formed by the electrochemical reduction are large enough for the contributions arising from the passivation layer to be too small to be clearly resolved. After scanning the potential several times between −0.3 and −1.2 V, however, the doublet associated with the Fe(OH$_2$) disappeared.

4.1.2. The Passive Film of Iron

The structure and properties of the passive film on iron may be regarded as of crucial importance to the further understanding of corrosion inhibition.[19] Despite the efforts of numerous research groups involving electrochemical and spectroscopic techniques, no consensus has yet been reached regarding such important aspects of the film as the oxidation state of the iron sites, the degree of long-range order, and the extent of hydration. Part of the controversy has centered around the use of *ex situ* spectroscopic methods for the acquisition of structural and compositional information. In particular, the relevance of the results obtained to the conditions that prevail in actual electrochemical environments has been regarded by many workers as highly questionable, since (a) the removal of the electrodes from the electrochemical cell results in a loss of potential control and (b) the exposure of specimens to either air or vacuum and photon, electron, or ion beams may be expected to modify considerably the nature of the film. Because of its high degree of specificity and the possibility of conducting *in situ* measurements, Mössbauer spectroscopy appears especially suited for the investigation of this particular system. This section will be based principally on the rather recent comprehensive studies of Eldridge *et al.*,[20] which may be regarded as an extension of the pioneering work of O'Grady published in 1980.[21]

Essentially all *in situ* Mössbauer studies of the passive film have been performed in borate buffer solutions of pH 8.4. The choice of this electrolyte has been motivated primarily by the work of Nagayama and Cohen,[22] who concluded based on electrochemical measurements that the passive layer in this medium could be reduced to metallic iron in a reproducible fashion. Except where otherwise noted, all the experiments to be presented in this section were conducted with thin ^{57}Fe-enriched film electrodes, vapor-deposited on a highly conducting gold on Melinex substrate.[6] Scanning electron micrographs provided evidence that this preparation procedure affords films of a much smoother character than those obtained by electrodeposition from a conventional plating bath. Although some carbon impurities were detected in such vapor-deposited iron films, it is highly unlikely that their presence would significantly affect the overall electrochemical characteristics of the pure metal.

The electrochemical cell employed in the experiments is illustrated in Figure 6b. A typical *in situ* Mössbauer spectrum obtained for an ^{57}Fe film (ca. 11 nm thickness) polarized at -0.4 V versus α-Pd/H in borate buffer, pH 8.4, is shown as curve A of Figure 14. The two peaks correspond to the inner components of the sextet characteristic of bulk metallic iron. Upon stepping the potential to 1.3 V, an additional doublet with parameters $\delta = 0.41$ mm s^{-1}, $\Delta = 1.09$ mm s^{-1}, and $\Gamma = 0.83$ mm s^{-1} attributed to the passive film was obtained after about 24 h of data acquisition (curve B, Figure 14). The electrode polarization was then interrupted, the electrolyte drained, and the film washed with distilled water and stored in a desiccator.

Two significant changes were observed in the *ex situ* spectrum of this dried film (curve C, Figure 14) compared to those obtained *in situ*: an increase in the intensity of the passive film features with respect to those of bulk iron and a decrease in the value of Γ (Table VI). After this measurement was completed, the cell was reassembled and filled with electrolyte and the same

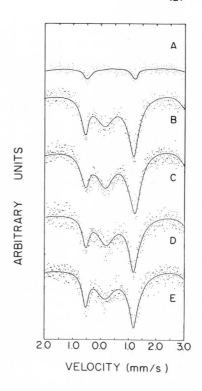

FIGURE 14. *In situ* Mössbauer transmission spectra for 11-nm [57]Fe film in borate buffer (pH 8.4) at −0.4 V (metallic Fe) (curve A) and after passivation at +1.3 V versus α-Pd/H (curve B). Curves C and D were obtained *ex situ* after drying the film and *in situ* (+1.3 V) after reintroducing the passive film in the same electrolyte. Spectrum E was recorded at +1.3 V after two reduction–passivation cycles. See text for additional details.

VELOCITY (mm/s)

electrode was polarized at 1.3 V. Although no change was observed in the magnitude of Γ for the passive layer feature as compared to that of the dry film *ex situ*, the relative peak amplitude was found to be about a third larger than that observed in the original *in situ* spectrum at the same potential (curve

TABLE VI. Transmission Mössbauer Data for a [57]Fe Film Passivated at +1.3 V versus α-Pd/H in Borate Buffer (pH 8.4)

Specimen	Isomer shift (δ/mm s^{-1} versus α-Fe)	Quadrupole splitting (Δ/mm s^{-1})	Width (Γ/mm s^{-1})	A_{pf}/A_{Fe} [a]	Figure
Initial passivation *in situ*	0.41	1.09	0.89	0.89	14b
Dry	0.38	1.07	0.75	1.11	14c
Repassivation, no reduction *in situ*	0.40	1.05	0.73	1.04	14d
Two-step cycle, repassivations *in situ*	0.37	1.06	0.79	0.88	14c

[a] This quantity corresponds to the fraction of total resonant area due to the passive film contribution.

D, Figure 14). The electrode was then reduced and passivated twice and a new *in situ* Mössbauer spectrum recorded (curve E, Figure 14), which was almost indistinguishable from the spectrum shown as curve B in Figure 14. In particular, the relative passive film/iron substrate resonant absorption contribution, A_{pf}/A_{Fe}, was found to be the same as that observed in the original passivated specimen. This indicates that within the sensitivity of this technique, the amount of iron lost in the electrolyte during the whole procedure was negligible and thus that most of the iron in the passive film was reduced back to the metal state. Additional evidence in support of this view was provided by the striking similarities between the conversion electron Mössbauer spectra of films that had been subjected to a single compared to multiple reduction-passivation cycles.

The dependence of the film characteristics on the rate at which the passivation is performed was investigated also. As shown in Table VII, slow scan rates invariably yield spectra with much larger passive film contribution and substantially lower quadrupole splittings than films formed by a fast sweep or a step. It is interesting to note in this regard that the parameters reported by O'Grady[21] for the passive film measured *in situ* after multiple reduction-passivation cycles are very similar to those of films obtained by slow scan rates. This provides rather conclusive evidence that the current distribution, which is largely determined by the cell geometry, may play a significant role in controlling certain structural and electronic properties of the film formed on the iron surface.

No spectral differences were found between films formed at 1.3 V and later polarized in sequence at 0.5 and 0.35 V. In contrast, films formed originally at 0.35 V and subsequently stepped to 0.5, 0.9, and 1.3 V exhibited a systematic increase in the relative passive film contribution, a decrease in the isomer shift, and an increase in the quadrupole splitting, to yield at the final potential of 1.3 V, parameters very similar to those of films obtained by direct passivation at 1.3 V. It was found also that films formed at lower potentials underwent spectral changes upon drying, an effect that was not observed for films prepared

TABLE VII. *In Situ* Transmission Mössbauer Data for a ^{57}Fe Film Passivated at +1.3 V versus α-Pd/H in Borate Buffer (pH 8.4) at Different Scan Rates

Specimen[a]	Isomer shift (δ/mm s^{-1} versus α-Fe)	Quadrupole splitting (Δ/mm s^{-1})	Width (Γ/mm s^{-1})	A_{pf}/A_{Fe}
Stepped	0.38	1.14	0.77	0.84
Scanned over 4.5 min	0.40	1.14	0.78	0.87
Scanned over 10.25 min	0.40	1.00	0.79	1.30

[a] The potential was always set initially at -0.4 V versus α-Pd/H.

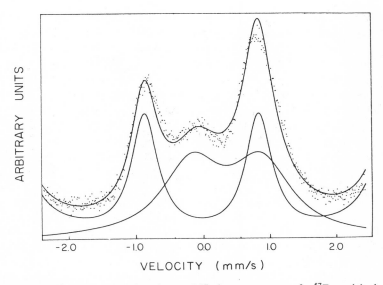

FIGURE 15. Quasi *in situ* conversion electron Mössbauer spectrum of a [57]Fe-enriched rotating iron film electrode polarized at +1.3 V versus α-Pd/H in borate buffer (pH 8.4). The narrow doublet corresponds to the inner lines of metallic Fe, whereas the broader doublet is attributed to the passive film.

at high potentials. This suggests that the amount of water incorporated in the film decreases as the potential is made more positive.

The Mössbauer parameters for the passive film obtained in quasi *in situ* conversion electron measurements conducted in the same borate buffer medium (Figure 15) were found to yield good agreement with those obtained in the transmission mode ($\delta = 0.38$, $\Delta = 1.03$). The sharper doublet in Figure 15 corresponds to the inner lines of metallic iron.

Although important information can be obtained by comparing the values of the Mössbauer parameters obtained for the passive film with those of known oxides and oxyhydroxides, extreme care must be exercised in the assignment of spectral features based solely on such data. This is due primarily to modifications in the magnitude of the hyperfine interactions associated with a given material induced by changes in the particle size (see above). If such effects are not assumed to play a major role, however, the value of the isomer shift may be regarded as characteristic of a ferric species in high spin, for which δ lies typically in the range between 0.35 and 0.75. Furthermore, the magnitude of Δ is much larger than it is for iron oxyhydroxides and crystalline oxides. This would be consistent with the presence of large geometric distortions, such as those expected in noncrystalline lattices, indicating, as originally suggested by O'Grady, that the passive film consists of a highly disordered ferric oxyhydroxide.

An approach that can provide considerable insight into the nature of highly dispersed or amorphous materials, as was briefly mentioned in the previous section, involves a careful examination of the temperature dependence of the Mössbauer spectra. Unfortunately, the temperature range in which *in situ* experiments of this type would be feasible is very restricted. Nevertheless, *ex situ* Mössbauer spectra recorded in the X-ray fluorescence backscattering detection mode[8] have indicated that the doublet associated with the passive film, prepared in the same fashion as in the experiments described earlier, essentially disappears at 80 K. Similar experiments in which the thickness of the enriched iron layer was decreased so as to reduce the contribution due to the iron substrate and thus improve the overall resolution resulted in a broad magnetically split spectrum at liquid nitrogen temperature. This was attributed either to the onset of magnetic ordering or a blocking of the superparamagnetic particles. Although neither of these explanations appears to be entirely satisfactory, the results provide evidence for the presence of a multiplicity of iron sites in the film, which would be consistent with a highly disordered structure. *In situ* extended X-ray absorption fine structure (EXAFS) measurements[23] are expected to provide much needed insight into this specific issue.

4.2. Mixed Ni–Fe Oxyhydroxides as Electrocatalysts for Oxygen Evolution

The presence of iron in nickel oxyhydroxide electrodes has been found to reduce considerably the overpotential for oxygen evolution in alkaline media associated with the iron-free material.[24] An *in situ* Mössbauer study of a composite Ni–Fe oxyhydroxide was undertaken by Corrigan *et al.* in order to gain insight into the nature of the species responsible for the electrocatalytic activity.[25] This specific system appeared particularly interesting as it offered a unique opportunity for determining whether redox reactions involving the host lattice sites can alter the structural and/or electronic characteristics of other species present in the material.

Thin films of a composite nickel–iron oxyhydroxide (9:1 Ni/Fe ratio) and iron-free Ni oxyhydroxide were deposited onto Ni foils by electroprecipitation at constant current density from metal nitrate solutions. A comparison of the cyclic voltammetry of such films in 1 M KOH at room temperature (Figure 16) shows that the incorporation of iron in the lattice shifts the potentials associated formally with the $NiOOH/Ni(OH)_2$ redox processes in the negative direction and decreases considerably the onset potential for oxygen evolution. It may be noted that the oxidation peak is much larger than the reduction counterpart, providing evidence that within the timescale of the cyclic voltammetry, a fraction of the nickel sites remain in the oxidized state at potentials more negative than the reduction peak.

The *in situ* Mössbauer experiments were conducted with 90% ^{57}Fe-enriched 9:1 Ni/Fe oxyhydroxide films which were deposited in the fashion described above onto gold on Melinex supports in a conventional

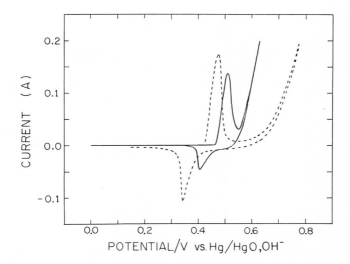

FIGURE 16. Cyclic voltammograms for a composite Fe–Ni oxyhydroxide (Fe/Ni 1:9) on a Ni foil substrate in 1 *M* KOH (solid curve). Scan rate: 10 mV s^{-2}. The dashed curve was obtained for an iron-free Ni oxyhydroxide film under the same experimental conditions.

electrochemical cell. Prior to their transfer into the *in situ* Mössbauer cell, the electrodes were cycled twice between 0 and 0.6 V versus Hg/HgO, OH$^-$ in 1 *M* KOH. Two such films were used in the actual Mössbauer measurements in order to reduce the counting time. A description of the *in situ* Mössbauer cell involved in these experiments may be found in the original literature.[25]

The *in situ* spectrum obtained at 0.5 V versus Hg/HgO, OH$^-$ (oxidized state) is shown as curve A in Figure 17. Following this measurement, the potential was swept to 0.0 V (reduced state), and a new *in situ* spectrum was recorded after the current had dropped to a very small value (curve B, Figure 17). Essentially identical results were obtained when the films were examined first in the reduced and then in the oxidized state.

The spectrum of the oxidized form was successfully fitted with a singlet yielding an isomer shift of 0.22. For the spectrum in the reduced state, a satisfactory fit could be achieved with two singlets, which, when regarded as the components of an asymmetric doublet, yielded $\delta = 0.34$ and $\Delta = 0.43$. In view of the fact that the cyclic voltammetry indicated a slow reduction of the oxidized state, a statistical analysis of the data in curve B of Figure 17 was attempted with a symmetric doublet and a singlet to account for a possible contribution due to the oxidized phase. This approach afforded excellent results, yielding an isomer shift for the singlet very similar to that obtained for the oxidized species from curve A of Figure 17. Furthermore, the values of δ and Δ for the symmetric doublet were found to be nearly the same as those in the fit involving the asymmetric doublet. These are listed in Table VIII.

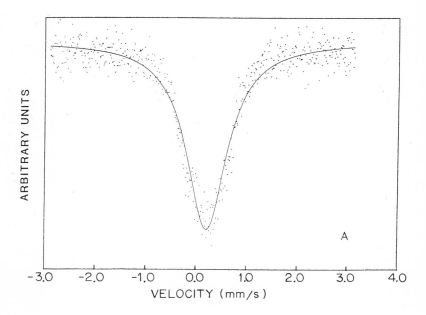

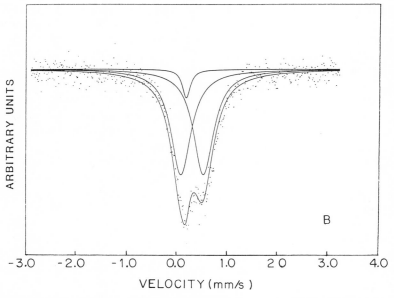

FIGURE 17. *In situ* Mössbauer spectrum of a composite Fe-Ni oxyhydroxide (Fe/Ni 1:9) polarized (A) at 0.5 V (oxidized state) and (B) at 0.0 V versus Hg/HgO, OH⁻ (reduced state) in 1 *M* KOH.

TABLE VIII. *In Situ* Mössbauer Parameters for an Iron–Nickel Mixed Oxyhydroxide in 1 *M* KOH

Potential (V versus Hg/HgO, OH⁻)	Isomer shift (δ/mm s^{-1} versus α-Fe)	Quadrupole splitting (Δ/mm s^{-1})	Width (Γ/mm s^{-1})	Figure
0.5	0.22		0.97	17a
0.0	0.32 (95)^a	0.44	0.47	17b
	0.19 (5)		0.19	

^{a} Values in parentheses represent the fraction of the total resonant absorption associated with each peak, assuming a common recoilless fraction for both species.

The isomer shift of 0.32 associated with the reduced state of the composite material is similar to that reported for various Fe(III) oxyhydroxides and indicates that iron is present in the ferric form. The much smaller quadrupole splitting, however, provides evidence that the crystal environment of such ferric sites is different from that in common forms of ferric oxyhydroxides. The lower value of the isomer shift observed upon oxidation of the composite film indicates a partial transfer of electron density away from the Fe(III) sites, which could result indirectly from the oxidation of the Ni(II) sites to yield a highly oxidized iron species. It is interesting to note that recent Raman measurements have provided evidence for the presence of highly symmetrical Ni sites in oxidized nickel hydroxide films.[26] Based on such information, it was postulated that the film structure could be better represented as NiO_2 rather than as $NiOOH$. It seems thus reasonable that the oxidized composite oxhydroxide could contain symmetrical sites which might be occupied by Fe ions. This would be consistent with the presence of a singlet, rather than a doublet, in the spectrum shown as curve A in Figure 17. Based on these arguments, Corrigan *et al.*[25] concluded that the composite metal oxyhydroxide may be regarded as a single phase involving distinct iron and nickel sites as opposed to a physical mixture of $Ni(OH)_2$ and FeOOH particles. This is not surprising since the composite hydroxide is a better catalyst than either of the individual hydroxides.

In summary, the results of this investigation indicated that the formal oxidation of the nickel sites in a composite nickel–iron oxyhydroxide modifies the electronic and structural properties of the ferric sites, yielding a more *d*-electron-deficient iron species. Although it may be reasonable to suggest that the electrocatalytic activity of this composite oxide for oxygen evolution may be related to the presence of such highly oxidized iron sites, additional *in situ* spectroscopic measurements such as EXAFS may be necessary in order to support this view.

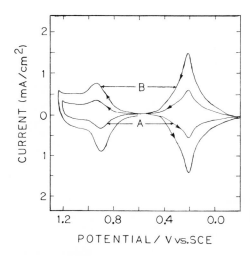

FIGURE 18. Cyclic voltammetry of a Prussian blue film electrodeposited on a glassy carbon electrode ($6\,mC\,cm^{-2}$ PB) at (A) 10 and (B) $20\,mV\,s^{-1}$ in deaerated $1\,M$ KCl (pH 4.0).

POTENTIAL / V vs.SCE

4.3. Prussian Blue

The addition of ferric ions to an aqueous solution of Fe(II) hexacyanide results in the formation of a highly colored colloidal precipitate known as Prussian blue (PB), a material regarded as the oldest coordination compound reported in the scientific literature.† It has been found recently that films of PB deposited on electronically conducting substrates are capable of undergoing a reversible blue-to-transparent color transition when the electrode potential is changed between two appropriate values.[27] This is an illustration of what has been referred to as electrochromism, a phenomenon that may find wide application in connection with electronically controlled color display devices.

The voltammetric behavior of PB films is characterized by the presence of two prominent reversible peaks. This is shown in Figure 18. The nature of the redox processes associated with the electrochromic effect have been investigated using *in situ* Mössbauer spectroscopy by Itaya and coworkers,[28] in what may be regarded as a classical example of the use of this technique in the study of electrochemical systems.

Films of PB were formed in this case on glassy carbon surfaces by electrodeposition from a mixture of $K_3Fe(CN)_6$ and highly enriched $^{57}FeCl_3$. The *in situ* experiments were conducted in the same electrolyte as that for the voltammogram of Figure 18, using a cell which is shown schematically in Figure 19. The *in situ* Mössbauer spectra were obtained at a potential of 0.6 V (Prussian blue) and at −0.2 V versus SCE, which corresponds to the reduced or transparent form of PB. These are given as curves A and B, respectively, in Figure 20. The parameters associated with PB, $\delta = 0.37$ and $\Delta = 0.41$, are

† For a review of the electrochemical and other physicochemical properties of thin films of Prussian blue and related polynuclear transition metal cyanides, see Ref. 27.

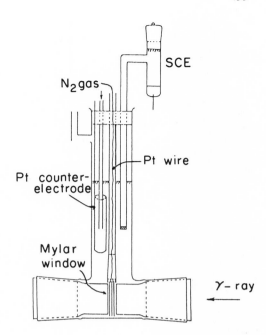

FIGURE 19. In situ Mössbauer-electrochemical cell.

characteristic of a Fe^{3+} ion in high spin (Table IX). As noted by Itaya *et al.*, a similar spectrum has been reported by Maer *et al.*[29] at liquid nitrogen temperature for a precipitate formed by mixing $^{57}Fe(SO_4)_3$ and $K_4Fe(CN)_6$ solutions. For the spectra recorded at -0.2 V, the isomer shift and quadrupole splitting are typical of a ferrous species in high spin. This provides evidence that (i) the redox process responsible for the electrochromic behavior involves a drastic change in the electron density about the iron sites not coordinated to the cyanide ligands and (ii) that no chemical reaction such as ligand exchange between the hydrated ferric ions and $Fe(III)(CN)_6^{3-}$ takes place during the electrochemical deposition of thin PB films.

Further work in this area may be required in order to determine the nature of the redox process associated with the additional redox peak at 0.9 versus

TABLE IX. *In Situ* Mössbauer Parameters for Prussian Blue in 1 *M* KCl

Potential (V versus SCE)	Isomer shift (δ/mm s^{-1} versus α-Fe)	Quadrupole splitting (Δ/mm s^{-1})	Figure
0.6	0.37	0.41	20a
−0.2	1.14	1.31[a]	20b

[a] This value is listed as 1.13 in the original paper. Inspection of the reported spectra, however, has indicated that it is actually close to 1.31.

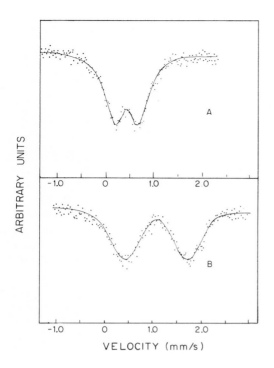

ARBITRARY UNITS

VELOCITY (mm/s)

FIGURE 20. *In situ* Mössbauer spectra of a Prussian blue film electrodeposited on a glassy carbon electrode in deaerated 1 *M* KCl (pH 4.0) at (A) 0.6 V and (B) −0.2 V versus SCE.

SCE. In addition, it would be of utmost interest to establish whether or not a change in the oxidation state of one of the sites modifies the electron density at the nucleus of the other redox center, in analogy with the phenomenon observed in the case of the mixed Fe–Ni oxyhydroxides films discussed in the previous section.

4.4. Transition Metal Macrocycles as Catalysts for the Electrochemical Reduction of Dioxygen

A number of transition metal macrocycles have been shown to promote the rates of oxygen reduction when adsorbed on a variety of carbon surfaces.[30] Attention has been mainly focused on phthalocyanines and porphyrins containing iron and cobalt centers, as their activity in certain cases has been found to be comparable to that of platinum. Essential to the understanding of the mechanism by which these compounds catalyze the reduction of O_2 is the description of the interactions, not only with the reactant, but also with the substrate. *In situ* techniques can provide much of this needed information, and indeed a number of such methods have been used in connection with this type of system.[31] One of the first illustrations of the use of Mössbauer spectroscopy for the *in situ* investigation of molecules adsorbed on electrode surfaces involved iron phthalocyanine (FePc) adsorbed on high-area carbon.[7] It was later found,[15] however, that this compound undergoes thermal decomposition at surprisingly low temperature in an inert atmosphere, and thus the features observed in the early work were not due to individual molecules of FePc adsorbed on the substrate but rather to small particles of a ferric form of iron oxyhydroxide. Shortly thereafter, Blomquist *et al.*[9] reported for the first time *in situ* Mössbauer spectra of a fuel cell type electrode containing a polymeric form of iron phthalocyanine, p-FePc, dispersed on a high-area carbon. This work was particularly interesting as the measurements were conducted with the electrode operating as an oxygen cathode in the constant-current mode.

A schematic diagram of the cell used in the *in situ* Mössbauer studies is shown in Figure 21. The key feature in this design is a hollow glass tube with a plastic window attached to one of its ends, placed very close to the working electrode so as to minimize the absorption or radiation associated with the electrolyte. During the Mössbauer measurements, the electrode was polarized in 2.3 M H_2SO_4 at a constant current density of 10 mA cm^{-2}, and several hours were required to obtain an acceptable signal-to-noise ratio. Three *in situ* spectra were recorded consecutively over a period of about three days, involving in each case increasing intervals of time due to losses in the signal strength (see below). These are given in Figure 22.

Some difficulties were encountered in the interpretation of the spectra due primarily to two factors: (i) the electrode potential was found to vary in a linear fashion during the measurements (inset, Figure 22), and hence, each of the recorded spectra was then an average of the state of the electrode over the specified period of time, and (ii) because of the way in which these

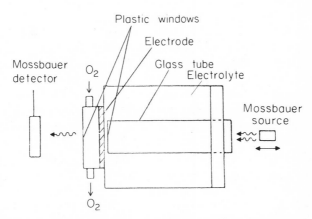

FIGURE 21. *In situ* Mössbauer cell for measurements involving a fuel cell type electrode operating as an oxygen cathode.

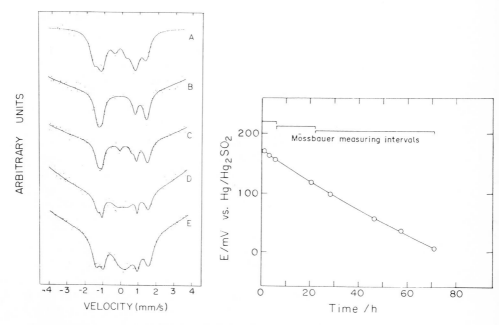

FIGURE 22. *In situ* Mössbauer spectra of a fuel cell type electrode containing a polymeric form of FePc operating as an oxygen cathode in 2.3 M H_2SO_4. Curves A and E were obtained before and after conducting the actual experiments. The other spectra were acquired between (B) 0 and 6 h, (C) 6 and 22 h, and (D) 22 and 71 h.

electrodes were prepared, some of the FePc detected by the Mössbauer measurement may not have been in contact with the electrolyte and thus would not contribute to the electrochemical activity.

A satisfactory statistical fit to the data could be obtained, however, with three symmetric doublets, which were attributed to central, peripheral, and oxidized iron sites in the polymer. The associated Mössbauer parameters are listed in Table X.

Based on this information and the results of additional experiments involving the analytical determination of iron in the electrode and in the electrolyte as a function of the time of operation of the oxygen-fed cathode, it was concluded that:

(i) the activity of the electrode was directly correlated with the amount of total iron in the electrode; i.e., as the observed potential decreases, increased amounts of iron are found in the electrolyte,

(ii) the initial loss of resonant absorption area is associated primarily with oxidized sites, suggesting that the species associated with these sites undergoes facile dissolution in the electrolyte, and

TABLE X. *In situ* Mössbauer Parameters for a Gas-Fed Electrode Containing a Polymeric Form of Iron Phthalocyanine Dispersed on High-Area Carbon[a]

Time	Isomer shift (δ/mm s^{-1} versus α-Fe)	Quadrupole splitting (Δ/mm s^{-1})	A (%)	Assignment[b]	Figure
Before[c]	0.27	2.79	40	C	22a
	0.14	1.95	32	P	
	0.33	0.80	28	O	
0–6 h	0.33	2.74	65	C	22b
	0.12	1.86	35	P	
6–22 h	0.33	2.77	62	C	22c
	0.12	1.93	29	P	
	0.44	0.75	9	O	
22–71 h	0.32	2.82	49	C	22d
	0.13	1.96	15	P	
	0.30	0.57	36	O	
After[c]	0.37	2.95	40	C	22e
	0.13	1.92	16	P	
	0.32	0.45	44	O	

[a] The experiments were performed in the constant-current mode in 2.3 M H$_2$SO$_4$.
[b] C: central site; P: peripheral site; O: oxidized site.
[c] These refer to measurements conducted before and after polarization.

(iii) FePc species in which the iron center is formally in the ferrous state are responsible for most of the electrocatalytic activity for O$_2$ reduction. These sites, however, undergo oxidation during operation, most likely as a result of a buildup in the concentration of hydrogen peroxide, generating labile Fe(III) species. This would be consistent with the observed gradual deterioration of the overall performance of the electrode.

4.5. Tin

Despite the fact that iron has been the subject of the large majority of Mössbauer studies reported in the literature, it is rather interesting to note that the first *in situ* application of this technique involved tin as the active nuclide. In that work, Bowles and Cranshaw[32] examined the emission spectra of ^{119m}Sn adsorbed on a high-area platinum electrode. Although some of the conclusions made by the authors may be open to question, the results obtained clearly indicated that the tin, in whatever form was present at the interface, was bound sufficiently strongly to the host lattice to give a Mössbauer spectrum. Thus, these types of measurements were indeed feasible.

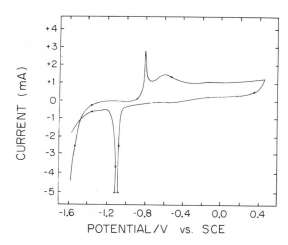

FIGURE 23. Cyclic voltammetry of a Sn electrode in borate buffer (pH 8.4). Scan rate: 60 mV s^{-1}.

More recently, Vertes and coworkers[33] have investigated the spectral properties of the passive film on tin in borate buffer medium (pH 8.4) over a wide potential range. A typical voltammetric curve for Sn in this specific electrolyte is shown in Figure 23. At -0.9 V, the transmission Mössbauer spectrum of a ^{119}Sn-enriched tin film, electrodeposited on an aluminum substrate (curve A, Figure 24), was found to have clear contributions due to β-Sn, $\delta = 2.5$, and SnO_2 or $Sn(OH)_4$, $\delta = 0.03$ (Table XI). The small absorption peak at about 4.2 mm s^{-1} was attributed to the high-velocity component of a

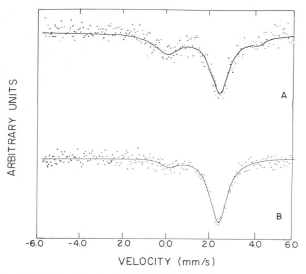

FIGURE 24. *In situ* Mössbauer spectra of a ^{119}Sn-enriched Sn electrode electrodeposited on an aluminum substrate in borate buffer (pH 8.4) at (A) -0.9 V and (B) $+0.2$ V versus SCE.

doublet with parameters consistent with a Sn(II) species, most probably present in an amorphous form. No Sn(II) species could be detected, however, with *in situ* measurements obtained at -0.78, -0.75, and $+0.2$ V, as illustrated for this last potential in Figure 24 (curve B). Based on the electrochemical and spectroscopic results, these authors concluded that in borate buffer (pH 8.4), the passive film in the potential region between -1.18 and -0.78 V consists of highly amorphous $Sn(OH)_2$, or hydrated stannous oxide and SnO_2, or $Sn(OH)_4$, whereas at much more positive potential only Sn(IV) oxide or hydroxide species are present in the film.

In a separate study, the surface structure of tin layers obtained by the electroless deposition in a strongly alkaline solution was examined by van Noort *et al.*,[34] employing the quasi *in situ* conversion electron Mössbauer technique introduced originally by Kordesch *et al.*[11] The spectrum obtained in the metallation bath with the electrode under rotation (curve A, Figure 25) was best fitted with three singlets. Two of these features could be unambiguously ascribed to β-Sn and SnO_2, whereas the third singlet could not be assigned to any known tin compound. This specific peak, however, was not observed when the spectrum was recorded with the same electrode stationary (curve B, Figure 25), leading these authors to conclude that this species was most probably a Sn(II) intermediate in the overall electroless deposition process. A third peak exhibiting the same rotation-dependent characteristics, although different Mössbauer parameters, was also detected in experiments involving tin layers prepared by electroless deposition but polarized in a slightly acidic medium containing no tin ions.

A series of experiments involving either stationary disks and different solutions or dry electrodes were conducted to determine whether the effects observed could be associated with mechanical motion or the presence of a liquid film on the surface. In all cases examined, however, the spectrum remained unchanged, leading van Noort *et al.* to conclude that the third peak was not due to an instrumental artifact but rather to the presence of transient Sn(II) species on the electrode surface.

TABLE XI. *In situ* Mössbauer Parameters for Tin in Borate Buffer (pH 8.4)

Potential (V versus SCE)	Isomer shift (δ/mm s^{-1} versus CaSnO$_3$)
-0.9	0.03
	2.50
	4.17[a]

[a] This feature has been regarded as the high-velocity component of a doublet with an isomer shift of about 2.8 mm s^{-1} and quadrupole splitting of 2.2 mm s^{-1}.

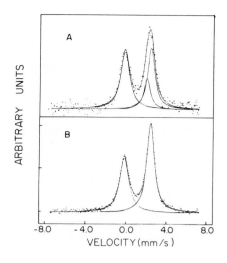

FIGURE 25. ^{119}Sn conversion electron Mössbauer spectrum of a rotating disk with its lower half in the metallization solution (curve A). Curve B was obtained with the same disk stationary.

4.6. In Situ Emission Mössbauer

Besides the early study of Bowles and Cranshaw on the interactions of Sn with Pt surfaces, referred to in the previous section, the most significant illustrations of the use of *in situ* Mössbauer in the emission mode appear to be found in the work of Simmons and coworkers,[35] who conducted a rather detailed study of passivation and anodic oxidation of Co surfaces.

The emission Mössbauer spectra of Fe in a rather thick (100–200 Å) cobalt film deposited by electrochemical means onto a Co specimen after preparation and during polarization at −1.1 V are shown in Figure 26. It should be stressed

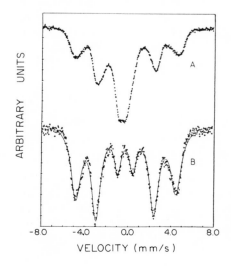

FIGURE 26. Emission spectrum of ^{57}Fe in cobalt recorded after deposition of a 100 to 200-Å ^{57}Co layer (curve A). Curve B shows the spectrum of the same specimen during cathodic polarization at −1.1 V versus SCE in a deaerated borate buffer solution (pH 8.5).

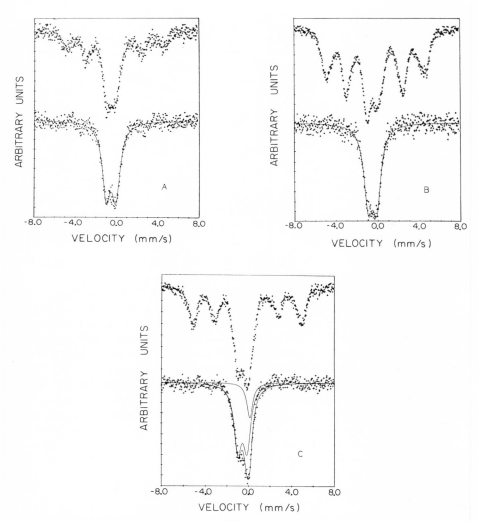

FIGURE 27. Emission Mössbauer spectra of ^{57}Fe in cobalt polarized at (A) +0.2, (B) +0.5, and (C) +0.8 V versus SCE in the same electrolyte as for Figure 29 before (upper spectra) and after (lower spectra) subtracting the contributions due to the unreacted metal.

that even though the original material is Co, the observed spectrum is that of the ^{57}Fe daughter nucleus. The intense lines in curve A in this figure are attributed to the presence of a layer consisting presumably of cobalt oxyhydroxides formed after the film was prepared. The latter undergoes reduction during cathodic polarization at −1.1 V, yielding the characteristic six-line spectrum of Fe in a Co matrix (curve B, Figure 26). Much broader lines were observed in similar experiments involving a film of only 20 to 50 Å thickness,

444 Daniel A. Scherson

TABLE XII. *In Situ* Mössbauer Parameters for ^{57}Fe in Anodically Polarized
Cobalt Surfaces Doped with ^{57}Co in Borate Buffer (pH 8.5)

Potential (V versus SCE)	Isomer shift[a] (δ/mm s^{-1} versus α-Fe)	Quadrupole splitting (Δ/mm s^{-1})	Figure
-0.1^b	-1.15	2.83	27a
	-0.43	0.78	
0.2	-0.39	0.94	
0.5	-0.38	0.81	27b
0.8	-0.39^c	0.85^c	27c
	-0.02		

[a] According to the convention, isomer shifts in the emission mode are reported with opposite sign.
[b] Data obtained at liquid nitrogen temperature.
[c] Values constrained during fitting.

a phenomenon attributed to hyperfine relaxation associated with the presence of small-size particles or magnetic dilution effects. The spectra obtained at 0.2, 0.5, and 0.8 V before and after subtracting the metal contribution are given in Figure 27. Based on the analysis of the Mössbauer parameters (Table XII), it was concluded that polarization at increasingly positive potentials leads to spectra consistent with the presence of more oxidized species. It is interesting to note that no Mössbauer signals could be observed when the electrode was polarized at 0.1 V at room temperature. The spectra obtained for this specimen at 77 K, however, indicated species in the +2 and +3 states oxidation states. As was pointed out in Section 3.1, chemical and Auger after effects can generate ^{57}Fe ionic states which may differ from those of the parent ^{57}Co and thus introduce uncertainties in the spectral assignments. Nevertheless, some of these ambiguities could, in principle, be removed by examining the emission Mössbauer spectra of known species.

Acknowledgments

The author would like to express his appreciation to Prof. Ernest Yeager for his guidance and support, and to Dr. Jeff Eldridge for providing invaluable comments during the preparation of this manuscript. Financial support was provided by the IBM Corp. through a Faculty Development Award and by the Gas Research Institute.

References

1. (a) R. L. Mössbauer, *Z. Phys.* **151**, 124 (1958); (b) R. L. Mössbauer, *Naturwissenschaften* **45**, 538 (1958).

2. (a) V. I. Goldanskii and R. Herber (eds.), *Chemical Applications of Mössbauer Spectroscopy*, Academic Press, New York (1968); (b) P. Guttlich, R. Link, and A. Trautwein, *Mössbauer Spectroscopy and Transition Metal Chemistry*, Inorganic Chemistry Concepts 3, Springer Verlag, Berlin (1978); (c) D. P. E. Dickson and F. J. Berry (eds.), *Mössbauer Spectroscopy*, Cambridge University Press, Cambridge (1986); (d) I. J. Gruverman (ed.), *Mössbauer Effect Methodology*, Vols. I–IX, Plenum Press, New York.

3. A. Hudson and H. J. Whitfield, *Inorg. Chem.* 6, 1120 (1967).

4. (a) D. Scherson, S. B. Yao, E. B. Yeager, J. Eldridge, M. E. Kordesch, and R. W. Hoffman, *J. Electroanal. Chem.* 150, 535 (1983); (b) A. Vertes and I. Czako-Nagy, *Izv. Khim.* 19, 380 (1986).

5. G. J. Long, *Mössbauer Effect Ref. Data J.* 6, 42 (1983).

6. M. E. Kordesch and R. W. Hoffman, *Thin Solid Films* 107, 365 (1983).

7. D. Scherson, S. B. Yao, E. B. Yeager, J. Eldridge, M. E. Kordesch, and R. W. Hoffman, *J. Phys. Chem.* 87, 932 (1983).

8. J. Eldridge, Ph.D. dissertation, Case Western Reserve University, Cleveland, Ohio (1984).

9. J. Blomquist, U. Helgeson, L. C. Moberg, L. Y. Johansson, and R. Larson, *Electrochim. Acta* 27, 1453 (1982).

10. D. Scherson, C. Fierro, E. B. Yeager, M. E. Kordesch, J. Eldridge, and R. W. Hoffman, *J. Electroanal. Chem.* 169, 287 (1984).

11. M. E. Kordesch, J. Eldridge, D. Scherson, and R. W. Hoffman, *J. Electroanal. Chem.* 164, 393 (1984).

12. D. L. Rath and D. M. Kolb, *Surface Sci.* 109, 641 (1981).

13. U. Frese, T. Iwasita, W. Schmickler, and U. Stimming, *J. Phys. Chem.* 89, 1059 (1985) and references therein.

14. Y. Geronov, T. Tomov, and S. Georgiev, *J. Appl. Electrochem.* 5, 351 (1975).

15. D. Scherson, C. A. Fierro, D. Tryk, S. L. Gupta, E. B. Yeager, J. Eldridge, and R. W. Hoffman, *J. Electroanal. Chem.* 184, 419 (1985).

16. (a) L. D. Burke and O. J. Murphy, *J. Electroanal. Chem.* 109, 379 (1980); (b) V. A. Macagno, J. R. Vilche, and A. J. Arvia, *J. Appl. Electrochem.* 11, 417 (1981).

17. K. L. Hassett, L. C. Stecher, and D. N. Hendrickson, *Inorg. Chem.* 19, 416 (1980).

18. C. Fierro, R. Carbonio, D. Scherson, and E. B. Yeager, *J. Phys. Chem.* 91, 6579 (1987).

19. See, e.g., M. Froment (ed.), *Passivity of Metals and Semiconductors*, Elsevier, Amsterdam (1983); R. P. Frankenthal and J. Kruger (eds.), *Passivity of Metals*, Electrochemical Society, Princeton (1978).

20. J. Eldridge, M. E. Kordesch, and R. W. Hoffman, *J. Vac. Sci. Technol.* 20, 934 (1982).

21. W. O'Grady, *J. Electrochem. Soc.* 127, 555 (1980).

22. M. Nagayama and M. Cohen, *J. Electrochem. Soc.* 109, 781 (1962).

23. J. M. Fine, J. J. Rusek, J. Eldridge, M. E. Kordesch, J. A. Mann, R. W. Hoffman, and D. S. Sandstrom, *J. Vac. Sci. Technol.* 1A, 1036 (1983).

24. (a) S. I. Cordoba, R. E. Carbonio, M. Lopez-Teijelo, and V. A. Macagno, *Electrochim. Acta* 31, 1321 (1986); (b) D. A. Corrigan, *J. Electrochem. Soc.* 134, 377 (1987).

25. D. Corrigan, R. S. Conell, C. Fierro, and D. Scherson, *J. Phys. Chem.* 91, 5009 (1987).

26. J. Desilvestro, D. A. Corrigan, and M. J. Weaver, *J. Phys. Chem.* 90, 6408 (1986).

27. K. Itaya, I. Uchida, and V. Neff, *Acc. Chem. Res.* 19, 162 (1986).

28. K. Itakaya, T. Ataka, S. Toshima, and T. Shinohara, *J. Phys. Chem.* 86, 2415 (1982).

29. K. Maer, Jr., M. L. Beasley, R. L. Collins, and W. O. Milligan, *J. Am. Chem. Soc.* 90, 3201 (1968).

30. (a) F. van der Brink, E. Barendrecht, and W. Visscher, *J. Royal Neth. Chem.* 99, 253 (1980); (b) M. R. Tarasevich and K. A. Radyushkina, *Russ. Chem. Rev.* 49, 718 (1980).

31. E. B. Yeager, C. Fierro, and D. Scherson, in: *Catalyst Characterization Science* (M. Deviney and J. Gland, eds.), ACS Symposium Series No. 288, American Chemical Society, Washington, D.C. (1985), p. 535.

32. B. J. Bowles and T. E. Cranshaw, *Phys. Lett.* 17, 258 (1965).

33. (a) A. Vertes, H. Leidheiser, Jr., M. L. Varsanyi, G. W. Simmons, and L. Kiss, *J. Electrochem. Soc.* **125**, 1946 (1978); (b) M. L. Varsanyi, J. Jaen, A. Vertes, and L. Kiss, *Electrochim. Acta* **30**, 529 (1985).
34. H. M. van Noort, B. C. M. Meenderink, and A. Molenaar, *J. Electrochem. Soc.* **133**, 263 (1986).
35. G. W. Simmons, E. Kellerman, and H. Leidheiser, Jr., *J. Electrochem. Soc.* **123**, 1276 (1976); G. W. Simmons and H. Leidheiser, in: *Mössbauer Effect Methodology* (I. J. Gruverman and C. Seidel, eds.), Vol. 10, Plenum Press, New York (1976), p. 215.

Index